T0310442

Foundation Engineering for Expansive Soils

Foundation Engineering for Expansive Soils

John D. Nelson
Kuo Chieh (Geoff) Chao
Daniel D. Overton
Erik J. Nelson

WILEY

Library of Congress Cataloging-in-Publication Data:

Nelson, John D.
    Foundation engineering for expansive soils / John D. Nelson [and 3 others].
        1 online resource.
    Includes index.
        Description based on print version record and CIP data provided by publisher;
resource not viewed.
    ISBN 978-1-118-41799-7 (pdf) – ISBN 978-1-118-41529-0 (epub) –
ISBN 978-0-470-58152-0 (hardback)
1. Soil-structure interaction. 2. Swelling soils. 3. Foundations. I. Title.
    TA711.5
    624.1'51–dc23

                                                        2014043883

Contents

Preface

The practice of foundation engineering was first developed to address problems associated with settlement due to saturated soils that were prevalent in areas with soft coastal and deltaic deposits. As population and business centers moved into areas with more arid climates, problems with other types of soils became evident. Some soils that were capable of supporting a load in a natural unsaturated state were observed to either expand or collapse when wetted. These soils did not conform to the classical theories of soil mechanics and foundation engineering, and more research began to focus on the behavior of unsaturated soils.

Within the general category of unsaturated soils, the expansive soils posed the greatest problems, and created the most financial burden. In response to major infrastructure development in the late 1950s and 1960s there was an upswing in research regarding the identification of expansive soils and factors influencing their behavior. Engineers became more cognizant of the need for special attention to the unique nature of expansive soils.

The general curricula taught at universities did not specifically address the design of foundations for these soils, and engineers did not become aware of expansive soils unless they began to practice in areas where those soils existed. Therefore, the practice of foundation engineering for expansive soils developed around experience and empirical methods.

Few books have been written specifically on the subject of design of foundations on expansive soils. Fu Hua Chen wrote a book entitled *Foundations on Expansive Soils* that was published in 1975. A second edition of that book was published in 1988. Those books were based to a large extent on Mr. Chen's personal experiences along the Front Range of Colorado. The Department of the Army published a technical manual in 1983 titled, *Foundations for Expansive Soils*. That manual served as the basis for the design of structures on military bases, and was available to the civilian engineering community as well.

At about that same time the US National Science Foundation funded a research project at Colorado State University (CSU) dealing with expansive soils. The scope of that project included a survey of the practices followed by engineers throughout the United States and Canada, as well as individuals from other countries. On the basis of that survey, and research that had been conducted by that time, Nelson and Miller (1992) published a book entitled *Expansive Soils: Problems and Practices in Foundation and Pavement Engineering*.

In 1993 Fredlund and Rahardjo published their text, *Soil Mechanics for Unsaturated Soil*. That book extended the framework of classical soil mechanics to incorporate soil suction as an independent stress state variable, and provided the rigor needed for a theoretical understanding of unsaturated soils. A part of that book was devoted to the mechanics of expansive soil.

In the 20 years since the publication of Nelson and Miller (1992), the authors of this book have worked together and have performed hundreds of forensic investigations on expansive soils. In the course of that work, many new ideas have emerged, additional research has been conducted, and methods of analyses were developed that have been applied to foundation design. This book reflects the authors' experiences over the period since the book by Nelson and Miller was written. It incorporates a broader scope of analysis and a greater degree of rigor than the earlier work.

In a presentation at the 18th International Conference on Soil Mechanics and Geotechnical Engineering in which he introduced his most recent book, *Unsaturated Soil Mechanics in Engineering Practice*, Dr. Fredlund noted the need for practitioners to continue to publish works that will extend the application of the concepts of unsaturated soil mechanics to the solution of practical geotechnical engineering problems. It is believed that this book responds to that call and will provide a sound basis on which to establish a practice of foundation engineering for expansive soils.

Many people have contributed to the completion of this book, most notably Ms. Georgia A. Doyle. She has read the entire manuscript, provided necessary and valuable editing and coordination, and queried the authors where material was not clear. Many valuable comments were received from Dr. Donald D. Runnells after his review of chapter 2, and Dr. Anand J. Puppala after his review of chapter 10. Professor Erik. G. Thompson developed the FEM analysis for the APEX program presented in chapter 12.

In addition, many current and former staff of the authors' company, Engineering Analytics, Inc., have contributed in one way or another. Special recognition goes to Kristle Beaudet, Todd Bloch, Denise Garcia, Debbie Hernbloom, Jong Beom Kang, Lauren Meyer, Ronald Pacella, and Rob Schaut. Their help along the way is much appreciated.

John D. Nelson
Kuo Chieh (Geoff) Chao
Daniel D. Overton
Erik J. Nelson

List of Symbols

A_c	clay activity
B	width of footing
B	slope of matric suction vs water content curve
c	cohesion of soil
C	molar concentration
C_c	compression index
C_{DA}	"Department of Army" heave index
C_H	heave index
C_h	suction compression index
C_m	matric suction index
C_p	peak cohesion
C_r	residual cohesion
C_s	swelling index
C_t	compression index with respect to net normal stress
C_w	CLOD index
$C(\psi)$	correction function
d	distance between particles
d	pier shaft diameter
D	constrained modulus of soil
D_0	depth of nonexpansive fill
e	void ratio
e	base of natural logarithm, 2.71828
e_0	initial void ratio
E_s	Young's modulus of the soil
E_A	Young's modulus of the soil in units of bars
f	lateral restraint factor
f_s	anchorage skin friction
f_u	uplift skin friction
F_t	nodal force tangent to pier

F_t	maximum interior tensile force
g	gravitational acceleration
G_s	specific gravity of solids
h	pressure head
h_d	displacement pressure head
h_m	matric suction head
h_o	osmotic pressure head
H	thickness of a layer of soil
ΔH	change in thickness of that layer due to heave
H_t	total hydraulic head
i	hydraulic head gradient
I_{pt}	instability index
I_{ss}	shrink swell index
k	spring constant in APEX
K_a	active earth pressure
$K(h)$	coefficient of unsaturated hydraulic conductivity
K_0	coefficient of earth pressure at rest
K_p	passive earth pressure
K_s	coefficient of permeability
L	length of pier
L_{reqd}	required pier length
LL	liquid limit
$\Delta L/\Delta L_D$	linear strain relative to dry dimensions
m	molality
m	factor for swelling pressure correlation
n	porosity of the soil
p_a	active earth pressure
pF	unit for soil suction
p_0	equivalent fluid pressure
p_p	passive earth pressure
P	partial pressure of pore water vapor
P	load per linear dimension
P	footing load
P/P_0	relative humidity
P_{dl}	dead load on footing
PI	plasticity index
PL	plastic limit
P_{max}	maximum tensile force
P_0	saturation pressure of water vapor over a flat surface of pure water at the same temperature
P_0	total load due to earth pressure

q	flow rate of water
q	distributed load
q_a	allowable bearing pressure
q_u	unconfined compressive strength
q_m	mean rate of infiltration at the ground surface
r	radius
r^2	correlation coefficient
r_w	sources or sinks of water
R	universal gas constant
R	resistance force
RF	risk factor
RF_w	weighted risk factor
R_p	pullout capacity of helical bearing plate
s	coefficient for load effect on heave
S	degree of saturation
$\%SP$	percent of swelling pressure that is applied by the total applied stress on the soil
T	absolute temperature
T_s	surface tension
u_a	pore air pressure
$(u_a)_d$	displacement air entry pressure
u_w	pore water pressure
U	total uplift force
U_t	nodal displacement tangent to pier
V	molar volume of a solution
V	total volume of soil
$\Delta V/V$	volumetric strain
V_w	volume of water in an element of soil
w	gravimetric water content
w_{aev}	air entry gravimetric water content
w_u	gravimetric water content corresponding to a suction of 1 kPa
w_e	weight of sample at equilibrium
w_s	weight of oven-dry sample
y_{max}	differential soil movement
y_s	net ground surface movement
z	depth
z	elevation head
z	soil layer thickness
z_A	depth of active zone
z_{AD}	depth of design active zone

z_p	depth of potential heave
z_s	zone of seasonal moisture fluctuation
z_w	zone (depth) of wetting
z_{wt}	height above the water table
Δz	thickness of soil layer
α	compressibility factor
α	soil to pier adhesion factor
α	drainage slope
α_1	coefficient of uplift between the pier and the soil
α_2	coefficient of anchorage between the pier and the soil
β	contact angle with tube
β	reduction factor for expansive earth pressure
γ	unit weight of soil
γ_d	dry density of soil
γ_{sat}	saturated unit weight of soil
γ_t	total unit weight of soil
γ_w	unit weight of water
$\gamma_{\psi o}$	osmotic suction volumetric compression index
γ_σ	mean principal stress volumetric compression index
$\gamma_{\psi m}$	matric suction volumetric compression index
δ	interface friction angle
δ_{max}	differential heave
ε	strain
ε_{iso}	isotropic swelling strain
ε_s	strain
ε_{sn}	shrinking strain
ε_{sw}	swelling strain
ε_T	total range of strain
$\varepsilon_{s\%}$	percent swell
$\varepsilon_{s\%N}$	normalized percent swell
$\varepsilon_{s\%vo}$	percent swell measured from a sample inundated at the overburden stress in the consolidation-swell test
θ	volumetric water content
θ_f	volumetric water content above the wetting front
θ_r	residual volumetric water content
θ_s	saturated volumetric water content
λ	pore size distribution index
ν	number of ions from one molecule of salt
ν	Poisson's ratio
ρ	heave

$\Delta\rho$	differential movement
ρ_{max}	maximum heave
ρ_0	free-field heave
ρ_p	pier heave
ρ_s	solute mass/density
ρ_{ult}	total heave
$\sigma = (\sigma' + u_w)$	total stress, normal stress
$\sigma' = (\sigma - u_w)$	effective stress
$\sigma'' = (\sigma - u_a)$	net normal stress
σ''_{cs}	consolidation-swell swelling pressure
σ''_{cv}	constant volume swelling pressure
σ''_{cvN}	reduced swelling pressure
σ_{ext}	external stress
σ''_f	final vertical stress
σ''_h	lateral stress
σ''_i	inundation stress
σ_{int}	internal stress between particles
$\Delta\sigma''_v$	increment of applied stress
$\sigma_{vo}, \sigma'_{vo}, \sigma''_{vo}$	vertical overburden stress in terms of total, effective, or net normal stress
τ	shear stress
ϕ	osmotic coefficient
ϕ	angle of internal friction
ϕ_p	peak angle of internal friction
ϕ_r	residual angle of internal friction
χ	chi parameter
ψ	total suction
ψ_{ae}	air entry soil suction
$\psi_m = (u_a - u_w)$	matric suction
ψ_o	osmotic suction
ψ_r	residual suction

List of Abbreviations

APEX	analysis of piers in expansive soils
ANN	artificial neural network
CEAc = CEC/clay content	cation exchange capacity activity
CEC	cation exchange capacity
CNS	cohesive non-swelling
COLE	coefficient of linear extensibility
CS	consolidation-swell
CV	constant volume
DTA	differential thermal analysis
EI	expansion index
ET	evapotranspiration
FSI	free swell index
LE	linear extensibility
LMO	lime modification optimum
PVC	potential volume change
PVC	polyvinyl chloride
PVR	potential vertical rise
SAMC	standard absorption moisture content
SI	shrinkage index
SL	shrinkage limit
SSA	specific surface area
SWCC	soil water characteristic curve
SWCR	soil water characteristic relationship
TP	total potassium
UV	ultraviolet
XRD	X-ray diffraction

1

Introduction

The design of foundations for structures constructed on expansive soils is a major challenge for geotechnical engineers practicing in areas where such soils are prevalent. The forces exerted by expansive soils and the movements that they cause to even heavily loaded structures can be well in excess of those experienced by ordinary soils. Also, the costs associated with development of expansive soil sites are much higher than those for nonexpansive soil sites. The site investigation and design phase requires more extensive testing and analyses, and the construction phase requires more inspection and attention to detail. Special considerations must be addressed during their occupancy with regard to the maintenance of facilities constructed on expansive soils. Furthermore, the cost to repair the problems caused by expansive soils may be prohibitive. There are many examples where repair costs exceed the original cost of construction.

The nature of expansive soils and the magnitude of costs associated with shortcomings in design, construction, and operation are such that there exists little margin for error in any phase of a project. In that regard, the following quote is appropriate (Krazynski 1979):

> To come even remotely close to a satisfactory situation, trained and experienced professional geotechnical engineers must be retained to evaluate soil conditions. The simple truth is that it costs more to build on expansive soils and part of the cost is for the professional skill and judgment needed. Experience also clearly indicates that the cost of repairs is very much higher than the cost of a proper initial design, and the results are much less satisfactory.

In the initial phases of a project, the owner or developer is faced with costs that may be significantly higher than initially estimated. They generally are intolerant of shortcomings and demand that the foundations be designed and constructed such that the movements are within tolerable limits. At the same time, they are reluctant to undertake the required additional cost for something that exists below ground and cannot be seen. One large foundation contractor, Hayward Baker, has as part of its motto, "You never see our best work." An important task of the engineer is to convince the client that the additional cost is not merely justified, but is critical. This is especially true for critical structures such as hospitals and public buildings, where failure could have serious consequences.

Expansive soils problems exist on every continent, with the exception perhaps of Antarctica. Expansive soils have been encountered in almost every state and province of the United States and Canada, but they are more troublesome in the western and southwestern areas because areas of low precipitation often tend to be more problematic.

In spite of the fact that expansive soils have been designated a geologic hazard, public awareness is lacking. Few universities offer formal courses relating to geotechnical applications for expansive soils. There is a shortage of continuing education courses in the subject, and research is limited to a relatively small number of institutions. As a result, few practicing foundation engineers have received formal education in this area. It is intended that this book will provide a service for awareness, education, and technical reference in this important area.

1.1 PURPOSE

This book is intended to provide the background and principles necessary for the design of foundations for expansive soils. The nature of expansive soil is described from an engineering perspective to develop an appreciation as to how the microscopic and macroscopic aspects of soil interact to affect expansive behavior. Tools that are necessary to use in the practice of expansive soil foundation design are developed in a fashion that can be easily implemented. The application of these tools to the design of foundations is demonstrated.

An important underlying theme of the book is the ability to predict ground heave and structural movement caused by expansive soil. This is a fundamental part of foundation design. Rigorous calculations of slab heave and potential movement of deep foundations should be a part of every design. Several chapters in this book are devoted to that important subject.

1.2 ORGANIZATION

The organization of this book is designed to first present the fundamental nature of expansive soil and then address the factors that influence expansion. Those tools provide the means for the design of foundations based on the concept of minimizing structural movement. The first eight chapters present the nature of expansive soil and the tools needed to perform the analyses for foundation design. The remaining chapters apply these tools.

Chapter 2 begins with a microscopic view of the molecular structure of the clay particle and the chemistry of the surrounding water. The concept of a clay micelle is introduced and used to explain the nature of expansive soil and the formation of an expansive soil deposit. That concept is extended to show the manner in which macroscopic factors such as density and water content influence the expansion potential. The distribution of expansive soil throughout the world and representative expansive soil profiles are discussed.

Chapter 3 concentrates on the factors of a site investigation for an expansive soil site that may not be included in the investigation of a non-expansive soil site. Chapter 4 is devoted to a discussion of soil suction and its role in defining the state of stress. Soil suction is an important parameter that relates to negative pore water pressure and is a major factor influencing the behavior of expansive soil. The state of stress and the stress state variables for unsaturated soils are presented in chapter 5. The nature in which they relate to the classical effective stress concept for saturated soil is discussed, and important constitutive relationships for expansive soils are presented.

Chapter 6 is devoted to the oedometer test. This is the principal method for measuring the expansion potential of a soil and is used extensively in predicting heave. The migration of water through soil is presented in chapter 7. Methods of analysis are presented for the determination of water content profiles to be used in computation of heave and the design of foundations. Chapter 8 discusses methods for computing predicted heave. Two basic methods of predicting heave are presented. One method is based on the application of oedometer test results and the other uses measurements of soil suction.

Chapter 9 introduces the design of foundations and other structural elements for expansive soil sites. Shrink-swell of soils is also considered. This chapter presents general considerations for foundation design and discusses those factors that are unique to expansive soils.

One approach to foundation design on expansive soil is to mitigate their effects by treating the soil or by controlling the water content regime around the structure. Methods of accomplishing those measures are presented in chapter 10. Chapters 11 and 12 present methods of design for shallow and deep foundations to accommodate the forces and movement of expansive soils. Centered on those designs is the computation of predicted foundation movement. This is the fundamental parameter that must guide a successful design. A discussion of several foundation repair options is also provided in these chapters to guide the reader to a successful rehabilitation of distressed structures.

A part of the foundation design must consider the floors and slabs and their interaction with foundation elements. This is addressed in chapter 13. The use of slab-on-grade floors is discouraged in lieu of structural floor systems. Consideration is given to the effect of soil heave on exterior flatwork. Again, repair options for slab-on-grade floors are also provided at the end of this chapter.

Finally, chapter 14 addresses lateral loads that are exerted on foundations and retaining walls by expansive soils. Lateral earth pressure from expansive backfill is generally of such a magnitude as to preclude the use of expansive soil as backfill.

The concepts and design methods that are presented in the previous chapters are demonstrated by several examples that are provided to help guide the reader through the calculations. Case studies of actual sites investigated by the authors have been presented as well. The case studies show applications of the principles presented in this book, not just from a study of the failure to properly apply expansive soil theory, but also from a research perspective that has been gained during development of many of the design methods provided herein.

This book presents design methods in both English and SI units. SI units are presented within the parentheses shown immediately after the English units. There are two exceptions to this. Chapter 4 on soil suction uses SI units primarily. Because most of the data that are used in the examples have been collected from projects in the United States, examples and case studies use only English units.

1.3 TERMINOLOGY

Many of the terms used in geotechnical practice with expansive soils are either new or have been used in different contexts by different engineers around the world. The following definitions and discussion explain the different terms used in this book. It is suggested that the engineering community adopt this terminology.

Expansive soil is generally defined as any soil or rock material that has a potential to increase in volume under increasing water content. In some cases, it is necessary to present a quantitative description of expansive soil. In those cases, expansive soil is described in terms of the following parameters: (1) the percent swell that a soil exhibits when inundated under a prescribed vertical stress and (2) the swelling pressure of the soil. These parameters are typically measured by one-dimensional consolidation-swell tests or constant volume tests conducted in the laboratory on representative samples of soil.

Soil suction is the magnitude of the tensile stress in the pore water of an unsaturated soil. It consists of two components, matric suction and osmotic suction. For purposes of general engineering practice, changes in matric suction are most important.

Oedometer test is a one-dimensional test in which a soil sample is confined laterally and subjected to vertical stress while being wetted. In the consolidation-swell (CS) test, the sample is wetted under a prescribed inundation stress and allowed to swell. In the constant volume (CV) test the sample is restrained from swelling while it is being wetted.

Consolidation-swell swelling pressure, σ''_{cs}, is the load required to compress the soil to its original thickness after it has been inundated and allowed to swell in a consolidation-swell (CS) test.

Constant volume swelling pressure, σ''_{cv}, is the load required to prevent swell, and thus maintain a constant volume of the soil after it has been inundated in a constant volume (CV) test.

Design life of a foundation is the useful lifespan of a foundation assumed for design purposes. Different elements of a structure are designed for different design lives. For example, the design life of a shingle roof is much less than that of a foundation. The design life for a foundation generally ranges from about 50 years to 200 years. A typical design life for a residential structure is 100 years, but for commercial structures it could be less.

Zone of seasonal moisture fluctuation, z_s, is the depth to which the water content fluctuates in the soil due to changes in climatic conditions at the ground surface. The zone of seasonal moisture fluctuation does not take into consideration the effect of water being introduced to the soil profile by other sources such as landscape irrigation or water introduced below the ground surface by leaking pipes or drains, or off-site sources of groundwater.

Active zone, z_A, is the zone of soil that contributes to heave due to soil expansion at any particular point in time (Nelson, Overton, and Durkee 2001). Thus, the depth of the active zone can vary with time. Historically, this zone has been considered to be the depth to which climatic changes can influence the change in water content in the soil (i.e., the "zone of seasonal moisture fluctuation"). That concept, however, is correct only if external influences such as grading, surface drainage, or irrigation are not present.

Zone (depth) of wetting, z_w, is the zone of soil in which water contents have increased due to the introduction of water from external sources, or due to capillarity after the elimination of evapotranspiration. The zone of wetting was originally defined in Nelson, Overton, and Durkee (2001). In the case where a wetting front is moving downward from the surface, the depth of this zone is the *zone of wetting*. The soil above the wetting front represents the zone in which heave is taking place, and hence, represents a time-varying active zone.

Depth of potential heave, z_p, is the greatest depth to which the overburden vertical stress equals or exceeds the swelling pressure of the soil. This represents the maximum depth to which the soil can heave, and, thus, is the deepest depth possible for the active zone.

Design active zone, z_{AD}, is the zone of soil that is expected to become wetted by the end of the design life. This is the active zone for which the foundation is to be designed.

Free-field heave is the amount of heave that the ground surface will experience due to wetting of the soils with no surface load applied. The surface load applied by slabs-on-grade and pavements is very small relative to the swelling pressure. Therefore, the heave of slabs and pavements is essentially the same as the free-field heave. The term *free-field heave* has been used for many years by various researchers and practitioners worldwide in the fields of expansive soil and frozen soil (O'Neill 1988; Rajani and Morgenstern 1992, 1993, and 1994; Ferregut and Picornell 1994; Venkataramana 2003; Miao, Wang, and Cui 2010; Ismail and Shahin 2011 and 2012). This term is also commonly called *free-field soil movement* (Poulos 1989; Ong 2004; Ong, Leung, and Chow 2006; Wang, Vasquez, and Reese 2008; and Ti et al. 2009). The free-field heave is also the heave to be expected for infrastructure, such as slabs-on-grade and pavements, where very low surface loads are applied.

Current heave is the heave that has occurred at the time being considered. This is the heave that has been produced by the current degree of wetting that has occurred in the active zone, taking into account whether the soil has been fully wetted or partially wetted.

Ultimate heave is the maximum amount of heave that a soil profile can exhibit if it were to become fully wetted throughout the entire zone above the depth of potential heave.

Future maximum heave is the amount of future heave expected to occur since the time of investigation. Generally this calculation assumes that the entire depth of potential heave will become fully wetted.

Design heave is the amount of heave that will be experienced during the design life of the foundation. Design heave is calculated based on the change in the subsurface water content profile in the design active zone. Water migration modeling can be used to predict the final water content profile at the end of the design life of the foundation. Calculations of the design free-field heave should also take into account the degree of wetting in the design active zone. It is also the amount of heave that the foundation must be designed to tolerate within its design life.

References

Ferregut, C., and M. Picornell. 1994. "Calibration of Safety Factors for the Design of Piers in Expansive Soils." In *Risk and Reliability in Ground Engineering*, Institution of Civil Engineers, edited by B. O. Skipp, 277–290, London: Telford.

Ismail, M. A., and M. A. Shahin. 2011. "Finite Element Modeling of Innovative Shallow Foundation System for Reactive Soils." *International Journal of Geomat* 1(1): 78–82.

———. 2012. "Numerical Modeling of Granular Pile-Anchor Foundations (GPAF) in Reactive Soils." *International Journal of Geotechnical Engineering* 6, 149–156.

Krazynski, L. M. 1979. "Site Investigation. The Design and Construction of Residential Slabs-on-Ground: State-of-the-Art." *Proceedings of Workshop, Building Research Advisory Board (BRAB), Commission of Sociotechnical Systems*, Washington, DC, 133–148.

Miao, L., F. Wang, and Y. Cui. 2010. "Improvement and Controlling Deformation of the Expansive Soil Ground." *Proceedings of the 5th International Conference on Unsaturated Soils*, Barcelona, Spain, 1321–1324.

Nelson, J. D., D. D. Overton, and D. B. Durkee. 2001. "Depth of Wetting and the Active Zone." *Expansive Clay Soils and Vegetative Influence on Shallow Foundations*, ASCE, Houston, TX, 95–109.

O'Neill, M. W. 1988. "Adaptive Model for Drilled Shafts in Expansive Clay." In *Special Topics in Foundations*, Geotechnical Special Publication No. 16, edited by B. M. Doas, ASCE, 1-20.

Ong, D. E. L. 2004. "Pile Behaviour Subject to Excavation-Induced Soil Movement in Clay." PhD dissertation, National University of Singapore, Singapore.

Ong, D. E. L., C. E. Leung, and Y. K. Chow. 2006. "Pile Behavior Due to Excavation-Induced Soil Movement in Clay. I: Stable Wall. " *Journal of Geotechnical and Geoenvironmental Engineering*, ASCE 132(1): 36–44.

Poulos, H. G. 1989. "Program PIES—Axial Response of Piles in Expansive Soils, Users' Guide." Centre for Geotechnical Research, University of Sydney, Australia.

Rajani, B. B., and N. R. Morgenstern. 1992. "Behavior of a Semi-Infinite Beam in a Creeping Medium." *Canadian Geotechnical Journal* 29(5): 779–788.

————. 1993. "Uplift of Model Steel Pipelines Embedded in Polycrystalline Ice." *Canadian Geotechnical Journal* 30(3): 441–454.

————. 1994. "Comparison of Predicted and Observed Responses of Pipeline to Differential Frost Heave." *Canadian Geotechnical Journal* 31(6): 803–816.

Ti, K. S., B. B. K. Huat, J. Noorzaei, M. S. Jaafar, and G. S. Sew. 2009. "Modeling of Passive Piles—An Overview." *Electronic Journal of Geotechnical Engineering*, 14.

Venkataramana, K. 2003. "Building on Expansive Clays with Special Reference to Trinidad." *West Indian Journal of Engineering* 25(2): 43–53.

Wang, S. T., L. Vasquez, and L. C. Reese. 2008. "Study of the Behavior of Pile Groups in Liquefied Soils." *The 14th World Conference on Earthquake Engineering*, Beijing, China.

2

Nature of Expansive Soils

To develop at least a conceptual understanding of expansive soils, it is important to take into consideration microscale and macroscale factors. Microscale factors include mineralogy, pore fluid chemistry, and soil structure. These microscale factors influence macroscale physical factors such as plasticity, density, and water content to dictate the engineering behavior of a soil.

This chapter will discuss the general physicochemical factors that comprise the microscale aspects and how they influence soil behavior and expansion potential. The manner in which macroscale effects influence expansion is discussed next. It is shown how those factors are used to identify expansive soils. Some of the identification methods just identify the presence of potentially expansive minerals in a naturally occurring soil, whereas others quantify the expansion potential. The chapter finishes with a discussion of the geographic distribution of expansive soils around the world and presents characteristics of some expansive soil profiles.

2.1 MICROSCALE ASPECTS OF EXPANSIVE SOIL BEHAVIOR

The microscale aspects of expansive soil consider the mineral composition of the clay particles, and the manner in which they react with the chemistry of the soil water. This section begins with a description of the mineral that forms the solid clay particle. It then describes the cations that are attracted to the clay particle by electrical forces and the interaction of the cations with water of hydration and other surrounding water molecules. The solid clay particle, the cations, and the bound water form a unit termed a *micelle,* which is discussed in detail in this chapter. The nature of the micelles of different minerals influence the soil behavior and its expansive characteristics.

2.1.1   The Clay Particle

2.1.1.1   Mineral Composition

Chemically, clay minerals are silicates of aluminum and/or iron and magnesium (Grim 1959). Most of the clay minerals have sheet or layered structures and can have various shapes. A typical clay particle of expansive soil consists of a microscopic platelet having negative electrical charges on its flat surfaces and positive electrical charges on its edges. The mineral composition of clay can be depicted as being made up of combinations of two simple structural units. In the description of the minerals that is presented in this chapter, the structural units are represented by conceptual *building blocks*. A detailed description of these units is presented in Mitchell and Soga (2005).

The two basic elemental units of the building blocks are the silicon tetrahedron and the alumino-magnesium octahedron. They are depicted schematically in Figure 2.1. The silicon tetrahedron is made up of silicon and oxygen atoms. Because the valence of silicon is 4+, it can bond with negatively charged ions such as oxygen (O^{2-}) or hydroxyl (OH^-), as shown in Figure 2.1a. The relative sizes of the silicon and oxygen atoms cause this structural unit to assume the shape of a tetrahedron. The alumino-magnesium octahedron consists of aluminum or magnesium atoms surrounded by hydroxyls, as shown in Figure 2.1d. These atoms are arranged such that they can be thought of as forming an octahedral shape.

In the silicon tetrahedron shown in Figure 2.1a, the oxygen atoms each have an unsatisfied chemical bond. The oxygen atoms at the base of a tetrahedron are shared with adjacent tetrahedra, and the resulting arrangement of tetrahedra forms sheets, as shown in Figure 2.1b. Sharing of oxygen atoms between the tetrahedra satisfies the oxygen atoms at the bases of the tetrahedra, but the oxygen atoms at the apexes still have unsatisfied bonds, as shown in Figure 2.1b. Therefore, the upper face of the silica sheet is capable of forming chemical bonds with positively charged cations.

The octahedral units share hydroxyls to form a sheet structure, as shown in Figure 2.1e. The arrangement of the octahedral units is such that the hydroxyls in the sheet structure do not have unsatisfied chemical bonds. The central cation in the octahedral sheet can vary as will be discussed in Section 2.1.1.3.

Figures 2.1c and 2.1f show schematic symbols that represent the building blocks that are used to depict the crystalline structure of the different clay minerals. By varying the manner in which these two building blocks are arranged, a variety of different clay minerals can be created. A number of different minerals are depicted in Mitchell and

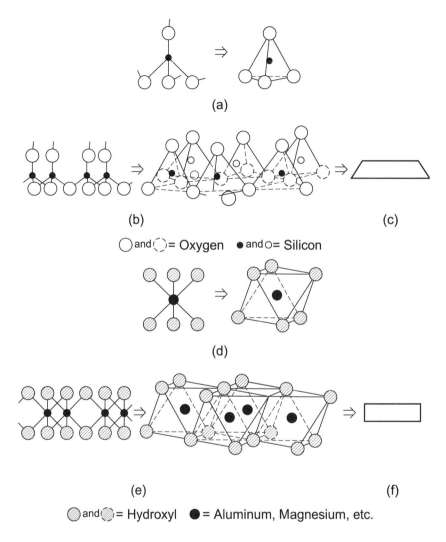

FIGURE 2.1. Atomic structure of silicon tetrahedra and alumino-magnesium octa-hedra: (a) silicon tetrahedron; (b) silica sheet; (c) symbolic structure for silica sheet; (d) alumino-magnesium octahedron; (e) octahedral sheet; (f) symbolic structure for octa-hedral sheet (after Lambe and Whitman, 1969; Mitchell and Soga, 2005).

Soga (2005). The various minerals are classified into groups according to the stacking sequence of the sheets. For purposes of this book, it is sufficient to consider only three basic minerals:

1. Kaolinite
2. Illite
3. Montmorillonite

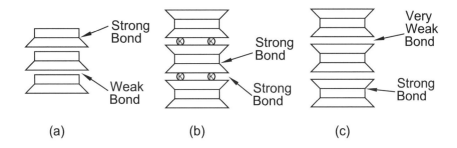

FIGURE 2.2. Schematic diagrams of the structure: (a) kaolinite; (b) illite; (c) montmorillonite.

Figure 2.2 shows schematic diagrams of idealized structures of these three minerals. The bonding between the different building blocks plays a very important part in the behavior of the different minerals as will be discussed in Section 2.1.1.2.

The term *bentonite* is often used in reference to expansive soils. This term refers to clays that are rich in montmorillonite. Bentonite is a highly plastic, swelling clay material containing primarily montmorillonite. It is mined commercially and is used for a variety of purposes such as drilling fluids, slurry trenches, cosmetics, paint thickeners, and many others. Not all expansive soils are bentonite, but they are frequently referred to simply as bentonite, usually by nonengineer laymen.

2.1.1.2   Interlayer Bonding

The size of the mineral particles in a soil is influenced by the nature of the bonds between the silica and octahedral sheets. The structure of the kaolinite as depicted in Figure 2.2 shows that there is a strong bond between the top of the silica sheet and the octahedral sheet. This is because of the arrangement of the atoms and the fact that the oxygen atoms at the top of the silica tetrahedra can replace some of the hydroxyls in the octahedral sheets. Thus, there is a strong chemical bond between the silica sheet and the octahedral sheet (Grim 1959).

In Figure 2.2a, the chemical bonds at the bottom of the silica sheet are satisfied, and consequently, the bond between the bottom of the silica sheet and the octahedral sheet is formed primarily by weaker hydrogen bonds. That is labeled a weak bond. Thus, when cleavage of the mineral occurs, the sheets will separate at the weak bond. Nevertheless, the bonds are sufficiently strong that the clay particles can sustain a number of building blocks in each particular clay particle. As a result, the kaolinite particles tend to be relatively large with a lateral dimension as great as 1 micron (μm) or more, and a thickness of $^1/_3$ to $^1/_{10}$ of the lateral dimension (Lambe and Whitman 1969).

Chemical bonds form between atoms and are effective at small distances. At larger distances where chemical bonds are not effective, van der Waals bonding may be effective. They are not nearly as strong as chemical bonds. There are several types of forces that contribute to van der Waals bonding, each of which is of different strength (Companion 1964).

For the montmorillonite, the bond between the bases of the two silica sheets is formed by weaker van der Waals forces, and in Figure 2.2c it is labeled as "very weak," in contrast to just "weak." As a result, montmorillonite particles may be only one or two sets of building blocks thick. Thus, the thickness of the clay particles may be as small as about 10 Ångström[1] (Lambe 1958; Grim 1968).

The structure of the illite particle is basically the same as that of montmorillonite. However, in illite the bond between the bases of the silica sheets is formed by potassium cations that are shared by adjacent sheets. The size of the potassium ions is such that they fit into the spaces in the bases of the silica sheet formed by the arrangement of the tetrahedra. Sharing of the potassium ions between the silica sheets produces a strong bond (Grim 1959). Thus, the kaolinite and illite minerals exhibit much less expansive behavior than does the montmorillonite.

Figure 2.3 shows scanning electron micrographs of kaolinite, illite, and montmorillonite.

2.1.1.3 *Isomorphous Substitution and Surface Charges*

When the octahedral sheets are formed in nature, some of the central positions that would normally be occupied by Al^{3+} atoms are occupied by other elements instead. This is a process termed *isomorphous substitution*. *Isomorphous* means "same form" and refers to the substitution of one kind of atom for another in the crystal lattice. The type of central cation in the octahedral units can vary widely. Commonly, it is either aluminum (Al^{3+}) or magnesium (Mg^{2+}), but it could be other cations such as iron (Fe^{2+}, Fe^{3+}), manganese (Mn^{2+}), or others. If the predominant cation in the octahedron is aluminum, the sheet is referred to as a gibbsite sheet. If it is magnesium, the sheet is referred to as a brucite sheet. This discussion of expansive minerals will consider the sheets to

---

[1]One Ångström = 10^{-10} m. The Ångström unit (Å) is not in conformity with SI units, but is in common usage for atomic dimensions. The unit is named after the Swedish physicist Anders Jonas Ångström (1818–1874). The letter Å is pronounced "oh" as in "boat."

(a)

(b)

FIGURE 2.3. Scanning electron micrographs: (a) kaolinite; (b) illite; (c) montmorillonite (reproduced by permission of OMNI/Weatherford Laboratories).

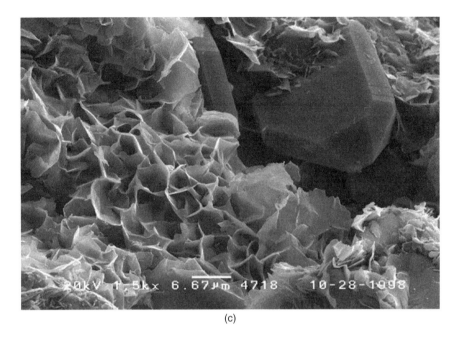

(c)

FIGURE 2.3. (*continued*)

be gibbsite sheets because that is the predominant building block of montmorillonite, which is one of the most expansive clay minerals.

The substitution of an atom having a lower valence than Al^{3+} has two effects. One effect is that it results in a charge deficiency at that location. The second is that it distorts the crystal structure since it is not the same size as the atom it is replacing. The end result is that the face of the structural unit, and hence the clay particle, has a net negative charge. Other factors related to the atoms in the crystal lattice also contribute to the negative charge on the clay particle, but isomorphous substitution is the most important. A more detailed discussion of clay mineralogy is presented by Grim (1959), Mitchell and Soga (2005), or others.

2.1.2 Adsorbed Cations and Cation Hydration

The negative charges on the faces of the clay particles are balanced by positively charged cations. The type of cations in the environment will influence the nature of the clay soil. For purposes of this discussion, it is convenient to consider sodium (Na^+) as the cation that is present, since it is monovalent, and its presence usually results in the greatest expansion potential.

The cations in a soil can exist in different stages. For a completely dry soil, the cations would be at a very low stage of hydration. As water becomes available to the soil, it will bond with the sodium cations by the process of hydration. The mechanism of hydration of cations is outside the scope of this book. Frank and Wen (1957) present a well-accepted concept of hydration. They depicted a simple model of a hydrated ion as shown in Figure 2.4. Here, region A is termed the *region of immobilization*. In that region, the water molecules are strongly held in the field of the ion and are immobile. In region B, the water molecules have less structure but are held to the ion. Region C contains water with normal structure but the water molecules are polarized by the weak ionic field (Frank and Wen 1957). The water held in regions A, B, and C is termed *water of hydration*. For purposes of considering the role of ion hydration in soil expansion, the water in region A can be assumed to be "fixed" and forming a permanent part of the ion. Region C can be considered

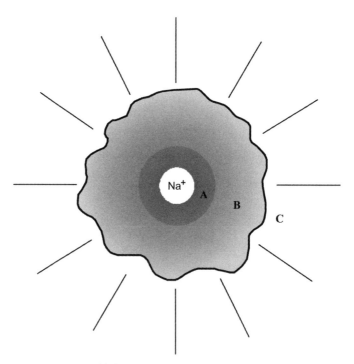

A) Region of Immobilization
B) Region of Structure Breaking
C) Structurally "Normal" Water

FIGURE 2.4. Ion water interaction (modified from Frank and Wen 1957).

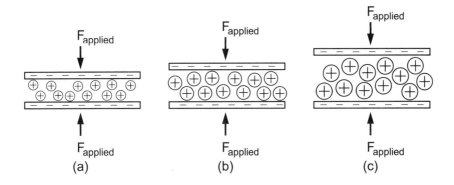

FIGURE 2.5. Role of cation hydration on soil expansion: (a) low hydration; (b) partial hydration; (c) full hydration.

as part of the water that is adsorbed to the clay particle and that can be removed easily by air drying. Region B is considered to be that part of the water of hydration that is removed by heat or desiccation of the soil.

A conceptual model to depict the role of cation hydration on soil expansion is shown in Figure 2.5. In Figure 2.5a, two parallel clay particles are separated by cations that have a very low level of hydration. These would be cations that have water in region A and perhaps some of region B. The energy of hydration is quite large such that if water is added to the soil, it is drawn into region B. Figure 2.5b shows the cations as having been hydrated to the stage where region B is completed. As more water is added, it goes into region C, as shown in Figure 2.5c.

Lambe and Whitman (1969) give the radii of unhydrated and hydrated sodium ions as 0.98 Å and 7.8 Å, respectively. Thus, in going from an unhydrated state to a fully hydrated state the ion grows in size by more than sevenfold. The adsorbed cations shown in Figure 2.5a are not completely dehydrated, and the difference between Figures 2.5a and 2.5c does not represent a sevenfold change in cation size. However, if the hydration of the cations were to cause even a 10 percent increase in the diameter of cations, this would represent a 10 percent change in spacing between particles, which would correlate to a significant amount of soil expansion.

2.1.3 The Clay Micelle

The mineral particle and the adsorbed cations act together as a single unit. In a system comprising soil and water, a higher concentration of cations at one location over that in another location will cause the migration of water molecules toward the location of higher concentration. This is a phenomenon termed *osmosis,* and will be discussed in detail in chapter 4. Some of the water close to the clay particle will exist as water of hydration and some will be that water that is drawn towards the soil particles and held there by surface forces. Lambe (1958) terms this water *adsorbed water.* For purposes of this discussion, water that is attracted to the clay particles because of higher cation concentrations will be referred to as *osmotic water.* As already noted, the transition between the hydration water in region C and the osmotic water is probably not a precise, well-defined boundary.

To discuss the nature of soil, it is necessary to consider three components—the mineral, the cations, and the associated water. To consider these three components together as a unit, it is convenient to introduce the concept of a *clay micelle.* A clay micelle is depicted in Figure 2.6. For clay minerals this can be thought of as a negatively charged inner mineral core surrounded by positively charged cations that neutralize the mineral charge. The clay particle and the cations, along with water of hydration and osmotic water that is held closely to the inner mineral core, form the micelle (Lambe 1958). The following discussion will describe the nature of the clay micelle.

Without the presence of the adsorbed cations, the electrical charges on the faces of the clay particles give rise to repulsive forces between

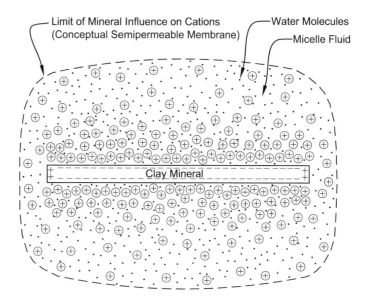

FIGURE 2.6. A clay micelle showing the concentration of cations near the surface of a clay particle.

individual particles. The cations in the micelle that are adsorbed tightly on the surface of the clay mineral balance these electrical surface forces. The surface forces are greatest near the surface of the clay particle and decrease with distance. The cations in solution that are closest to the clay particle will be constrained more tightly from migrating away by the surface forces. This results in a gradient in the cation concentration that is highest near the surface and decreases with distance away from the surfaces as shown in Figure 2.6. At a sufficiently large distance from the surface, the effect of the surface forces is negligible, and the cation concentration is the same as that in the free water in the soil. That distance is shown by the outer dashed line in Figure 2.6. The dashed line in Figure 2.6 does not represent a precise and definite distance but, rather, a decrease in influence from the particle surface. According to that model, the micelle can be defined as a discrete unit comprising a single clay particle surrounded by an aqueous solution. The aqueous solution within the area of influence of the micelle will be termed the *micelle fluid*.

The thickness of the water and cations in the micelle (i.e., the micelle fluid) is influenced primarily by the nature of the adsorbed cations and the electrical surface charges on the particles. For two different minerals, both of which have the same type of adsorbed cations, the thickness of the micelle fluid will be influenced mainly by the outermost few structural units, and will not be influenced greatly by the overall thickness of the particles. Micelles for montmorillonite and kaolinite are depicted in Figure 2.7a. Lambe (1958) shows the thickness of the micelle fluid as being 200 Å for montmorillonite and 400 Å for kaolinite. Thus, given that kaolinite particles are about 100 times thicker than montmorillonite particles, the ratio of micelle fluid to mineral thickness is about 50 times greater in montmorillonite than in kaolinite.

2.1.4 Crystalline and Osmotic Expansion

The model of expansion potential presented in Figure 2.5 is related to hydration of the adsorbed cations and the osmotic attraction of water into the micelle. Experiments by Norrish (1954) suggested that swelling of montmorillonite takes place in two distinct ways. At close spacings (<22 Å), the expansion was dependent on the exchangeable cation and the amount of water taken up was directly related to the hydration energy of the cation. At a spacing greater than 35 Å, the montmorillonite was thought to develop the rest of the micelle fluid and the swelling was essentially osmotic.

This concept helps to explain the nature of swelling in expansive soil. Because the energy of hydration is high, the first water that is introduced

KAOLINITE   MONTMORILLONITE

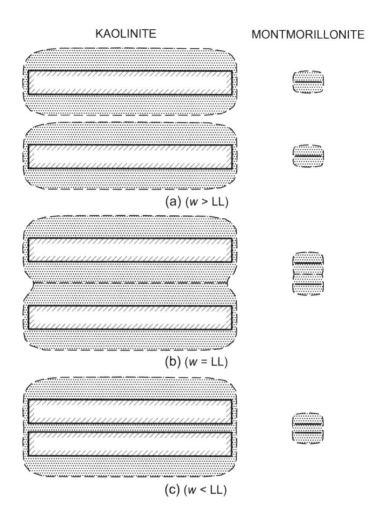

FIGURE 2.7. Kaolinite and montmorillonite micelles at water content: (a) above; (b) at; (c) below the liquid limit.

into a dry soil goes into hydrating the cations. Subsequently added water goes into satisfying the osmotic forces. Thus, in simple terms the initial phase of swelling is "crystalline" and is due to hydration of the cations as previously described (Slade, Quirk, and Norrish 1991). After that phase, the concentration of cations between the soil particles causes water to be drawn into the interparticle spacing due to osmotic forces. This second phase of swelling is termed *osmotic*.

It is unlikely that the two phases of swelling are clearly distinct, and there likely exists a transition zone between the crystalline and osmotic phases. It can be postulated that the development of regions A and B as shown in Figure 2.4 relates to the crystalline phase, and development of region C relates to both crystalline and osmotic phases.

Drying of a wetted soil can cause removal of much of the osmotic water and some of the crystalline water. This will cause the micelles to decrease in size, and cause the particles to move closer together. Thus, expansive soils may also decrease in volume when the water content decreases. For this reason, those soils are frequently referred to as *shrink-swell* soils. The shrinkage phase of a soil in a natural environment most likely involves removal of the osmotic water first and perhaps some of the water in region C. Desiccation of the soil by heat such as sun-baking or drying in the laboratory begins to remove more of the water of hydration.

2.1.5   Effect of Mineralogy on Plasticity of Soil

As a consequence of the difference between micelles as depicted in Figure 2.7, the more expansive minerals tend to have higher plasticity and a higher liquid limit. A conceptual model to explain the differences in liquid limit for different clay types is shown in Figure 2.7. In a saturated soil having water content greater than the liquid limit, there would be no interaction between the micelles, as shown in Figure 2.7a. In this condition, the micelles would act as a colloid and the suspensions would have no shear strength. The point where the micelles just begin to interact, shown in Figure 2.7b, defines the water content at which the soil just begins to develop some shear strength due to interaction between the particles. This is the point at which the water content of the soil is equal to the liquid limit. At lower water contents, as shown in Figure 2.7c, the micelles intersect and the soil particles share cations in the micelle. At this point, the soil will develop shear strength. At values of water content that are normally encountered in practice, all clay particles are sharing micelles with adjacent particles (Lambe 1958).

In Figure 2.7b, which depicts a soil having a water content equal to the liquid limit, it is evident that the water of the montmorillonite makes up a much greater percentage of two micelles than does that of the kaolinite. Thus, the water content, and hence the liquid limit, of the montmorillonite is much greater than that of the kaolinite. This demonstrates why the liquid limit of montmorillonite is so much greater than that of kaolinite.

The previous discussion has shown that the mechanism of swell takes place primarily in the spacing between particles (i.e., in the micelle fluid). In the montmorillonite, a greater percentage of the overall soil is occupied by the micelle fluid and the montmorillonite exhibits more expansion potential than the kaolinite. Thus, the montmorillonite, having a higher micelle fluid content relative to the solid particle, exhibits both a higher liquid limit and a higher expansion potential.

2.1.6   Effect of Mineralogy on Expansion Potential

The differences between the micelles shown in Figure 2.7 demonstrate the effect that mineralogy will have on expansion potential. If the physicochemical effects previously discussed were to cause a change in the size of the micelle fluid by a given amount, this would cause a much larger change in the relative amount of fluid for the montmorillonite than it would for the kaolinite. The difference in expansion potential between kaolinite and montmorillonite is demonstrated in Figure 2.8. That figure shows two compacted specimens of soil, before and after wetting. One specimen is pure kaolinite and the other pure montmorillonite. The samples were both wetted and allowed to swell. It is seen that the montmorillonite swelled by about 30 percent, whereas the kaolinite swelled very little.

2.1.7   Effect of Type of Cation on Expansion Potential

The discussion so far has related primarily to soil having monovalent adsorbed cations such as sodium (Na^+), as shown in Figure 2.6. Such an ideal model seldom exists in nature and it would be expected that both monovalent and multivalent cations would be present. The expansive nature of soil having an abundance of multivalent cations is much less than that of a soil having only monovalent cations. For example, if the sodium cations in the micelle shown in Figure 2.6 are replaced by divalent calcium cations (Ca^{2+}), there would need to be only half as many cations present to balance the negative charges on the clay particles, and the size of the micelle would shrink. Similarly, in the schematic diagram of soil expansion shown in Figure 2.5, if the adsorbed cations were divalent, there would be half as many present, and the expansion potential would be less. In addition, the bonding strength of the divalent cations can be greater than that of the monovalent cations. Benson and Meer (2009) showed that clays having an abundance of monovalent cations had a much higher swell index than those with an abundance of divalent cations.

This will also apply to the shrinkage of a soil. As would be expected, the shrink-swell potential will also be greater for clays with monovalent

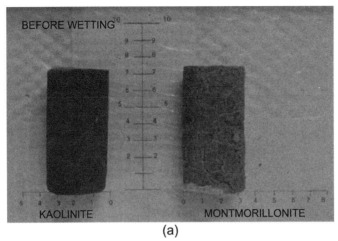

(a)

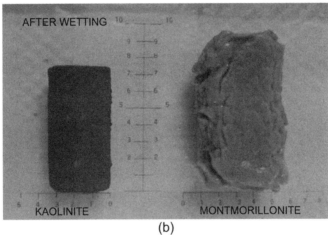

(b)

FIGURE 2.8. Compacted samples of kaolinite and montmorillonite: (a) before wetting; (b) after wetting (photographs taken at Colorado State University, Geotechnical Engineering Laboratory, scale in cm).

cations than for those with divalent cations. The amount of shrinkage may not equal the amount of swelling, depending on the nature of the cations and the hydration energy of the cations. Thus, there is a certain amount of hysteresis in the swelling and shrinking process. The different mechanisms of crystalline swelling and osmotic swelling and the nature of the associated volume change can have a pronounced effect on the performance of clay liners, which depend on the swelling characteristics of the clay to maintain a low permeability (Benson and Meer 2009).

2.2   MACROSCALE ASPECTS OF EXPANSIVE SOIL BEHAVIOR

This section discusses the macroscale properties, such as density and water content and how they are influenced by the interaction of the micelles and the microscale aspects that were already discussed.

2.2.1   Development of Natural Soil Deposits

The nature of a clay soil deposit depends on the nature of the millions of micelles that compose the soil. To envisage the development of expansion potential in a natural soil deposit, it is convenient to consider the formation of the soil. Many highly expansive soils were deposited millions of years ago under inland seas and lakes. Figure 2.9 illustrates the formation of a sedimentary deposit at the floor of a body of water. Sediment is initially introduced in the form of muddy water. Because this suspension has no shear strength, it can be thought of as a soil that has a water content that is well above the liquid limit. Water content is shown as a function of depth in Figure 2.9b. In the upper portions of the sediment where the water content is greater than the liquid limit, the shear strength is zero. As more sediment is deposited, the soil settles out and becomes more dense. When the weight of the sediment being deposited causes the soil particles to become close enough so that the water content has decreased to the liquid limit, the sediment begins to develop shear strength. At that point, the micelles are close enough to interact

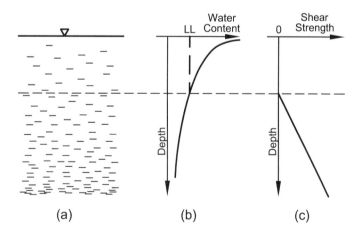

FIGURE 2.9. Sediment deposit beneath a body of water: (a) deposition; (b) water content; (c) shear strength.

with each other and the electrical forces in the micelle layer balance the weight of the material above. As more sediment is deposited, the particles are forced closer together and the soil becomes denser, stiffer, and stronger.

Over geologic time, soil that was deposited in this way can become overconsolidated. This can occur as the water recedes over time and the free water between particles drains out. External loads can be applied by processes such as aeolian, colluvial, or alluvial deposition. Another source of external loading could be applied by glacial or volcanic activity. These external forces cause the micelles to be compressed such that the micelle fluid is shared by neighboring micelles. As time progresses, some of the overburden material can be removed by erosion, melting of glaciers, or other forces. The end result is the formation of highly overconsolidated claystones and clayshales.

The compression of the soil under high loads, together with the mineralogical aspects, contribute to the expansion potential of the soil. If two parallel clay particles are brought into close contact, a combination of different surface forces results in repulsive forces between particles except at very close spacings, where attractive forces may exist. In expansive soil, only the repulsive forces are of concern. The variation of internal stress between particles, σ_{int}, with distance between particles, d, is shown in Figure 2.10. If an external stress, σ_{ext}, is applied, the

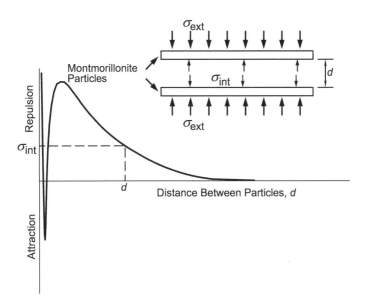

FIGURE 2.10. Internal stress, σ_{int}, between parallel clay particles.

spacing between micelles must decrease, such that the internal surface forces between particles balance the externally applied loads. Under large applied loads such as those applied by glaciers or thousands of feet of overlying soil, the particles become very closely packed.

In a natural deposit, the particles are not aligned in a perfectly parallel fashion. Thus, the edges of some particles may be in contact with the faces of others. Under very high stresses that occur at those contact locations the crystal lattices of adjacent particles can actually be forced together such that they form bonds (Bjerrum 1967). Such bonds are termed *diagenetic bonds*. When the load is removed by erosion or glacial melting, the diagenetic bonds and the cations being shared between micelles prevent the particles from returning to the spacing shown in Figure 2.10 where the internal stress, σ_{int}, would be in equilibrium with the externally applied stress, σ_{ext}. In addition, drying of the soil results in depression of the size of the micelle and dehydration of the adsorbed cations. The result is a soil in which the clay particles are closely packed with partially dehydrated adsorbed cations between the particles. Dehydration of the soil without the externally applied load can also cause similar effects, but the particles may not be as tightly packed as in the case of the highly overconsolidated soil. The result is that strain energy is stored both in the diagenetic bonds and the dehydrated shared cations.

The strain energy stored in the diagenetic bonds, together with the dehydrated cations, act to resist the internal repulsive forces. If changes occur in the environment that cause the cations to be hydrated, the diagenetic bonds are broken, and water is drawn into the micelle by osmotic forces. This causes an imbalance between the internal and external forces. When the internal repulsive forces become greater than the external ones, an increase in the interparticle spacing results, which causes expansion.

2.2.2 Effect of Plasticity on Expansion Potential

The previous discussion has shown that the higher the montmorillonite content in the clay, the more expansive will be the soil. The predominant clay mineral of the soil is reflected in the plasticity. The clay mineral also influences the expansion potential of the clay. Thus, it is to be expected that there is a correlation between the plasticity of a soil and its expansion potential. Figure 2.11 shows zones for different minerals superimposed on a plasticity chart as presented by Holtz, Kovacs, and Sheahan (2011). The zone for montmorillonite clays is located close to the U-line. One would expect, therefore, that the plasticity of expansive

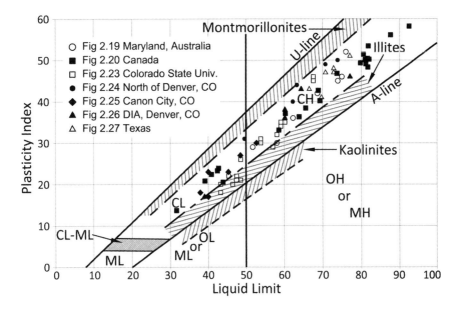

FIGURE 2.11. Plasticity characteristics of clay minerals (shaded mineral areas as identified in Holtz, Kovacs, and Sheahan 2011).

soil would plot close to the zone shown for montmorillonite. This, in fact, is in agreement with actual data. It has been observed that even if the soils are classified as CL soils, the plasticity of the more expansive soils plot close to the U-line. Data from a number of expansive soil sites that will be discussed in Section 2.4 are plotted together on the plasticity chart in Figure 2.11. As one would expect, the values tend to plot well above the A-line, and tend toward the U-line.

2.2.3 Effect of Soil Structure, Water Content, and Density on Expansion Potential

The orientation of the soil particles in the soil mass and the spacing between particles will influence the manner in which the particles interact. Depending on the conditions that existed during deposition, the particles may achieve varying degrees of orientation. Figure 2.12 depicts the particle orientation for flocculated and dispersed soil structures. The depiction of the two different structures shown in Figure 2.12 represents the extremes from fully flocculated to fully dispersed. In most soils, the orientation of particles would be somewhere between those two extremes. To consider the effect of density and soil structure on expansion potential, it is convenient to consider the structures that are shown.

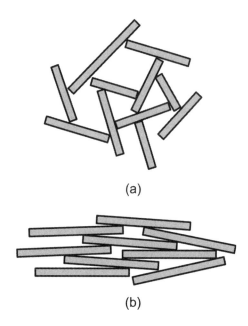

FIGURE 2.12. Sediment structures: (a) "flocculated" or "random" orientation; (b) "dispersed" or "oriented" (modified from Lambe and Whitman 1969).

For a flocculated structure, such as that shown in Figure 2.12a, it is clear that the interaction between micelles is influenced primarily by the contacts between the ends of the particles and the faces of adjacent ones. The spacing between particles is larger in the flocculated structure than in the dispersed structure. It is evident, therefore, that crystalline and osmotic swelling would be less effective in the flocculated structure than for the dispersed structure, which is depicted in Figure 2.12b. For highly overconsolidated clays that have been subjected to high overburden stresses the soil structure would tend more to the dispersed structure.

In addition to particle orientation, the spacing between particles and the hydration states of the cations will influence the expansion potential. Thus, a soil with a high dry density and low initial water content would be expected to exhibit a higher expansion potential than a less dense soil with higher initial water content. Chen (1973 and 1988) performed oedometer tests on samples that had been prepared to the same initial density but different initial water contents. His results are shown in Figure 2.13a. It is clear that the initial water content had a pronounced effect on the percent swell.

Chen (1973 and 1988) also conducted oedometer tests on samples with the same initial water content and different dry density. Those results are shown in Figures 2.13b and 2.13c, respectively. They show

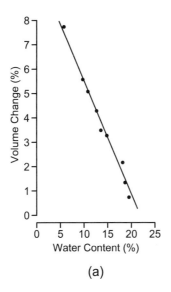

(a)

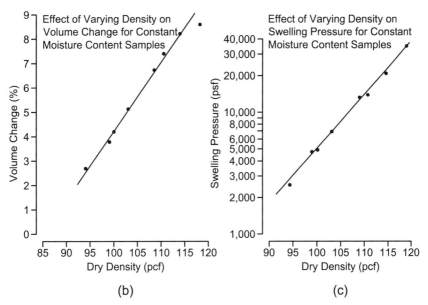

(b)   (c)

FIGURE 2.13. Effect of initial water content and dry density on expansion potential: (a) water content versus volume change; (b) dry density versus volume change; (c) dry density versus swelling pressure (Chen 1973).

quite clearly that the initial density had a pronounced effect on both the percent swell and the swelling pressure. Thus, the drier and denser the initial state of the soil, the greater the expansion potential.

2.3 IDENTIFICATION OF EXPANSIVE SOILS

The nature of the soil and the factors previously discussed can be employed to identify whether a soil has the potential to be expansive. Such identification is important during the reconnaissance and preliminary stages of a site investigation to indicate appropriate sampling and testing methods to be used. The identification methods used to identify the swell potential of expansive soils can generally be grouped into two categories. The first category mainly involves measurement of physical properties of soils, such as Atterberg limits, free swell, and potential volume change. The second category involves measurement of mineralogical and chemical properties of soils, such as clay content, cation exchange capacity, and specific surface area. Practicing geotechnical engineers typically use only the measurement of physical properties to identify expansive soils. However, the measurement of mineralogical and chemical properties is used routinely by agricultural and geological practitioners and should not be disregarded by the engineering community.

Many of the methods of identification only identify if minerals with the potential to produce expansion are present. They do not consider the physical properties such as in situ water content and density. Thus, they do not necessarily identify if the natural soil deposit is actually expansive, nor do they quantify the actual expansion potential. Nevertheless, they are useful indicators of the need to explore the potential expansivity in more detail.

2.3.1 Methods Based on Physical Properties

2.3.1.1 *Methods Based on Plasticity*
Atterberg limits are commonly used to characterize soils and are used in some methods to identify expansive soils. Two indices defined on the basis of the Atterberg limits are the plasticity index, PI, and the liquidity index, LI. Many expansive soil identification methods make use of one or both of these indices. As shown in Figure 2.11, the more expansive minerals tend to exhibit higher plasticity.

Peck, Hanson, and Thornburn (1974) suggested that there is a general relationship between the plasticity index of a soil and the potential for expansion, as shown in Table 2.1. However, Zapata et al. (2006)

TABLE 2.1   Expansion Potential of Soils and Plasticity Index
(Peck, Hanson, and Thornburn 1974)

Plasticity Index (%)	Expansion Potential
0–15	Low
0–35	Medium
20–55	High
> 35	Very high

showed that the expansion potential for remolded expansive soils correlates poorly to the plasticity index alone. They concluded that the correlation is significantly improved by correlating expansion potential with the product of plasticity index and percentage passing the No. 200 (75 µm) sieve. It is important to keep in mind that although the plasticity of a soil may be an indicator of expansive minerals, that in itself is not definitive identification of an expansive soil.

Atterberg limits and clay content can be combined into a parameter called activity, A_c. This term was defined by Skempton (1953) as

$$\text{Activity}\,(A_c) = \frac{\text{Plasticity Index}}{\%\,\text{by weight finer than}\,2\,\mu m} \qquad (2\text{-}1)$$

Skempton suggested three classes of clays according to activity. The suggested classes are "inactive" for activities less than 0.75, "normal" for activities between 0.75 and 1.25, and "active" for activities greater than 1.25. Active clays provide the most potential for expansion. Typical values of activities for various clay minerals are shown in Table 2.2. Sodium montmorillonite has the most expansion potential, which is reflected by the extraordinarily high value of activity in Table 2.2.

2.3.1.2 *Free Swell Test*
The free swell test consists of placing a known volume of dry soil passing the No. 40 (425 µm) sieve into a graduated cylinder filled with water and measuring the swelled volume after it has completely settled. The free

TABLE 2.2   Typical Activity Values for Clay Minerals
(Skempton 1953)

Mineral	Activity
Kaolinite	0.33–0.46
Illite	0.9
Montmorillonite (Ca)	1.5
Montmorillonite (Na)	7.2

swell of the soil is determined as the ratio of the change in volume from the dry state to the wet state over the initial volume, expressed as a percentage. A high-grade commercial bentonite (sodium montmorillonite) will have a free swell value from 1,200 to 2,000 percent. Holtz and Gibbs (1956) stated that soils having free swell values as low as 100 percent may exhibit considerable expansion in the field when wetted under light loading. Also, Dawson (1953) reported that several Texas clays with free swell values in the range of 50 percent have caused considerable damage due to expansion. This was due to extreme climatic conditions in combination with the expansion characteristics of the soil.

Because of its simplicity and ease of operation, the free swell test is used as the sole swell potential index in the Chinese Technical Code for Building in Expansive Soil Areas (CMC 2003). However, even for the same type of soil, the results of the test can be influenced significantly by many factors, such as the amount of soil tested, degree of soil grinding, and drop height of the soil sample (Chen et al. 2006).

The Bureau of Indian Standards (1997) 2720 Part 40 uses the free swell index (FSI) method to indirectly estimate swell potential of expansive soils. In this test, two oven-dried soil specimens are each poured into a graduated cylinder. One cylinder is filled with kerosene oil and the other with distilled water. Both of the samples are stirred and left undisturbed for a minimum of 24 hours after which the final volumes of soils in the cylinders are noted. The FSI is calculated as,

$$\text{FSI} = \frac{(soil\ volume\ in\ water - soil\ volume\ in\ kerosene)}{soil\ volume\ in\ kerosene} \times 100\% \quad (2\text{-}2)$$

The expansion potential of the soil as classified according to the FSI is shown in Table 2.3.

2.3.1.3   Potential Volume Change (PVC)

The potential volume change (PVC) method was developed by T. W. Lambe (1960) for the Federal Housing Administration. The PVC apparatus is shown in Figure 2.19a. It has been used by many State

TABLE 2.3   Expansion Potential Based on Free Swell Index

Free Swell Index (FSI)	Expansion Potential
< 20	Low
20–35	Medium
35–50	High
> 50	Very high

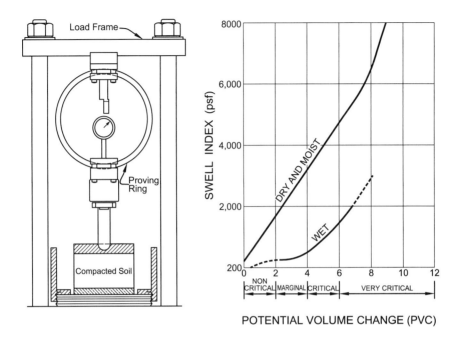

FIGURE 2.14. (a) Potential volume change (PVC) apparatus; (b) swell index vs. PVC.

Highway Departments as well as some geotechnical engineers. The PVC test consists of placing a remolded soil sample into an oedometer ring. The sample is then wetted in the device and allowed to swell against a proving ring. The swell index is reported as the pressure on the ring and is correlated to qualitative ranges of potential volume change using the chart shown in Figure 2.14b (Lambe 1960). The advantage of the test is its simplicity. The disadvantage is that the stiffness of the proving ring is not standardized, thereby allowing for different amounts of swelling to take place, depending on the stiffness of the proving ring. The swelling pressure that is developed will vary with the amount of swelling that the proving ring allows. Because the test uses remolded samples, the swell index and PVC values are more useful for identification of potential expansive behavior and should not be used as design parameters for undisturbed in situ soils.

2.3.1.4 Expansion Index (EI) Test
The expansion index test was developed in southern California in the late 1960s in response to requests from several local agencies for the standardization of testing methods in that area. The method was evaluated statistically by five different testing laboratories in California, and

TABLE 2.4   **Expansion Potential Based on Expansion Index**

Expansion Index (EI)	Expansion Potential
0–20	Very low
21–50	Low
51–90	Medium
91–130	High
> 130	Very high

was adopted as a standard by many California government agencies and the Uniform Building Code (1997) (UBC Standard 18-2). ASTM International has published a standard method of test for the EI test (ASTM D4829). The test is basically a consolidation-swell oedometer test, which will be described in chapter 6.

The test consists of compacting a soil at a degree of saturation of 50% ± 2% under standard conditions. A vertical stress of 144 psf (7 kPa) is applied to the sample, and the sample is inundated with distilled water. The expansion index, reported to the nearest whole number, is calculated by equation (2-3).

$$EI = \frac{(final\ thickness - initial\ thickness)}{initial\ thickness} \times 1,000 \qquad (2\text{-}3)$$

The expansion potential of the soil is classified according to the expansion index, as shown in Table 2.4. The 1997 Uniform Building Code states the following:

Foundations for structures resting on soils with an expansion index greater than 20, as determined by UBC Standard 18-2, shall require special design consideration. If the soil expansion index varies with depth, the variation is to be included in the engineering analysis of the expansive soil effect upon the structure.

The International Building Code (2012) and the International Residential Code (2012) both adopted the expansion index test for identification of an expansive soil. Both of the codes state the following:

Soils meeting all four of the following provisions shall be considered expansive, except that tests to show compliance with Items 1, 2, and 3 shall not be required if the test prescribed in Item 4 is conducted:

1. Plasticity Index (PI) of 15 or greater, determined in accordance with ASTM D 4318.
2. More than 10 percent of the soil particles pass a No. 200 sieve, determined in accordance with ASTM D 422.

3. More than 10 percent of the soil particles are less than 5 micrometers in size, determined in accordance with ASTM D 422.
4. Expansion Index greater than 20, determined in accordance with ASTM D 4829.

2.3.1.5   Coefficient of Linear Extensibility (COLE)

The COLE test determines the linear strain of an undisturbed, unconfined sample on drying from 5 psi (34 kPa) suction to oven-dry suction (150,000 psi = 1,000 MPa). The procedure involves coating undisturbed soil samples with a flexible plastic resin. The resin is impermeable to liquid water, but permeable to water vapor. Natural clods of soil are brought to a soil suction of 5 psi (34 kPa) in a pressure vessel. They are weighed in air and in water to measure their weight and volume using Archimedes' principle. The samples are then oven-dried and another weight and volume measurement is performed in the same manner.

COLE is a measure of the change in a sample dimension from the moist to dry state. The value of COLE is given by:

$$\text{COLE} = \Delta L / \Delta L_D = (\gamma_{dD}/\gamma_{dM})^{0.33} - 1 \qquad (2\text{-}4)$$

where:

$\Delta L / \Delta L_D$ = linear strain relative to dry dimensions,
γ_{dD} = dry density of oven-dry sample, and
γ_{dM} = dry density of sample at 5 psi (34 kPa) suction.

The value of COLE is sometimes expressed as a percentage. Whether it is a percentage or dimensionless is evident from its magnitude. COLE has been related to swell index from the PVC test and other indicative parameters (Franzmeier and Ross 1968; Anderson, Fadul, and O'Connor 1972; McCormick and Wilding 1975; Schafer and Singer, 1976; Parker, Amos, and Kaster 1977). The linear extensibility, LE, can be used as an estimator of clay mineralogy. The LE of a soil layer is the product of the thickness, in centimeters, multiplied by the COLE of the layer in question. The LE of a soil is defined as the sum of these products for all soil horizons (USDA-NRCS 2010). The ratio of LE to clay content is related to mineralogy as shown in Table 2.5.

TABLE 2.5   Ratio of Linear Extensibility (LE) to Percent Clay

LE/Percent Clay	Mineralogy
< 0.05	Kaolinite
0.05–0.15	Illite
> 0.15	Montmorillonite

2.3.1.6 *Standard Absorption Moisture Content (SAMC)*

The SAMC test was originally proposed by Yao et al. (2004) for identification of expansive soils. It was recommended in China's Specifications for Design of Highway Subgrades (CMC 2004). The advantage of the SAMC test is its simplicity. The SAMC is the equilibrium water content that a soil will attain under standardized conditions.

The test consists of placing an undisturbed soil sample on a porous plate within a constant humidity container over a saturated solution of sodium bromide. The weight of the soil sample at equilibrium is measured, after which it is oven-dried. The SAMC is calculated as

$$\text{SAMC (\%)} = \frac{W_e - W_s}{W_s} \qquad (2\text{-}5)$$

where:

W_e = weight of sample at equilibrium (77 °F and 60% relative humidity) and

W_s = weight of oven-dry sample.

Zheng, Zhang, and Yang (2008) developed a correlation between SAMC and the montmorillonite content.

China's Specifications for Design of Highway Subgrades (CMC 2004) presents a method for classifying expansive soils based on the standard absorption moisture content, plasticity index, and free-swell values, as shown in Table 2.6.

2.3.2 Mineralogical Methods

The identification of the presence of montmorillonite in a soil is one means of identifying the soil as being potentially expansive. The mineralogy of a clay can be identified on the basis of crystal structure or by means of chemical analyses.

Popular mineralogical methods of identification include X-ray diffraction (XRD), differential thermal analysis (DTA), and electron microscopy. XRD operates on the principle of measuring basal plane

TABLE 2.6 Classification Standard for Expansive Soils (CMC 2004)

Standard Absorption Moisture Content (%)	Plasticity Index (%)	Free-Swell Value (%)	Swell Potential Class
< 2.5	< 15	< 40	Nonexpansive
2.5–4.8	15–28	40–60	Low
4.8–6.8	28–40	60–90	Medium
> 6.8	> 40	> 90	High

spacing by the amount by which X-rays are diffracted around crystals. DTA consists of simultaneously heating a sample of clay and an inert substance. The resulting thermograms, which are plots of the temperature difference versus applied heat, are compared to those for pure minerals. Each mineral shows characteristic endothermic and exothermic reactions on the thermograms. Electron microscopy provides a means of directly observing the clay particles. Qualitative identification is possible based on size and shape of the particles.

Other mineralogical methods include X-ray absorption spectroscopy, petrographic microscopy, soil micromorphology, digital image analysis, atomic force microscopy, and diffuse reflectance spectroscopy (Ulery and Drees 2008). Mineralogical methods are seldom used in engineering practice. They are most useful for research purposes.

2.3.3 Chemical Methods

The most common chemical methods that are used to identify clay minerals include measurement of cation exchange capacity (CEC), specific surface area (SSA), and total potassium (TP). These methods are described in the following sections.

2.3.3.1   Cation Exchange Capacity (CEC)
The CEC is the total number of exchangeable cations required to balance the negative charge on the surface of the clay particles. CEC is expressed in milliequivalents per 100 grams of dry clay. In the test procedure, excess salts in the soil are first removed and the adsorbed cations are replaced by saturating the soil exchange sites with a known cation. The amount of known cation needed to saturate the exchange sites is greater for a mineral with a greater unbalanced surface charge. The composition of the cation complex that was removed can be determined by chemical analysis of the extract.

CEC is related to clay mineralogy. A high CEC value indicates the presence of a more active clay mineral such as montmorillonite, whereas a low CEC indicates the presence of a nonexpansive clay mineral such as kaolinite. In general, expansion potential increases as the CEC increases. Typical values of CEC for the three basic clay minerals are presented in Table 2.7 (Mitchell and Soga 2005).

A number of different procedures can be used to determine the CEC of a soil (Rhoades 1982). The measurement of CEC requires detailed and precise testing procedures that are not commonly done in most soil mechanics laboratories. However, this test is routinely performed in many agricultural soils laboratories and is inexpensive.

TABLE 2.7   **Typical Values of CEC, SSA, and TP for Clay Minerals (Mitchell and Soga 2005)**

Clay Mineral	Cation Exchange Capacity (CEC) (meq/100 g)	Specific Surface Area (SSA) (m²/g)	Total Potassium (TP) (%)
Kaolinite	1–6	5–55	0
Illite	15–50	80–120	6
Montmorillonite	80–150	600–800	0

A classification system was developed by McKeen and Hamberg (1981) and Hamberg (1985) that combines engineering index properties with the CEC. The system extended the concepts of Pearring (1963) and Holt (1969) that designated mineralogical groups based on clay activity and CEC.

McKeen and Hamberg (1981) and Hamberg (1985) extended the Pearring-Holt mineralogical classification system by assigning COLE values to different regions on a plot of activity, A_c, versus cation exchange activity, CEA_c ($CEA_c = CEC$/clay content). The chart was developed based on data obtained from soil survey reports of the Natural Resources Conservation Service, for soils in California, Arizona, Texas, Wyoming, Minnesota, Wisconsin, Kansas, and Utah. Figure 2.15 was developed to be used as a general classification system using the CEA_c versus A_c chart to indicate potentially expansive soils.

2.3.3.2   Specific Surface Area (SSA)

The specific surface area (SSA) of a soil is defined as the total surface area of soil particles in a unit mass of soil. As was demonstrated by Figure 2.7, the SSA of montmorillonite is much greater than that of kaolinite. As shown in Figure 2.7, a clayey soil with a high SSA will have higher water holding capacity and greater expansion potential (Chittoori and Puppala 2011). However, a high SSA alone does not necessarily indicate an expansive soil. For example, if a soil has a high organic fraction, that fraction may have a highly reactive surface with property characteristics similar to that for a material with a high specific surface area (Jury, Gardner, and Gardner 1991).

Several methods have been developed to measure the specific surface area of a soil. The most commonly used method uses adsorption of polar molecules, such as ethylene glycol, on the surfaces of the clay minerals.

Table 2.7 shows typical values of SSA for the three basic clay minerals. The montmorillonite minerals have an SSA about 10 times higher than the kaolinite group. Although the range of typical SSA

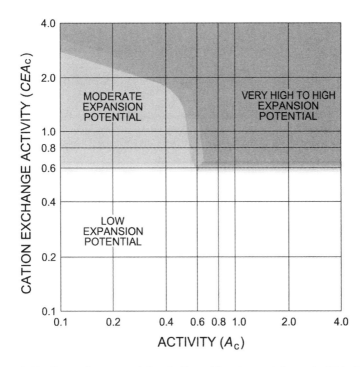

FIGURE 2.15. Expansion potential as indicated by clay activity and CEA_c (modified from Nelson and Miller 1992).

values shown in Table 2.7 may vary by 100 percent or more within the same group of minerals, the difference in SSA between groups, especially for montmorillonite, is so large that mineral identification is usually possible.

2.3.3.3   Total Potassium (TP)

The only clay mineral that includes potassium in its structure is illite. Therefore, the amount of potassium ions in a soil provides a direct indication of the presence of illite (Chittoori and Puppala 2011). Table 2.7 shows the differences in amount of total potassium between illite and the other two minerals. Thus, high potassium content is indicative of low expansion potential.

2.3.4   Comments on Identification Methods

Chittoori and Puppala (2011) have attempted to develop a rational and practical method to determine clay mineralogy distribution of a soil using the three chemical properties (CEC, SSA, and TP). These measurements were used to develop models based on an artificial neural network (ANN) to quantify and identify the dominant clay minerals in the fine fraction of the soil. They concluded that the measurements of CEC, SSA, and TP can be effectively used to identify clay minerals in a given soil. As noted previously, mineralogical methods are more commonly used for research purposes than by practicing geotechnical engineers due to the need for expensive and skill-oriented test devices. Also, it is imperative that the physical properties such as water content and density be taken into account, along with the presence of montmorillonite in a soil.

2.4   CHARACTERISTICS OF EXPANSIVE SOIL PROFILES

2.4.1   Geographic Distribution of Expansive Soils

Expansive soils deposits and problems associated with heaving soils have been reported on six continents and in more than 40 countries worldwide. Figure 2.16 shows the global distribution of reported expansive soil sites.

In North America, mapping of expansive soils has been carried out by a number of investigators using geologic and agricultural soil maps and soil reports (Hamilton 1968; Hart 1974; Snethen et al. 1975; Patrick and Snethen 1976; Snethen, Johnson, and Patrick 1977; Krohn and Slosson 1980; Ching and Fredlund 1984; Olive et al. 1989). Figure 2.17 shows a map of the general distribution of reported occurrences of expansive soils in the United States. The map shows areas where expansive soils have been encountered, but it must be emphasized that expansive soils are not limited to those areas. Expansive soils are more prominent in the western and southern part of North America. The most severe problems occur in the western parts of the United States and Canada, where problems are primarily attributed to highly overconsolidated claystone and clayshale.

2.4.2   Expansive Soil Profiles

Soil profiles for representative sites with reported expansion potential are discussed in the following sections. These soil profiles are presented to point out some important characteristics of expansive soil sites and to show the nature of the soil profiles in different areas of the world.

The nature of expansive soil is such that swelling is associated with an increase in water content of a dry soil having densely packed clay particles. In addition, the mineralogy must be such as to produce expansion potential. The preceding discussion has shown the relationship between plasticity and expansion potential.

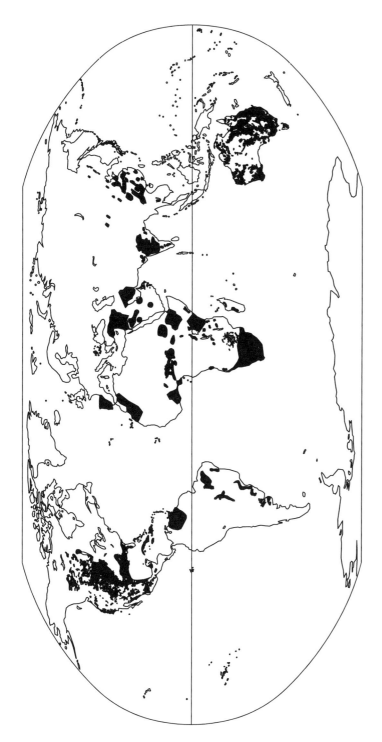

FIGURE 2.16. Global distribution of reported expansive soil sites (data from Dudal 1962; Hamilton 1968; Donaldson 1969; van der Merwe and Ahronovitz 1973; Richards, Peter, and Emerson 1983; Ching and Fredlund 1984; Popescu 1986; Agarwal and Rathee 1987; Ola 1987; Chen 1988; Olive et al. 1989; Al-Rawas and Woodrow 1992; Garrido, Rea, and Ochoa 1992; Juca, Gusmao, and Da Silva 1992; Ramana 1993; Shi et al. 2002; Al-Mhaidib 2006; Taboada and Lavado 2006; Suzuki, Fujii, and Konishi 2010; Sawangsuriya et al. 2011).

41

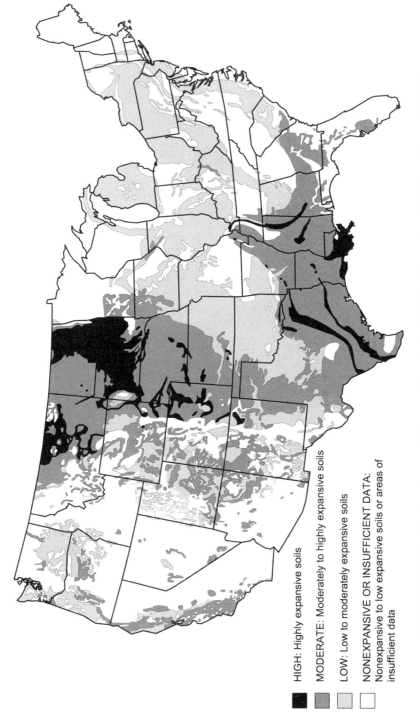

FIGURE 2.17. Distribution of potentially expansive soils in the United States (modified from Olive et al. 1989).

HIGH: Highly expansive soils

MODERATE: Moderately to highly expansive soils

LOW: Low to moderately expansive soils

NONEXPANSIVE OR INSUFFICIENT DATA:
Nonexpansive to low expansive soils or areas of
insufficient data

Another parameter that indicates high expansion potential is soil suction. The units commonly used for soil suction are kilopascals (kPa) or pF.[2] Soil suction will be discussed with respect to its effect on soil expansion in chapter 4. Parameters that one would expect at an expansive soil site would be relatively high plasticity, low natural water content, a relatively high density, and high values of soil suction.

2.4.2.1 Welkom, South Africa

Some of the earliest research on expansive soils was carried out in South Africa. In the 1950s, areas in South Africa observed heaving of foundations and roads at Leeuhof, Vereeniging, and Pretoria in the Transvaal. Figure 2.18 shows a soil profile from Welkom, South Africa, at which

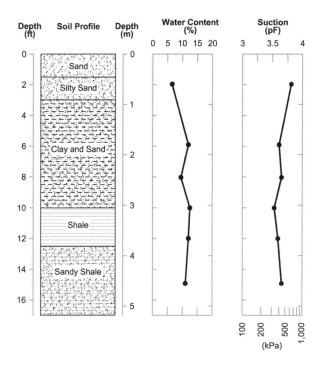

FIGURE 2.18. Profile of expansive soils at Welkom, South Africa (de Bruijn 1965).

---

[2]The unit pF expresses soil suction in terms of head. The pF of a soil is the soil suction in terms of the logarithm of the negative head of water expressed in centimeters. Although it does not conform to standard SI units, it is in common usage in geotechnical engineering.

expansive soils were excavated along with water content and soil suction values (de Bruijn 1965). Although density and expansive properties were not reported, the clay and shale shown in Figure 2.18 have low natural water content and high suction values. The suction values range from about 3.5 pF to 4 pF (300 kPa to 1,000 kPa). These values along with the low natural water content are typical values for sites with dry and dense expansive clay soils.

2.4.2.2   *Maryland, Australia*

Figure 2.19 shows an expansive soil profile from Maryland, Australia. Richards, Peters, and Emerson (1983) estimated that 20 percent of the surface soils of Australia can be classified as expansive. Seasonally induced ground surface movements can include swell and shrinkage extending over a range of several inches. The soil profile in Figure 2.19 consists of residual clays of high plasticity to medium plastic silty clay, underlain by weathered siltstone. The residual clay in this profile is the most expansive of the soils. It exhibits values of percent swell up to about 6 percent. The relatively high percent swell of the clay correlates with the high plasticity index.

An important feature to observe in Figure 2.19 is that the in situ water content is nearly equal to the plastic limit. This is a characteristic of dry, highly consolidated soils. It will be seen that this is also true for all of the soil profiles presented in the following figures.

Suction values below a depth of about 0.5 m are relatively high, with many values being at or above 4 pF (1,000 kPa). The lower values of percent swell at depth reflect the silty nature of the deeper soils, which is reflected in the lower values of liquid limit and plasticity index.

2.4.2.3   *Regina, Saskatchewan, Canada*

The area around Regina, Saskatchewan in Canada is extensively covered with preglacial, lacustrine clay sediments that have high plasticity. An expansive soil profile near Regina is shown in Figure 2.20. As was noted for most of the previously discussed soil profiles, the values of in situ water content are about equal to the plastic limit in this soil profile as well. The average liquid limit is 75 percent and the average plastic limit is 25 percent. Gilchrist (1963) performed oedometer tests on remolded samples of Regina clay. He measured values of percent swell as high as about 14 percent at an inundation stress of 800 psf (38 kPa) and swelling pressures as high as about 40,000 psf (2,000 kPa).

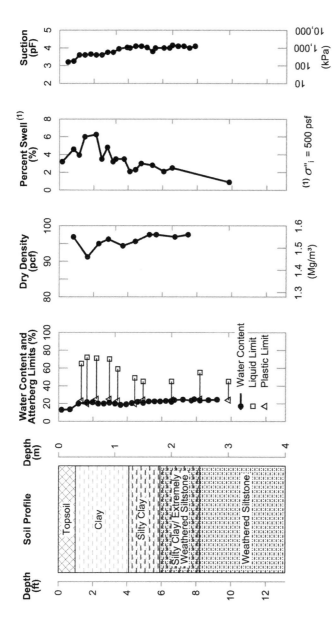

FIGURE 2.19. Profile of expansive soils from Maryland, Australia (Fityus, Cameron, and Walsh 2004).

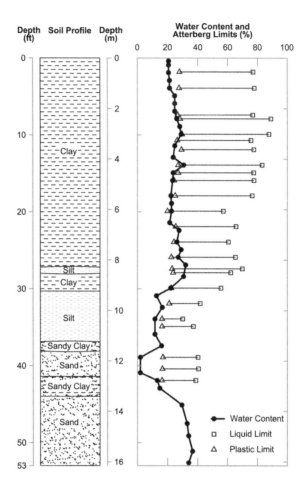

FIGURE 2.20. Profile of expansive soils from Regina, Saskatchewan, Canada (Fredlund, Rahardjo, and Fredlund 2012).

Regina is in a semiarid climate where the annual precipitation is approximately 14 in. (350 mm). Buildings founded on shallow foundations on the lacustrine clay have been observed to experience heave in amounts of 2 in. to 6 in. (50 mm to 150 mm) after construction (Fredlund, Rahardjo, and Fredlund 2012).

2.4.2.4 Front Range Area of Colorado, USA
The area of the USA immediately east of the easternmost range of the Rocky Mountains is known generally as the "Front Range" area.

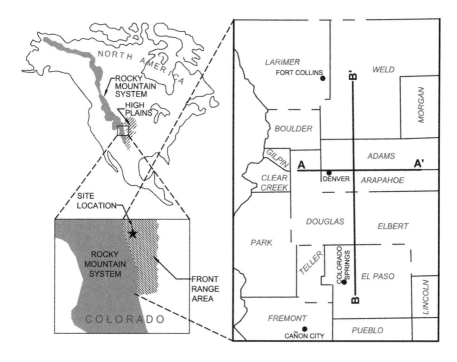

FIGURE 2.21. Map of the Front Range area of Colorado.

Figure 2.21 shows the general location of the Front Range area of Colorado. In this area, dense highly overconsolidated claystones and clayshales are common, and they can cause severe distress to light structures constructed on them. The bedrock in this area can belong to various geologic formations, depending on the location. Figure 2.22 shows geologic cross-sections that depict the relative ages of the formations. As will be seen in following figures, the expansion potential of the various formations can be similar. However, the existence of various features such as dip, presence of coal seams, etc. can cause a significant variation in the behavior of the bedrock at different locations.

Several soil profiles obtained from the Front Range area of Colorado are shown in Figures 2.23 to 2.26. As noted, the in situ water content is close to the plastic limit in all of these figures. The expansive soils in the Front Range area of Colorado typically have a liquid limit ranging from 35 to 75 percent and a plasticity index ranging from 15 to 50 percent. The expansive soils can swell by over 10 percent and exhibit swelling pressures in the general range of 10,000 psf to 30,000 psf

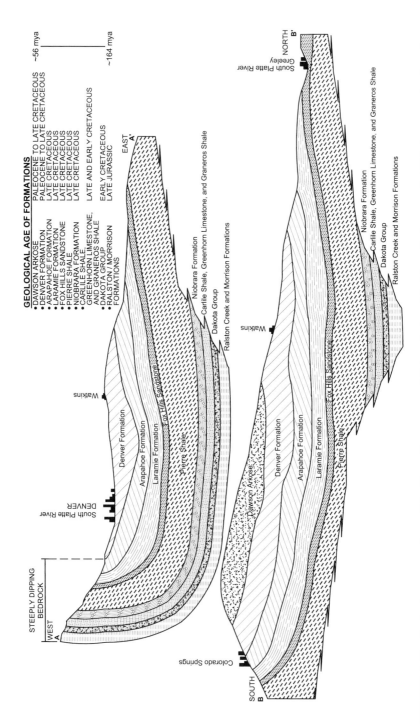

FIGURE 2.22. Geologic cross-sections of the Front Range area of Colorado (modified from Robson et al. 1981) (cross-section locations are shown on Figure 2.21).

48

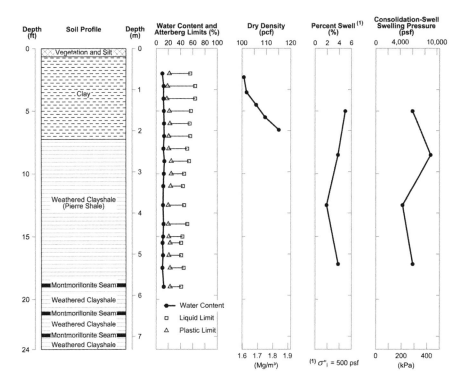

FIGURE 2.23. Profile of expansive soils from Colorado State University, Fort Collins, Colorado.

(480 kPa to 1,440 kPa). Values of swelling pressure as high as 58,000 psf (2,780 kPa) have been measured.

Figure 2.23 shows a soil profile at the Colorado State University Expansive Soils Field Test Site in Fort Collins, Colorado. This site is characterized by the older (lower) sediments that exist within the Pierre Shale formation. Montmorillonite seams are believed to be the result of volcanic ash that has been weathered. The dry density can be high. The swelling pressure shown in Figure 2.23 is not particularly high, but together with the percent swell, the expansion potential is very high.

Figure 2.24 shows a soil profile for a site north of Denver, Colorado. This site is located on the Laramie Formation. Suction values range from about 3.5 pF to 4.5 pF (310 kPa to 3,100 kPa). These values are

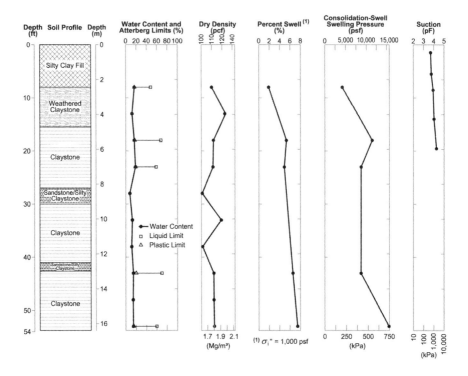

FIGURE 2.24. Profile of expansive soils from a site north of Denver, Colorado.

typical of the Front Range area. At this site, post-tensioned slab foundations experienced total heave in amounts up to 7.4 in. (190 mm) with differential heave in amounts of 4 in. (100 mm) or more within 8 years after construction. The remaining potential free-field heave for this site was calculated to range from 6 in. to 20 in. (165 mm to 520 mm). This was in addition to the heave already experienced by the buildings.

Figure 2.25 shows a profile at a site near Cañon City, Colorado, located on the Pierre Shale. At this site a heavy masonry building was constructed on spread footings. Floor slabs experienced heave in amounts as much as 9 in. (230 mm). Suction values ranged from about 4.0 pF to 4.5 pF (1,000 kPa to 3,000 kPa).

Figure 2.26 shows an expansive soil profile near the Denver International Airport (DIA). The sediments in this area belong to the Denver Formation and have high expansion potential. The Denver Formation is younger than the Pierre Shale or the Laramie Formation,

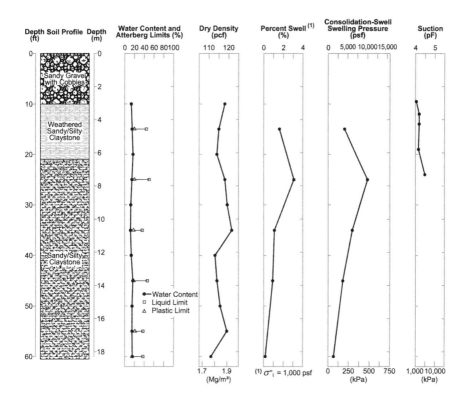

FIGURE 2.25. Profile of expansive soils near Cañon City, Colorado.

but the expansive characteristics of these formations in the Front Range area are similar. The building had been undergoing heave movement since the time of construction in 1991. Slab heave in amounts up to 6 in. (150 mm) from the time of construction have been measured (CSU 2004; Chao 2007). Figure 2.26 indicates that the expansive bedrock exhibited values of percent swell up to over 10 percent and a swelling pressure up to 25,000 psf (1,200 kPa). Of particular interest at this site was the water content profile. Irrigation had been discontinued and water content values were low in the upper several feet. However, deep wetting was occurring as a result of water being introduced through a deep-seated coal seam.

The similarities of the sites in the Front Range area for different formations of different ages reflect the highly overconsolidated nature of the claystone. The low values of initial water content and the high values

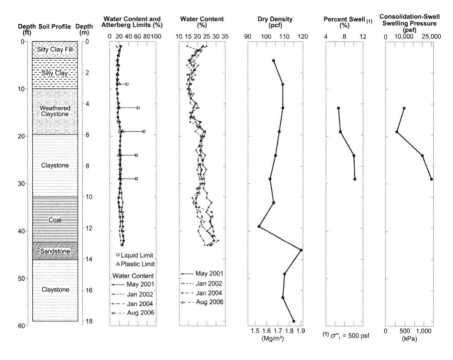

FIGURE 2.26. Profile of expansive soils from Denver International Airport, Denver, Colorado (CSU 2004; Chao 2007).

of dry density contribute to the high expansion potential that is typical of the soils in the Front Range area.

2.4.2.5   San Antonio, Texas, USA

In San Antonio, Texas, many of the foundations with heaving problems are found in areas located on the Taylor marl. An expansive soil profile from San Antonio is shown in Figure 2.27. The data shown in that figure were collected from an exploratory borehole drilled and sampled in the summer during a dry time of the year. The clay soil exhibited high swelling potential. It exhibited percent swell up to almost 12 percent and swelling pressures as high as about 40,000 psf (1,900 kPa). At the time the soil was sampled, the in situ water content was close to the plastic limit. During wetter time periods, the water content was higher. This soil is not highly overconsolidated and the soils exhibit shrink-swell over a fairly wide range due to seasonal change.

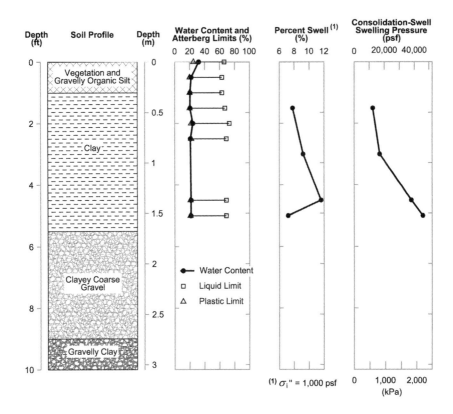

FIGURE 2.27. Profile of expansive soils from Fort Sam Houston, San Antonio, Texas.

References

Agarwal, K. B., and R. K. Rathee. 1987. "Bearing Capacity in Expansive Soils." *Proceedings of the 6th International Conference on Expansive Soils*, New Delhi, India, 1, 433–438.

Al-Mhaidib, A. I. 2006. "Swelling Behavior of Expansive Shale, a Case Study from Saudi Arabia." In *Expansive Soils—Recent Advances in Characterization and Treatment*, edited by A. A. Al-Rawas and M. F. A. Goosen. London: Taylor and Francis Group.

Al-Rawas, A., and L. K. R. Woodrow. 1992. "The Distribution of Expansive Soils in Oman." *Proceedings of the 7th International Conference on Expansive Soils*, 1, 432–437.

Anderson, J. U., K. E. Fadul, and G. A. O'Connor. 1972. "Factors Affecting the Coefficient of Linear Extensibility in Vertisols." *Soil Science Society of America Proceedings*, 27, 296–299.

ASTM D4829-11. 2011. "*Standard Test Method for Expansion Index of Soils*," ASTM International: West Conshohocken, PA.

Benson, C. H., and S. R. Meer. 2009. "Relative Abundance of Monovalent and Divalent Cations and the Impact of Desiccation on Geosynthetic Clay Liners." *Journal of Geotechnical and Geoenvironmental Engineering*, ASCE 135(3): 349–358.

Bjerrum, L. 1967. "Progressive Failure in Slopes of Overconsolidated Plastic Clay and Clay Shales." *Journal of Soil Mechanics and Foundation Engineering* 23(SM5): 3–47.

Bureau of Indian Standards. 1997. "Is 2720 Part 40. Determination of Free Swell Index of Soils." New Delhi: Bureau of Indian Standards.

Chao, K. C. 2007. Design Principles for Foundations on Expansive Soils." PhD dissertation, Colorado State University, Fort Collins, CO.

Chen, F. H. 1973. "The Basic Physical Property of Expansive Soils." *Proceedings of the 3rd International Conference on Expansive Soils*, Haifa, Israel, 1, 17–25.

———. 1988. *Foundations on Expansive Soils*. New York: Elsevier Science.

Chen, S. X., S. Song, Yu, Z. G. Liu, F. Yu, and X. C. Xu. 2006. "Effect of Soil Sample Preparation on Free Swelling Ratio and Its Improved Measures." *Rock and Soil Mechanics* 27(8): 1327–1330.

China Ministry of Construction (CMC). 2003. "Technical Code for Building in Expansive Soil Area." GBJ112-87, Beijing: Chinese Planning Press.

———. 2004. "Specifications for the Design of Highway Subgrades." JTG D 30, Beijing: Renmin Communication Press.

Ching, R. K. H., and D. C. Fredlund. 1984. "A Small Saskatchewan Town Copes with Swelling Clay Problems." *Proceedings of the 5th International Conference on Expansive Soils*, Adelaide, South Australia, 306–310.

Chittoori, B., and A. J. Puppala. 2011. "Quantitative Estimation of Clay Mineralogy in Fine-Grained Soils." *Journal of Geotechnical and Geoenvironmental Engineering*, ASCE 137(11): 997–1008.

Colorado State University (CSU). 2004. "Moisture Migration in Expansive Soils at the TRACON Building at Denver International Airport, Phase II. (January 2003–June 2004)." Fort Collins, CO.

Companion, A. L. 1964. *Chemical Bonding*. New York: McGraw-Hill.

Dawson, R. F. 1953. "Movement of Small Houses Erected on an Expansive Clay Soil." *Proceedings of the 3rd International Conference on Soil Mechanics and Foundation Engineering*, 1, 346–350.

De Bruijn, C. M. A. 1965. "Some Observations on Soil Moisture Conditions Beneath and Adjacent to Tarred Roads and Other Surface Treatments in South Africa." *Moisture Equilibria and Moisture Changes in Soils Beneath Covered Areas, A Symposium in Print*, edited by G.D. Aitchison, Australia: Butterworths, 135–142.

Donaldson, G. W. 1969. "The Occurrence of Problems of Heave and the Factors Affecting Its Nature." *Proceedings of the 2nd International Research and Engineers Conference on Expansive Soils*, College Station, TX, 25–36.

Dudal, R. 1962. "Dark Clay Soils of Tropical and Subtropical Regions." *Soil Science* 95(4): 264–270.

Fityus, S. G., D. A. Cameron, and P. F. Walsh. 2005. "The Shrink Swell Test." *Geotechnical Testing Journal*, ASTM 28(1): 1–10.

Frank, H. S., and W. Y. Wen. 1957. "Structural Aspects of Ion-Solvent Interaction in Aqueous Solutions: A Suggested Picture of Water Structures." *Faraday Society Discussions* 24, 133–140.

Franzmeier, D. P., and S. J. Ross. 1968. "Soil Swelling: Laboratory Measurement and Relation to Other Soil Properties." *Soil Science Society of America Journal* 32, 573–577.

Fredlund, D. G., H. Rahardjo, and M. D. Fredlund. 2012. *Unsaturated Soil Mechanics in Engineering Practice*. Hoboken, NJ: John Wiley and Sons.

Garrido, J. A. Z., M. L. P. Rea, and J. C. Ochoa. 1992. "Expansive Soil Properties in the Queretaro Valley." *Proceedings of the 7th International Conference on Expansive Soils*, Dallas, TX, 1, 45–50.

Gilchrist, H. G. 1963. "A Study of Volume Change in Highly Plastic Clay." Master's thesis, University of Saskatchewan, Saskatoon, SK (as presented in Fredlund, Rahardjo, and Fredlund 2012).

Grim, R. E. 1959. "Physico-chemical Properties of Soils." *Journal of the Soil Mechanics and Foundations Division*, ASCE 85(2): 1–70.

———. 1968. *Clay Mineralogy* (2nd ed.). New York: McGraw-Hill.

Hamberg, D. J. 1985. "A Simplified Method for Predicting Heave in Expansive Soils." Master's thesis, Colorado State University, Fort Collins, CO.

Hamilton, J. J. 1968. "Shallow Foundations on Swelling Clays in Western Canada." National Research Council of Canada, Department of Building Research, Technical Paper Number 263. Reprinted from *Proceedings of the International Research and Engineering Conference on Expansive Clay Soils*. College Station: Texas A & M University, 1965, 2, 183–207.

Hart, S. S. 1974. *Potentially Swelling Soil and Rock in the Front Range Urban Corridor, Colorado*. Denver, CO: Colorado Geological Survey.

Holtz, W. G., and H. J. Gibbs. 1956. "Engineering Properties of Expansive Clays." *Transactions ASCE* 121, 641–677.

Holtz, R. D., W. D. Kovacs, and T. C. Sheahan. 2011. *An Introduction to Geotechnical Engineering* (2nd ed.). Upper Saddle River, NJ: Prentice Hall.

International Building Code (IBC). 2012. *International Building Code*. Falls Church, VA: International Code Council.

International Residential Code (IRC). 2012. *International Residential Code for One- and Two-Family Dwellings*. Falls Church, VA: International Code Council.

Juca, J. F. T., J. A. Gusmao, and J. M. J. Da Silva. 1992. "Laboratory and Field Tests on an Expansive Soil in Brazil." *Proceedings of the 7th International Conference on Expansive Soils*, Dallas, TX, 1, 337–342.

Jury, W. A., W. R. Gardner, and W. H. Gardner. 1991. *Soil Physics* (5th ed.). New York: John Wiley and Sons.

Krohn, J. P., and J. E. Slosson. 1980. "Assessment of Expansive Soils in the United States." *Proceedings of the 4th International Conference Expansive Soils*, Denver, CO, 1, 596–608.

Lambe, T. W. 1958. "The Structure of Compacted Clay." *Journal of the Soil Mechanics and Foundations Division*, ASCE 84(SM2): 1654-1–1654-34.

———. 1960. "The Character and Identification of Expansive Soils, Soil PVC Meter." Federal Housing Administration, Technical Studies Program, FHA 701.

Lambe, T. W., and Whitman, R. V. 1969. *Soil Mechanics*. New York: John Wiley and Sons.

McCormick, D. E., and L. P. Wilding. 1975. "Soil Properties Influencing Swelling in Canfield and Geeburg Soils." *Soil Science Society of America Journal* 39, 496–502.

McKeen, R. G., and D. J. Hamberg. 1981. "Characterization of Expansive Soils." *Transportation Research Record* 790, Transportation Research Board, 73–78.

Mitchell, J. K., and K. Soga. 2005. *Fundamentals of Soil Behavior* (3rd ed.). Hoboken, NJ: John Wiley and Sons.

Nelson, J. D., and D. J. Miller. 1992. *Expansive Soils: Problems and Practice in Foundation and Pavement Engineering*. New York: John Wiley and Sons.

Norrish, K. 1954. "The Swelling of Montmorillonite." *Discussions of the Faraday Society* 18, 120–134.

Ola, S. A. 1987 "Nature and Properties of Expansive Soils in Nigeria." *Proceedings of the 6th International Conference on Expansive Soils*, New Delhi, India, 1, 11–16.

Olive, W., A. Chleborad, C. Frahme, J. Shlocker, R. Schneider, and R. Schuster. 1989. "Swelling Clays Map of the Conterminous United States." Map I-1940, US Geological Survey Miscellaneous Investigations Series.

Parker, J. C., D. F. Amos, and D. L. Kaster. 1977. "An Evaluation of Several Methods of Estimating Soil Volume Change." *Soil Science Society of America Journal* 41(6): 1059–1064.

Patrick, D. M., and D. R. Snethen. 1976. "An Occurrence and Distribution Survey of Expansive Materials in the United States by Physiographic Areas." US Army Engineer Waterways Experiment Station, Vicksburg, MS, Report No. FHWA-RD-76-82.

Peck, R. B., W. E. Hanson, and T. H. Thornburn. 1974. *Foundation Engineering* (2nd ed.). New York: John Wiley and Sons.

Popescu, M. E. 1986. "A Comparison between the Behavior of Swelling and of Collapsing Soils." *Engineering Geology* 23, 145–163.

Ramana, K. V. 1993. "Humid Tropical Expansive Soils of Trinidad: Their Geotechnical Properties and Areal Distribution." *Engineering Geology* 34, 27–44.

Rhoades, J. D. 1982. "Cation Exchange Capacity." In *Methods of Soil Analysis, Part 2: Chemical and Microbiological Properties*, Soil Science Society of Amcrica, Madison, WI, 149–157.

Richards, B. G., P. Peter, and W. W. Emerson. 1983. "The Effects of Vegetation on the Swelling and Shrinking of Soils in Australia." *Geotechnique* 33(2): 127–139.

Robson, S. G., A. Wacinski, S. Zawistowski, and J. C. Romero. 1981. "Geologic Structure, Hydrology, and Water Quality of the Laramie-Fox Hills Aquifer in the Denver Basin, Colorado." *Hydrologic Investigations Atlas HA-650*, US Geological Survey.

Sawangsuriya, A., A. Jotisankasa, B. Vadhanabhuti, and K. Lousuphap. 2011. "Identification of Potentially Expansive Soils Causing Longitudinal Cracks Along Pavement Shoulder in Central Thailand." *Proceedings of the 5th Asia-Pacific Conference on Unsaturated Soils*, Dusti Thani Pattaya, Thailand, in *Unsaturated Soils: Theory and Conference 2011*, edited by A. Jotisankasa, A. Sawangsuriya, S. Soralump, and W. Mairaing, 2, 693–705.

Schafer, W. M., and M. J. Singer. 1976. "Influence of Physical and Mineralogical Properties of Swelling of Soils in Yolo County, California." *Soil Science Society of America Journal* 40(4): 557–562.

Shi, B., H. Jiang, Z. Liu, and H. Y. Fang. 2002. "Engineering Geological Characteristics of Expansive Soils in China." *Engineering Geology* 67, 63–71.

Snethen, D. R., L. D. Johnson, and D. M. Patrick. 1977. "An Evaluation of Expedient Methodology for Identification of Potentially Expansive Soils." Soils and Pavements Laboratory, US Army Engineers Waterways Experiment Station, Vicksburg, MS, Report No. FHWA-RE-77-94.

Snethen, D. R., F. C. Townsend, L. D. Johnson, D. M. Patrick, and Vedros P. J. 1975. "A Review of Engineering Experiences with Expansive Soils in Highway Subgrades." Federal Highway Administration, USDOT, Report Number FHWA-RD-RD-75-48, NTUS Number AD A020309.

Skempton, A. W. 1953. "The Colloidal 'Activity' of Clays." *Proceedings of the 3rd International Conference on Soil Mechanics and Foundation Engineering*, Switzerland, 1, 57–61.

Slade, P. G., J. P. Quirk, and K. Norrish. 1991. "Crystalline Swelling of Smectite Samples in Concentrated NaCl Solutions in Relation to Layer Charge." *Clays and Clay Minerals* 39(3): 234–238.

Suzuki, M., K. Fujii, and J. Konishi. 2010. "Swelling Property and Its Anisotropy of Reconstituted and Undisturbed Samples." *Proceedings of the 4th Asia Pacific Conference on Unsaturated Soils, Newcastle, Australia, Unsaturated Soils: Experimental Studies in Unsaturated Soils and Expansive Soils*, edited by O. Buzzi, S. Fityus, and D. Sheng. Boca Raton, FL: CRC Press, 407–412.

Taboada, M. A., and R. S. Lavado. 2006. "Swelling in Non-Vertisolic Soils, Its Causes and Importance." In *Expansive Soils Recent Advances in Characterization and Treatment*, edited by A. A. Al-Rawas and M. F. A. Goosen. Leiden, The Netherlands: Taylor and Francis/Balkema, 55–76.

Ulery, A. L., and L. R. Drees. 2008. "Mineralogical Methods," In *Methods of Soil Analysis*, Part 5. Madison, WI: Soil Science Society of America.

Uniform Building Code (UBC). 1997. Uniform Building Code. Volume 1. Whittier, CA: International Conference of Building Officials.

van der Merwe, C. P., and M. Ahronovitz. 1973. "The Behaviour of Flexible Pavements on Expansive Soils in Rhodesia." *Proceedings of the 3rd International Conference on Expansive Soils*, Haifa, Israel, 1, 267–276.

Yao, H. L., Y. Yang, P. Cheng, and W. P. Wu. 2004. "Standard Moisture Absorption Water Content of Soil and Its Testing Standard." *Rock and Soil Mechanics* 25(6): 856–859.

Zapata, C. E., S. L. Houston, W. N. Houston, and H. Dye. 2006. "Expansion Index and Its Relationship with Other Index Properties." *Proceedings of the 4th International Conference on Unsaturated Soils*, Carefree, AZ, 2133–2137.

Zheng, J. L., R. Zhang, and H. P. Yang. 2008. "Validation of a Swelling Potential Index for Expansive Soils." In *Unsaturated Soils: Advances in Geo-Engineering*, edited by D. G. Toll, C. E. Augarde, D. Gallipoli, and S. J. Wheeler. London: Taylor and Francis Group.

3

Site Investigation

Soil investigation practices for expansive soil sites generally require more extensive sampling and specialized testing programs than sites with nonexpansive soils, even for small structures such as houses and one-story buildings. Site investigation for an expansive soil site must include not only those aspects common to all geotechnical investigations but also specialized methods designed to estimate the expansion potential of the soil. Additionally, because changes in water content are as important to the potential for expansion as are the physical properties of the soil, it is important to characterize the environmental conditions that may contribute to moisture change.

The site investigation should strive to obtain as complete a set of data as possible. However, where soil data may be incomplete, the geotechnical engineer must avoid the tendency to overanalyze an insufficient amount of data. Overanalysis of incomplete data can lead to inaccurate design assumptions. If a choice must be made between more accurate definition of the soil profile and precise measurement of soil properties by sophisticated laboratory techniques, the soil profile should receive primary attention.

3.1 PROGRAM OF EXPLORATION

The quality of the site investigation depends greatly on the experience and judgment of the field engineer, geologist, or technician. Soil exploration investigations are usually step-by-step processes that develop as information accumulates. The staged procedure involves the following steps:

1. Reconnaissance investigation—Review available information and perform an initial site visit.
2. Preliminary investigation—Conduct detailed surface mapping, preliminary borings, and initial laboratory testing and analysis for soil identification and classification.
3. Design-level investigation—Conduct soil borings for recovery of specialized samples, conduct specialized field and laboratory tests,

and perform analysis of data sufficient to determine soil parameters for use in design.

3.1.1 Reconnaissance Investigation

The reconnaissance site investigation should include review of as many available documents as possible related to the site history and an initial site visit. Existing documents might include topographic maps, geologic maps, historic reports, plans for existing structures, or previously prepared soil reports. Aerial photographs taken in different years prior to the investigation are particularly helpful in showing the history of the site and the locations of features such as ponds, swampy areas, roads, or other features that might influence the soil behavior.

The initial site visit can begin to identify likely problem areas. Topography and surficial geology can provide clues to the potential behavior of on-site soils, and aid in identifying areas that may be prone to moist or wet conditions, and existing disturbances on the site. Areas of exposed road cuts or rock outcrops can verify geology and indicate potential problem areas or expansive strata. Important things to document during the site visit include existing structures, exposed rock outcrops, slopes, on-site and off-site water sources, equipment/vehicle access, and so on.

Identification of potential sources of water is important for expansive soil sites. The site should be examined for drainage features, the steepness of valley slopes, slope problems such as landslides and mudflows, and the existence of nearby streams or rivers. Drainage features will govern the direction of surface water flow. Areas that restrict flow causing pooling or ponding of water should be noted.

3.1.2 Preliminary Investigation

For large sites, the site investigation should be conducted in stages to optimize the use of funding and to enhance the amount of pertinent data that can be obtained. A preliminary investigation may be used to supplement the reconnaissance investigation in development of the scope of the design-level investigation. The preliminary investigation is intended to provide a general knowledge of the soils present at a site and their potential for swelling and shrinking. This investigation may include some preliminary subsurface sampling and initial laboratory testing and analysis. The preliminary subsurface exploration program should emphasize sampling in areas where the reconnaissance indicates that problems might exist. It may include drilling several exploratory test holes across the site, particularly in and around the proposed building footprints. The purpose of drilling test holes is to define the general soil profile and the various strata that will influence the behavior of the foundation. Representative soil samples may be taken for preliminary laboratory tests. Soil samples from auger cuttings or large borings will aid in logging the subsurface profile, and provide samples for classification tests such as Atterberg limits, specific gravity, and grain size distribution.

The value of the preliminary program is to provide initial data to aid in planning the design-level investigation. It will aid in locating areas for detailed investigation, in determining depths and frequency of sampling, laboratory tests needed, and types of in situ measurements to be made. The preliminary investigation will provide an initial identification of other elements that are required for the detailed design-level exploration program. It also serves as a valuable part of the feasibility study to identify the suitability of a site for development and to provide the information to the developer on development issues and costs involved in further development. It is not considered to be of sufficient detail to be used for final design.

3.1.3   Design-Level Investigation

The design-level investigation is intended to provide the geotechnical engineering information and data necessary to develop the final design of the foundation. The design-level investigation requires a more thorough sampling and testing program than the preliminary investigation. The development of the design-level investigation must consider the following:

- Distribution of borings
- Depth of exploration
- Sampling frequency and depth

A design-level investigation includes a detailed definition of the soil profile, determination of soil properties, and quantification of the soil parameters and shrink-swell potential at the site. The distribution and depth of borings should be chosen to identify the soil profile and obtain samples for laboratory testing. In addition to the normal amount of laboratory testing that is required for a nonexpansive site, the sampling and testing for an expansive soil site must provide sufficient data for the analysis of the potential total and differential heave of the foundation soils as well as the bearing capacity and potential settlement.

Consequently, a greater number of samples will be required for expansive soil sites than for nonexpansive soil sites.

The design-level investigation should be carefully planned, but it is important to maintain flexibility so that the program can be modified to take advantage of data as they are gathered. For example, as the field investigation progresses, the program of drilling and sampling should be reviewed, and modified if necessary. The locations and depths of borings should be changed if necessary to collect additional samples as needed to fully define suspected problem areas. It is important not to be locked into a set drilling, sampling, and testing program that results in the site being inadequately defined.

The required intensity of site exploration will depend on the size and nature of the project, the geologic complexity, and the stage of progress in the investigation. The primary objective in any program of site characterization should be to acquire a sufficient amount of information so that all geotechnical engineering aspects of a particular project can be adequately defined.

Generally, a soils report is considered valid for one year after it has been issued. If the geotechnical engineer determines that the site conditions have not changed from when the soils report was issued, then the report should be reissued with the current date or a letter should be provided stating that the original report is still valid. In some cases, the reissued report may include supplemental data and analyses.

3.1.3.1   Distribution of Borings
The number and distribution of exploratory borings must consider spatial variability of the soils at the site and the depth to which expansion or collapse will influence potential foundation movement.

There are no specific standards regarding the distribution of borings. Judgment must be exercised both in the planning stage and as data are received in the field. Each site must be judged on its characteristics and geology. Neither the International Residential Code (2012) nor the International Building Code (2012) has any requirements for spacing or depth of borings or test pits. However, the following recommendations have been put forth by others.

- Chen (1988) states, "As a rule of thumb, test holes should be spaced at a distance of 50 to 100 ft. In no case should test holes be spaced more than 100 ft apart in an expansive soil area."
- The US Department of the Army (1983) states, "[S]pacing of 50 or 25 feet and occasionally to even less distance may be required when erratic surface conditions are encountered."

- Chen (2000) states, "Geotechnical engineers in the Rocky Mountain states, when conducting foundation investigation in a subdivision, insist on drilling one test hole for each lot. This is to ensure that the swelling potential and water table conditions are adequately covered." This is also in agreement with recommendations by the Colorado Association of Geotechnical Engineers (CAGE 1996).
- The Australian standard for residential slabs and footings (AS 2870-2011) specifies a minimum of one borehole or test pit per house site (Standards Australia 2011). In areas of deep movement where the soil profile is highly variable, they require a minimum of three boreholes per site.
- The European standard (prEN1997-2) provides guidance on spacing and depth of borings that is similar to that just stated (European Committee for Standardization 2002). However, their recommendations are not directed toward light structures on expansive soils.

Exceptions can be made to the above recommendations for sites where subsurface conditions are well defined and are relatively evenly distributed across the site, but only if the risks of distress due to expansive soil are well understood. In these cases, it may be reasonable to drill one boring for several lots or one boring between two adjacent houses. The Australian standard (AS 2870-2011) allows for that as well if soil profiling indicates predictable uniform soil conditions.

For an isolated single building site, where borings on adjacent lots are not available, the initial borings should be located close to the corners of the foundation. Unless subsurface conditions have been shown to be uniform, there should be at least one boring at each end of the building. For buildings with large footprints such as multifamily dwellings or commercial buildings, some borings interior to the footprint should be included. Additional borings will be dictated by the areal extent of the site and buildings, the location of proposed foundations, and soil conditions as they are encountered.

For large commercial projects, it is common to space boreholes on a 100 ft (30 m) grid. This gives general coverage over the entire area. For a single commercial building or large house, the general pattern of boreholes would be the same as that for ordinary soil sites. The primary governing criterion is that the site be adequately covered to define the general soil profile.

A common misconception is that drilling of test holes is the major cost of the soil investigation. Consequently, there is the tendency to drill as few holes as possible, oftentimes only one. The risk involved in such a policy can be quite large. Erratic soil conditions can exist between widely spaced test holes. This was demonstrated by one commercial project in which the engineer drilled one test hole at each corner of a proposed structure and found similar soil profiles in each hole. No further drilling appeared to be necessary. During construction, deep garbage fills were encountered at the middle of the site. The existence of these fills and the failure to identify them in the initial investigation not only necessitated further drilling but also voided the recommendations given in the report and caused long delays in construction.

Special care should be exercised when dealing with a site containing manmade fill. Many fill materials are expansive, and not uncommonly, the fill is placed at random. Often, it is not possible to delineate the extent of the fill, and extra sampling and testing is warranted. Such conditions should be documented clearly in the soil report.

If geologic conditions are predictable and uniform, the spacing of drill holes may be increased. One example of this was a site in southeastern Minnesota proposed for construction of 120 homes. The soil profile consisted of 3 ft to 30 ft (1 m to 10 m) of highly expansive shale overlying limestone, which, in turn, overlaid sound sandstone. The limestone was flat-lying and predictable, as evidenced by observations in road cuts and exploratory borings. Also, the engineering properties of the shale did not vary widely. Because of the predictability of the geology at this site, a total of less than 20 exploratory holes to the depth of the limestone were sufficient to characterize the entire site.

3.1.3.2   *Depth of Exploration*
The depth of exploration should extend to the entire depth that will influence the performance of the foundation. For an expansive soil site this will be significantly deeper than that for a nonexpansive site. This can be demonstrated by comparing two sites. A footing 3 ft or 4 ft (1 m or 1.2 m) in width will be considered. The depth of the zone of stress influence (i.e., to which the applied stress is less than 10 percent of the footing bearing stress) will be about 4 or 5 footing widths. On a nonexpansive soil site, borings to depths of 20 ft to 25 ft (6 m to 7 m) depth would be adequate. However, for an expansive soil site with even moderate swelling pressures, the depth to which heave can affect the foundation can extend to depths well in excess of 40 ft (12 m). Furthermore, on such sites it is not uncommon for deep foundations to be used. Thus, much deeper exploration is necessary for expansive soil sites.

A geotechnical engineer who is familiar with the geology of the area will have some indication of whether expansive soils should be expected, and the types of foundations that may be considered. The initial boreholes should be of sufficient depth to provide information pertinent to the range of possible foundation systems. The field engineer or technician should be given the responsibility to increase the initially planned depth of exploration if conditions so warrant. Good practice would be to advance the first borehole to a depth sufficient to indicate the probable foundation system that will be needed, and to add some deeper borings if the initial exploration indicates the desirability to do so.

In many areas, the depth to bedrock is the criterion used to determine the depth of the exploratory borings. The problem with that criterion is that in areas where claystone is expansive, the claystone "bedrock" is often the most expansive material in the soil profile. Thus, the depth of exploration, in at least a few holes, should extend to the depth beyond which heave is expected to occur. If bedrock of a nonexpansive nature, such as sandstone or limestone, is encountered, and if the bedrock is within reasonable reach, say within 40 ft (12 m), it is advisable to drill a few holes into the bedrock. In many areas, the depth to the top of the bedrock may be erratic, and the depth to bedrock can change by many feet within a short distance.

If deep foundations are being considered, the depth of exploration should extend sufficiently below the depth to which heave is expected to occur so as to define an adequate anchorage zone for the deep foundation elements. Soil samples should be taken to the expected depth of the anchorage zone, and laboratory testing should be done to that depth. For basement construction, the depth of the test holes should extend below the proposed elevation of the basement to a depth sufficient to investigate the possibility of groundwater becoming a problem in the basement.

3.1.3.3 Sampling Frequency and Depth

Soil sampling programs for different applications or projects will involve different strategies. For example, a sampling program for a building will involve more closely spaced samples at greater depths than that for a highway or airfield project. For expansive soil sites, most applications require that a greater number of samples be taken.

Sampling intervals should be frequent in each borehole. Sampling intervals may be increased for greater depths. Samples should also be taken if a significant change in soil consistency or strength is encountered. Samples should be taken down to depths below the depth of expected heave or to depths below which the loads applied by the structure have a significant effect, whichever is greater.

Continuous core samples of the soil between driven samples should be taken whenever possible. Continuous core allows for inspection of the entire soil column and is useful in identifying small but important

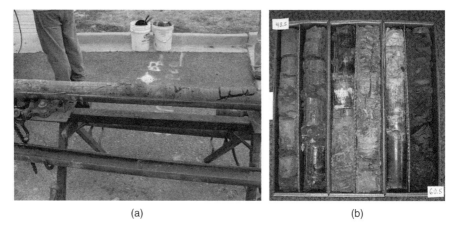

(a)   (b)

FIGURE 3.1. (a) Continuous core sampler; (b) soil sample in core box.

features that would be obliterated during augering. Continuous core samples are taken inside hollow-stem augers between where driven samples are taken. The continuous core is particularly useful for logging the soil profile in that it allows for inspection and visual identification of the in situ soil between other samples. This core is usually taken with a continuous core sampler that is advanced in the hollow-stem auger. Figure 3.1a shows the core barrel and the soil sample. The samples are placed in core boxes as shown in Figure 3.1b and taken to the laboratory for review.

In the laboratory, the core is inspected and the field log is reviewed by the project manager and the senior project engineer, along with the field engineer or technician who logged the hole. Refinements or changes to the log are made as necessary based on the detailed inspection of the core sample. The core is retained until the end of the project, and not uncommonly, it is re-reviewed one or more times if questions arise concerning the soil profile. The core is sufficiently undisturbed to permit detailed visual inspection of soil features, but it is not suitable for testing as an undisturbed sample. The ability to log the undisturbed nature of the soil profile between driven samples is invaluable in identifying small, but influential, features that are destroyed by the auger and cannot be detected in the cuttings.

The practice of taking continuous core samples increases the cost of the drilling, but the overall increase in cost of the geotechnical investigation is less than about 10 percent. The increase in cost is fully justified by the benefits of having the core to review.

Some foundation designs require that the upper few feet of expansive soil be excavated and recompacted in an attempt to decrease the expansion potential. If that is done, soil samples should be prepared to replicate the remolded and recompacted soil conditions and then tested for both compression response when loaded at placement conditions and expansion potential under overburden plus structural loads.

The following case study illustrates an example where the maximum depth of exploration and testing was inadequate.

CASE STUDY—DEPTH OF EXPLORATION

This case involved a geotechnical investigation and foundation recommendations for several one-story, single-family residences near Denver, Colorado (Nelson and Chao 2010). Two exploratory borings were drilled for each residence to depths of 10 and 25 ft. The soils at the site consisted of approximately 15 to 20 ft of silty clay overlying claystone bedrock. Three samples from depths of 2 ft, 6 ft, and 9 ft, respectively, were tested for expansion potential. It was reported that the silty clay exhibited moderate to high expansion potential. The samples of the claystone bedrock were not tested for expansion potential.

The engineering design report recommended that the expansive soils and bedrock be excavated and replaced with compacted fill to a minimum depth of 7 ft below the bottom of the foundation. Spread footings were recommended for the foundation. Figure 3.2 shows the configuration of the recommended excavation and depths of the borings and samples that were tested at one of the residences.

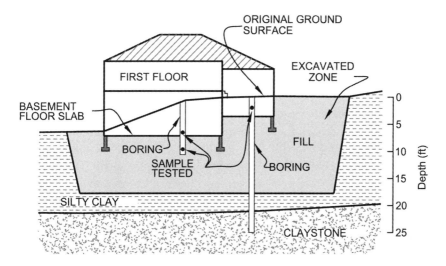

FIGURE 3.2. Soil profile and depth of exploration for a residence near Denver, Colorado.

Within six months after construction, distress to the houses began to be observed. After two years, significant cracks in the drywall and racked doorways were observed. After four years, the maximum out-of-levelness of the structures was reported to be about 3.5 in.

Although the depth of exploration at the site extended to the underlying claystone, as shown in Figure 3.2, only the silty clay was tested for expansion potential. However, all of the soil that was tested was removed during the excavation process, and none of the soil or underlying claystone bedrock in the foundation soils was tested for expansion potential. Thus, the foundation design was prepared with no geotechnical data on the foundation soils. No calculations of expected heave were reported. The forensic investigation indicated that the underlying claystone was highly expansive, and the remaining future free-field heave for the structures was calculated to be up to approximately 12 in.

3.2 FORENSIC INVESTIGATION

A fourth type of geotechnical investigation is a *forensic investigation*. This investigation is undertaken when problems have occurred. The purpose of the forensic investigation is to determine the nature of the geotechnical conditions at the site, to determine the cause of the distress, and to make preliminary recommendations for remediation.

Generally, the scope of such an investigation is similar to that of a preliminary investigation, except that it may be more focused on certain aspects. The forensic investigation is not intended to be comprehensive enough for final design of a remediation plan. The product of a forensic investigation may be a report of the results of the geotechnical investigation or, in the case of litigation, it may be an opinion report in which responsibility for various aspects of the distress is designated.

The resources available for the forensic investigation will are generally related to the severity of the problem and the magnitude of the remedial program. In any case, there are minimum data requirements to determine the cause and extent of the expansion. Because it is often difficult to determine if the structural distress is caused by settlement or heave, it is essential to determine the soil profile and the expansion potential of the foundation soils. Also, information such as sources of water and depth of the natural water table is important.

In many cases, a significant amount of this type of information can be obtained from visual inspection and experience with soil conditions of the area. In many other cases, especially if the damage is great and remedial measures will be extensive, or if litigation is involved, a detailed soil investigation including sampling, testing, and in situ measurements will be required.

Comparison of soil conditions and water contents adjacent to or below the structure or pavement with those at some location away from the structure or pavement will provide valuable information on the likely cause of expansion and the extent to which it may continue. The presence of the structure or pavement usually contributes to changes in the water content and state of stress in the soil.

Also, the investigation must determine the type of the existing foundation system, evaluate the quality of the construction, and determine the degree of distress that the foundation elements have suffered. This will generally require that test pits be excavated next to the grade beam and interior supports. Although the excavations and removal of slabs may be disruptive, they are frequently necessary in the course of stabilizing the foundation. They may or may not seriously impact the habitability of the structure depending on their extent and location and the tolerance of the inhabitants. It is necessary, however, to stress that such investigations are essential. If inadequate site investigations lead to wrong diagnoses and improper remedial measures, the resources that are saved actually represent a waste.

During soil testing, environmental conditions must be considered. For example, consolidation-swell tests conducted on soils that have already reached a maximum swell in the field will not indicate the expansion potential of an initially drier soil and may be misleading. Comparison of test results on samples taken under or near the structure with those for samples taken from a nearby uncovered area may give an indication of expansion that has taken place. Alternatively, the foundation soil can be air dried and rewetted. Comparison of the current moisture condition with that reported at the time of the initial site investigation or construction is an important indicator of potential expansion.

A site investigation can often indicate the nature of the foundation by observing the location and type of damage. Crack patterns can be analyzed with regard to the structural design and movement constraints. Movements can be transmitted to the roof framing, which should be checked. Crack patterns and magnitude of cracks indicate severity and cause of damage. However, care must be taken to evaluate relationships of crack patterns because not all cracking is due to soil volume changes and the cause of a particular crack is not always apparent.

Possible sources of water infiltration should be investigated. A rise in the groundwater table or presence of a perched water table should be observed. Utility lines and sprinkler systems should be checked for leaks.

The type of drainage control, such as gutters and downspouts, gives an indication of possible problems due to surface water intrusion. The location of water spigots and shrubbery and the amount of irrigation that is applied can also indicate possible causes of swelling. It should be noted that factors other than poor drainage control can contribute significantly to heaving, and repair of the drainage system alone usually will not solve the problem.

Elevation surveys are an additional part of forensic investigations that are normally not included in earlier site investigations. These surveys are used to measure differential movement of slabs and foundations. New structures generally exhibit an average out-of-levelness of approximately 0.5 in. (13 mm) (Noorany et al. 2005; Dye 2008). It is difficult to determine if out-of-levelness at some point in time is due to original construction or movement due to expansive soils. Although it is seldom done, good practice would be to perform an elevation survey to measure the foundation and floor elevation at the end of construction. If expansive soils movement is suspected, and a post-construction elevation survey was not conducted, a baseline survey should be established as soon as practical. Subsequent surveys taken over a 3- to 12-month interval will provide evidence of ongoing movement, help to delineate the movement pattern and rate, and provide a useful predictor of future movement and potential problems (Witherspoon 2005).

One particular complication that arises when making elevation surveys at expansive soil sites is the problem of establishing a stable benchmark. There has been more than one project in which the surveyor used a mark on a street curb as the benchmark. Obviously, the heave of the curb made subsequent surveys useless in assessing ongoing movement. To evaluate continuing heave of a structure, a stable benchmark must be used.

In areas with expansive soils, special techniques are necessary to provide benchmarks that are free from ground movements. To prevent the movement of benchmarks due to heaving of expansive soils, they must be anchored below the depths where these movements originate. The term *deep benchmarks* is used because the benchmarks are anchored at depths such that heave of the expansive soil will not cause movement of the benchmarks. The procedure for installation of a deep benchmark in expansive soil is detailed in Chao, Overton, and Nelson (2006) and illustrated in the following case study.

CASE STUDY—DEEP BENCHMARK INSTALLATION

This case study discusses the establishment of three deep benchmarks at an expansive soil site at the Denver International Airport (DIA) (Chao, Overton, and Nelson 2006). These benchmarks were constructed to provide a stable reference point that would not be influenced by expansion of the surface soils and bedrock. Benchmarks constructed using conventional methods are not generally installed to a sufficient depth to be stable in expansive soil or bedrock.

The soils at the DIA site consisted of a layer of silty/sandy clay fill, underlain by silty clay, weathered claystone, sandstone, and claystone bedrock with some coal seams. Extensive laboratory testing was conducted to measure soil properties. For use in predicting the depth of potential heave for design of the deep benchmark, a generalized soil profile was used. The depth of potential heave was calculated to be 89 ft. The required depths of anchorage for three proposed deep benchmarks were calculated to be approximately 120 ft, 98 ft, and 98 ft, respectively.

Samples were collected from the benchmark borings to confirm the expansion potential of the soils. The soil sampled in the benchmark borings varied somewhat from that encountered during the initial borings. This shows the importance for an experienced engineer or geologist to be on site to observe the core that is recovered and note any particular deviations from the assumptions used in determining the depth of the benchmark. Also, it is important to be conservative in selecting the depth of anchorage. It must be kept in mind that although a stratum of material with low expansive potential may be encountered, this may not be the stratum that controls the depth of potential heave. An underlying stratum of material with higher expansion potential may govern the depth of potential heave, and, therefore, the depth of anchorage.

The rods in the deep benchmarks were anchored in concrete at the bottom of the boreholes. For the deepest benchmark the depth was between 110 and 120 ft. For the other two the depths were between 85 and 98 ft. The installation procedure of the deep benchmarks was modified from the procedure for a class A rod benchmark established by the National Geodetic Survey (US Department of Commerce 1978) to account for the expansive soil conditions. The construction of the deep benchmarks is shown in Figure 3.3.

The deep benchmarks have been monitored using precision surveying methods since September 2000. The elevations of the two shallower benchmarks have been measured relative to the deepest benchmark. One of the two benchmarks indicated a relative elevation change of 0.04 in. during the monitoring period. The relative elevation of the other benchmark was the same as the reference. The monitoring results indicate that the deep benchmarks have remained stable during the monitoring period and are reliable as references for elevation monitoring.

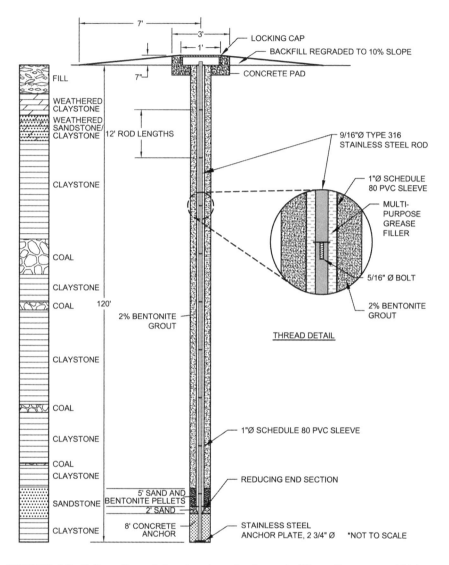

FIGURE 3.3. Soil profile and deep benchmark schematic (Chao, Overton, and Nelson 2006).

References

Chao, K. C., D. E. Overton, and J. D. Nelson. 2006 "Design and Installation of Deep Benchmarks in Expansive Soil." *Journal of Surveying Engineering*, ASCE 132(3): 124–131.

Chen, F. H. 1988. *Foundations on Expansive Soils*. New York: Elsevier Science.

———. 2000. *Soil Engineering: Testing, Design, and Remediation*. Boca Raton, FL: CRC Press.

Colorado Association of Geotechnical Engineers (CAGE). 1996. "Guideline for Slab Performance Risk Evaluation and Residential Basement Floor System Recommendations (Denver Metropolitan Area) plus Guideline Commentary." CAGE Professional Practice Committee, Denver, CO.

Dye, H. B. 2008. *Moisture Movement through Expansive Soil and Impact on Performance*. PhD dissertation, Arizona State University, Phoenix, AZ.

European Committee for Standardization. 2002. Eurocode 7—Geotechnical Design—Part 2: Ground Investigation and Testing, European Standard prEN1997-2, Brussels.

International Building Code (IBC). 2012. *International Building Code*. Falls Church, VA: International Code Council.

International Residential Code (IRC). 2012. *International Residential Code for One-and Two-Family Dwellings*. Falls Church, VA: International Code Council.

Nelson, J. D., and K. C. Chao. 2010. "Depth of Investigation for Foundation Soils." *Proceedings of the 17th Southeast Asian Geotechnical Conference*, Taipai, Taiwan, 184–187.

Noorany, I. M., E. D. Colbaugh, R. J. Lejman, and M. J. Miller. 2005. "Levelness of Newly Constructed Posttensioned Slabs for Residential Structures." *Journal of Performance of Constructed Facilities*, ASCE 19(1): 49–55.

Standards Australia. 2011. Australian Standard, Residential Slabs and Footings—Construction, AS 2870-2011. Standards Australia International Ltd., Sydney, NSW.

US Department of the Army (DA). 1983. "Technical Manual TM 5-818-7, Foundations in Expansive Soils." Washington, DC.

US Department of Commerce, National Oceanic and Atmospheric Administration. 1978. "Geodetic Bench Marks." NOAA Manual NOS NGS 1. Washington, DC.

Witherspoon, W. T. 2005. *Residential Foundation Performance*. Dallas, TX: Association of Drilled Shaft Contractors (ADSC), International Association of Foundation Drilling.

4

Soil Suction

If one were able to attach a stress gauge onto one end of a horizontal water column and soil onto the other end, the soil suction would be the tensile stress that is measured in the water. In soil physics, soil suction is generally referred to as the potential energy state of water in soil relative to some reference state (Jury, Gardner, and Gardner 1991). Because there is no absolute scale of energy, the potential energy state of water in soil is defined as the difference in energy per unit quantity of water compared to the state of pure, free water at a reference pressure, reference temperature, and reference elevation (Bolt 1976). The reference state is generally standard temperature and pressure and is arbitrarily given a value of zero.

Typical units of soil suction are summarized in Table 4.1. The second column in Table 4.1 shows units in pF. The pF value is expressed as the logarithm to the base 10 of the height, in centimeters, of a water column that the soil suction would support. Thus, it is a measure of the suction in units of pressure head. This unit is used fairly frequently in practice, but conventional units of stress (e.g., psi or kPa) are finding more widespread use.

4.1 SOIL SUCTION COMPONENTS

Total soil suction comprises two components, the matric suction and the osmotic suction. The matric suction represents the value of the tension in the water and represents a physical process. It can be thought of as the height to which water can be sucked up into an unsaturated soil, a process that is commonly referred to as *capillary rise*. Figure 4.1a illustrates the capillary rise in a small diameter tube. At the interface between the water and the air the water has a unique structure resulting in a distinct phase known as *surface tension*. The water surface will have a contact angle, β, with the tube. A free body diagram of the air–water interface having a contact angle $\beta = 0$ is shown in the expanded view in Figure 4.1a. This is similar to the tension at the air–water interface in an unsaturated soil. The surface tension is shown as being equal to T_s.

TABLE 4.1   Conversion of Various Soil Suction Units[1]

Height of Water Column (cm)	pF	psi	psf	kPa	bars	atmos.
1	0	0.0147	2.1	10^{-1}	10^{-3}	10^{-3}
10	1	0.147	21	1	10^{-2}	10^{-2}
10^2	2	1.47	210	10	10^{-1}	10^{-1}
10^3	3	14.7	2.1×10^3	10^2	1	1
10^4	4	147	2.1×10^4	10^3	10	10
10^5	5	1,470	2.1×10^5	10^4	10^2	10^2
10^6	6	14,700	2.1×10^6	10^5	10^3	10^3
10^7	7	147,000	2.1×10^7	10^6	10^4	10^4

[1] The values shown in Table 4.1 are rounded values and may not be exactly equal from one column to the other.

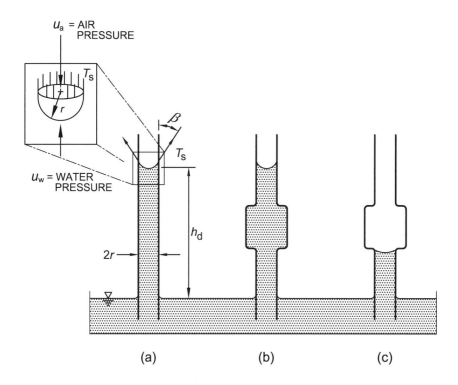

FIGURE 4.1. Capillary rise of water in thin tube (modified from Taylor 1948).

Considering equilibrium of this air–water interface gives

$$T_s(2\pi r) + u_w \pi r^2 - u_a \pi r^2 = 0 \qquad (4\text{-}1a)$$

or

$$(u_a - u_w) = \frac{2T_s}{r} \qquad (4\text{-}1b)$$

where:

r = radius of the tube.

The term $(u_a - u_w)$ is referred to as the *matric suction*. In practical applications in geotechnical engineering, the pore air pressure is taken as atmospheric pressure and the value of $(u_a - u_w)$ is the negative of the pore water pressure. For reasons that will be explained in Section 4.3, the pore water pressure in unsaturated soil is negative and, hence, the value of $(u_a - u_w)$ is positive.

Water can also be drawn into soil as a result of the dissolved salt concentration in the soil water. The salt cations have an affinity for water and if the salt concentration in the soil is greater than that in an external source of water, the external water will be pulled into the soil. If the water is restricted from entering the spaces between the soil particles, the attractive forces from the salt cations cause the water to be under tension. This type of soil suction is different from the matric suction. The forces caused by the imbalance of salt cation concentrations represent an osmotic process. Thus, this part of the soil suction is termed the *osmotic suction*.

This chapter will present the concepts of soil suction and will discuss the two components of suction. Methods of measuring soil suction and techniques that can be used to estimate soil suction will also be discussed.

4.1.1 Matric Suction

The capillary phenomenon in soils is generally presented in terms of the matric suction $(u_a - u_w)$ (McWhorter and Sunada 1977). This was demonstrated in Figure 4.1a by considering the rise of water in the capillary tube. Considering equilibrium of the water in the tube, the weight of the column of water below the air–water interface must equal the surface tension forces at the interface. Thus, assuming that the air pressure above the interface is the same as the air pressure acting in the pan at the bottom of the tube,

$$\gamma_w \pi r^2 h_d = 2T_s \pi r \cos \beta \qquad (4\text{-}2a)$$

or

$$h_d = \frac{2T_s \cos \beta}{\gamma_w r} \tag{4-2b}$$

where:

γ_w = unit weight of water,
h_d = capillary pressure head in the water, and
T_s = strength of the surface tension.

At this point we will introduce a concept known as the inkwell concept that was presented by Taylor (1948). This will be applied to soil at a later point. Two capillary tubes, shown in Figures 4.1b and 4.1c, have a wide spot part way up the tube. This wide spot is analogous to an "inkwell," hence the name *inkwell concept*. The tube shown in Figure 4.1b was initially filled to the top with water and allowed to drain down. It will come to equilibrium when the water has drained to the same height as shown in Figure 4.1a. That figure is labeled the "drainage" condition because water was draining out of the tube. The tube shown in Figure 4.1c was initially dry and was inserted into the water. Water rose in the tube until it reached the wide spot in the tube. Above that point, the diameter of the tube was too large to be able to support a column of water above that height. Consequently, water could not rise beyond that. That figure is labeled the *imbibition* condition because the tube was imbibing water.

Figure 4.2 applies the inkwell concept to capillarity in soil. Figure 4.2a shows a column of granular soil that was originally saturated and then placed in a pan of water. It is labeled the *drainage* condition. Water drained from the soil and the upper part of the soil became unsaturated. A meniscus will form between the soil grains at the air–water interface as shown in the magnified view of a small element of soil in Figure 4.2a. When the water had drained to the point where the surface tension in the water was able to support the weight of the column of water below it in the soil, further drainage did not occur. The height to which the water drained would be equal to h_d as given in Equation (4-2b). The degree of saturation will vary with height down to a point where the pore spaces are small enough to be able to support the column of water. Below that point, the soil will remain saturated. The degree of saturation in the soil column for the drainage condition is shown as a function of height in Figure 4.2c. The height to which the soil remains saturated, h_d, is referred to as the *displacement head* because above that height some of the pore water is displaced by air. Above the height h_d, the degree of saturation varies with height. The

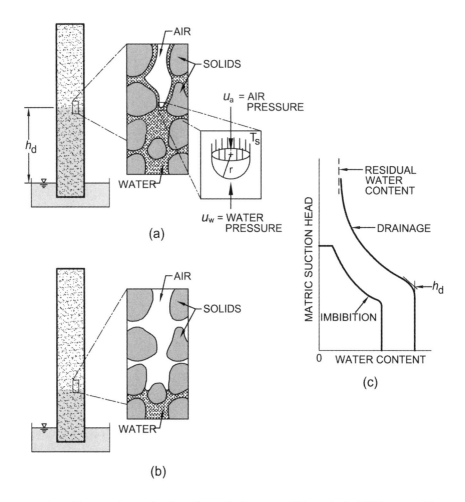

FIGURE 4.2. Matric suction in soil: (a) drainage condition; (b) imbibition condition; (c) soil water characteristic curve (modified from Nelson and Miller 1992).

curve shown in Figure 4.2c is sometimes referred to as the capillary retention curve but more commonly it is referred to as the soil water characteristic curve (SWCC).

Figure 4.2b shows a soil column in which the soil was dry prior to being placed in the pan of water. This figure is labeled *imbibition* because it represents a situation for which water is rising in the soil column. Not all of the pore spaces are small enough to cause water to rise by capillarity, analogous to the "inkwell" concept, and therefore, the soil may not become saturated even in the lower part of the column. The distribution of pore size dictates where and to what height water will rise

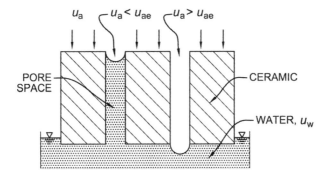

FIGURE 4.3. Illustration of air-entry value for porous stone.

up into the soil. There will be some height above which even the smallest pore size cannot support a water column to that height. The degree of saturation goes to zero at that point. The resulting water content profile is a curve such as the one labeled *imbibition* in Figure 4.2c.

For some applications, such as measuring or controlling soil suction, it is necessary to separate the soil water from a reservoir of water outside the soil. This is often accomplished by means of a fine-grained porous stone. In Figure 4.3, the pore spaces in a porous stone are depicted as a series of capillary tubes. The water in the porous stone can be maintained at a pressure equal to u_w, and the contact angle of the surface tension is taken as zero. The air pressure above the stone is equal to u_a.

If the water pressure is maintained at a value of zero, and u_a is increased to a value where the surface tension is not able to maintain the air–water interface as shown in the tube in the right side of Figure 4.3, the air pressure will displace the water from the porous stone and cause an air bubble to escape into the water below the stone. Considering equilibrium of the air–water interface at that point would show that the air pressure necessary to cause an air bubble to be forced out would be

$$(u_a)_d = \frac{2T_s}{r} \tag{4-3}$$

The similarity of equation (4-3) to equation (4-1b) is noted. The term $(u_a)_d$ is commonly referred to as the *displacement pressure* because that is the air pressure needed to displace the water. It has also been called the *air-entry pressure* because that is the pressure necessary to cause air to enter the stone, as shown in the tube on the left side of Figure 4.3. It has also been termed the *bubbling pressure* for obvious reasons. The air-entry pressure for a porous stone is commonly stated in units of bars.

The displacement head depicted by h_d in Figure 4.2c is more distinct for more ordinary soils than for expansive soils. This is because the soil volume, and hence the effective pore size, changes as the soil suction changes. Furthermore, in expansive soils there are macropores between soil particles and soil aggregates, and micropores within aggregates of clay particles.

4.1.2 Osmotic Suction

Osmotic suction, ψ_o, is caused by differences in salt concentration at different locations in the soil water. In Figure 4.4, a chamber of water on the left side containing a concentrated salt solution is separated by a semipermeable membrane from a chamber on the right side containing pure water with no salts. The semipermeable membrane allows water molecules to pass through it but not salt molecules. The concentration of the solution causes an attraction to water molecules, and hence a tendency for the pure water to flow into the salt solution through the semipermeable membrane. The pressure that the salt solution can exert on the pure water is related to the concentration of the salt in the solution. This pressure is called the osmotic pressure, as given by equation (4-4). If the water on the right side of the membrane is pure water and the salt solution on the left side has a molar concentration of C, the osmotic pressure would be given by the van't Hoff equation, which is shown in equation (4-4). This can be related to the osmotic pressure in terms of pressure head, as shown in Figure 4.4.

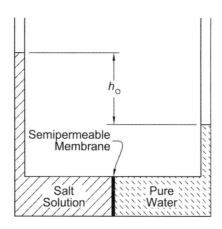

FIGURE 4.4. Osmotic pressure head across a semipermeable membrane.

$$\psi_o = RTC = \rho_s g h_o \tag{4-4}$$

where:

ψ_o = osmotic pressure/suction,
R = universal gas constant,
T = absolute temperature,
C = molar concentration of the solute,
ρ_s = solute mass density,
g = gravitational acceleration, and
h_o = osmotic pressure head.

The manner in which the osmotic suction manifests itself in soil is depicted in Figure 4.5. Two idealized clay particles are shown in close proximity to each other. The electrical charges on the clay particles hold salt cations in the clay micelle. As a result, the concentration of salt in the water in the space between and around the particles is higher than that outside of the micelle. Therefore, the electrical field around the clay particle can be thought of as creating a "pseudo-semipermeable" membrane, as shown in Figure 4.5. The relatively high concentration of salt between the particles causes pressure to be exerted on the water

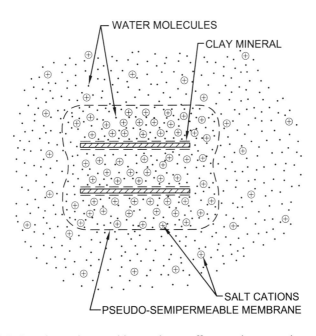

FIGURE 4.5. Pseudo-semipermeable membrane effect causing osmotic suction in clay.

molecules outside of this area of influence. This pressure is the osmotic suction of the soil.

4.1.3 Total Suction

Total suction, ψ, has been shown to be the sum of the matric suction, $(u_a - u_w)$, and the osmotic suction, ψ_o, of the soil (Miller 1996; Miller and Nelson 2006). Thus,

$$\psi = (u_a - u_w) + \psi_o \qquad (4\text{-}5)$$

As may be expected, the matric and osmotic suction will vary with the water content. However, within the range of water content changes that occur for most practical applications in geotechnical engineering, it has been shown that the osmotic suction changes little, if at all. Figure 4.6 shows experimental data presented in Krahn and Fredlund (1972). The data shown in Figure 4.6 were obtained using soil specimens compacted at various initial water contents. Each component of soil suction and the total suction was measured independently. Figure 4.6 shows that the matric suction showed wide variation with initial water content, but the osmotic suction remained fairly constant over the range of water content from 11 to 17 percent. It is noted that the curve for osmotic plus matric suction is nearly parallel to and about equal to that for measured total suction. Because of this, and the fact that total suction is easily measured, the total suction is often used as a surrogate for matric suction. The small difference between the measured value of total suction and the sum of measured matric and osmotic suction is likely the result of different measuring techniques being used for each curve.

4.2 SOIL WATER CHARACTERISTIC CURVE

The soil water characteristic curve (SWCC) defines the relationship between water content and soil suction. Although the SWCC will be influenced by applied stress, it is usually defined for conditions where the normal stress is small. The water content is generally quantified in terms of gravimetric water content, volumetric water content, or degree of saturation. Any one of these parameters can be used to define the SWCC, provided that the reference volume of the soil remains consistent.

Figure 4.2c defined features of the drying (drainage) and wetting (imbibition) portions of soil water characteristic curves. The displacement pressure head, or air-entry pressure, of the soil was shown to be that value of the matric suction, for drainage conditions, below which

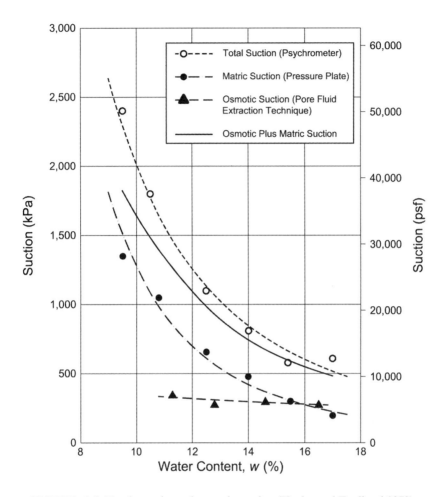

FIGURE 4.6. Total, matric, and osmotic suction (Krahn and Fredlund 1972).

the soil remains saturated. The residual water content is that value where a large suction change is required to remove additional water from the soil and where the SWCC becomes asymptotic to the soil suction axis (McWhorter and Sunada 1977; Corey 1994; Fredlund and Xing 1994).

Figure 4.7 shows SWCCs for wetting and drying conditions for ordinary soil. The format shown in Figure 4.7 is transposed from the one shown in Figure 4.2c. That is the format that is commonly used for expansive soil. The SWCCs shown in Figure 4.2 are ones that are typical for a soil matrix in which a change in suction produces little volume change and the residual water content is fairly well defined. In clayey soils, and especially expansive soils, there is not a well-defined value that defines a residual water content. Instead, there exists a point where

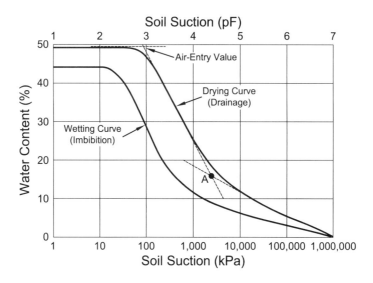

FIGURE 4.7. Soil water characteristic curves for ordinary soil (modified after Fredlund 2000).

a particular change in suction produces a smaller change in water content. That point is labeled A in Figure 4.7. Although this point does not define an actual "residual" water content as would exist for a rigid granular matrix, it does represent a point where the mechanism changes by which water is removed from the soil.

4.2.1   Mathematical Expressions for SWCC

A number of different equations have been proposed to express the functional relationship defined by the SWCC. Some commonly used equations are shown in Table 4.2. Others have been proposed by Gardner (1958), Williams et al. (1983), McKee and Bumb (1984), and Burger and Shackelford (2001), but they are not as widely used.

In all of the proposed equations there are a number of parameters that must be determined for each soil. These parameters, therefore, are actually soil properties that have to be measured by some method. The parameter λ in the Brooks and Corey equation takes into account the pore size distribution of the soil and is termed the *pore size distribution index*. Fredlund and Xing (1994) took into account the pore size distribution curve for the soil, and also included a correction function, $C(\psi)$, to force the soil water characteristic curve to pass through the value of soil suction equal to 10^6 kPa (7.0 pF) at zero water content.

TABLE 4.2   Proposed Soil Water Characteristic Relationships (Modified from Leong and Rahardjo 1997)

Reference	Equation	Unknown Parameters
Brooks and Corey (1964)	$\theta = \begin{cases} \theta_s; & \psi \leq \psi_d \\ \theta_r + (\theta_s - \theta_r)(\alpha\psi)^{-\lambda}; & \psi > \psi_d \end{cases}$	$\theta_r, \alpha, \lambda$
van Genuchten (1980)	$\theta = \theta_r + \dfrac{\theta_s - \theta_r}{[1 + a\psi^b]^c}$	θ_r, a, b, c
Fredlund and Xing (1994)	$\theta = C(\psi) \dfrac{\theta_s}{\left[\ln\left(e + \left(\dfrac{\psi}{a}\right)^b \right) \right]^c}$	$C(\psi), a, b, c$
Fredlund, Rahardjo, and Fredlund (2012)	$w = \begin{cases} = w_u - S_1 \log(\psi); & 1 \leq \psi < \psi_{aev} \\ = w_{aev} - S_2 \log\left(\dfrac{\psi}{\psi_{aev}}\right); & \psi_{aev} \leq \psi < \psi_r \\ = S_3 \log\left(\dfrac{10^6}{\psi}\right); & \psi_r \leq \psi < 10^6 kPa \end{cases}$	$S_1, S_2, S_3, w_{aev}, \psi_{aev}, \psi_r$

where

θ = volumetric water content,

θ_r = residual volumetric water content,

θ_s = saturated volumetric water content (measured in the laboratory, and hence, assumed to be a known parameter),

λ = pore size distribution index,

ψ = soil suction (i.e., matric suction at low suctions and total suction at high suctions),

ψ_d = displacement pressure,

ψ_{aev} = air-entry soil suction,

ψ_r = residual soil suction,

$C(\psi)$ = correction function that forces the volumetric water content to be zero at a soil suction of 10^6 kPa,

e = base of natural logarithm, 2.71828,

w = gravimetric water content,

w_u = gravimetric water content corresponding to a suction of 1 kPa,

w_{aev} = gravimetric water content corresponding to the air-entry suction, ψ_{ae}, and

$a, b, c, \alpha, S_1, S_2,$ and S_3 = fitting parameters.

Even though each equation has its own limitations, all of the proposed equations provide a reasonable fit for soil water characteristic data in the low and intermediate suction ranges.

Because of its simplicity and the fact for many years it was well known, the Brooks and Corey relationship is still used in many applications. However, in more recent applications the van

Genuchten equation and the Fredlund and Xing equation have become more widely used. The equation developed by Fredlund, Rahardjo, and Fredlund (2012) has some advantages for use in computer modeling.

4.2.2   Soil Water Characteristic Curves for Expansive Soils

The equations presented in Table 4.2 were generally developed for ordinary soils, in which the microscale and physicochemical effects discussed in chapter 2 were not a concern. In addition, the general platey shape of the clay particles in expansive soil differs markedly from the shape of the particles in silts and sands for which many of the equations in Table 4.2 were developed. Consequently, the SWCC for an expansive soil will have different characteristics, particularly the lack of a distinct value of displacement pressure head and the lack of a distinct residual water content.

The SWCCs for expansive soil are generally bilinear in nature as shown in Figure 4.8. That figure shows SWCCs for a highly overconsolidated claystone of the Pierre Shale from the Front Range area of Colorado, a normally consolidated black cotton expansive soil taken from a test site in San Antonio, Texas, and claystone of the Denver Formation. A distinct bilinear nature of the SWCC for all three soils is evident as is the hysteresis between the wetting and drying curves.

The bilinear nature of the SWCCs reflects the physicochemical nature of the water interaction between micelles and within the micelles as described above. The bifurcation point on the curves may be considered analogous to the air-entry pressure for granular soils. However, there is not an abrupt change to an unsaturated state at that point. It has also been suggested that this point of bifurcation represents a transition between micropore and macropore spaces (McKeen and Neilsen 1978; Marinho 1994; Miller 1996). The macropore spaces would represent the spaces between aggregates of particles and the micropore spaces would represent the spaces within the aggregates. A number of other researchers have also observed similar behavior for expansive soils (McKeen and Neilsen 1978; Marinho 1994; Chao 1995; Miller 1996; Chao et al. 1998, Fredlund and Pham 2006; Miao, Jing, and Houston 2006; Puppala, Punthutaecha, and Vanapalli 2006; Chao et al. 2008; and Pham and Fredlund 2008, etc.).

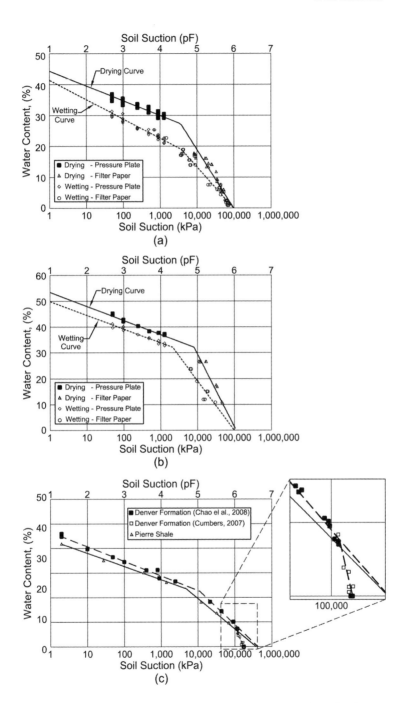

FIGURE 4.8. Bilinear SWCCs: (a) Pierre Shale; (b) Texas expansive soil; (c) Denver Formation and Pierre Shale (data from Chao et al. 1998 and 2008; Durkee 2000; Cumbers 2007).

The bifurcation point of the curves in Figure 4.8 occurs at a suction value between 1,000 kPa and 10,000 kPa (4 pF and 5 pF). Cumbers (2007) showed that there also exists a point at high values of suction at which the slope changes again. This is shown by the dotted line in the expanded part of Figure 4.8c. At this point the water content decreases more rapidly with an increase in suction. It is believed that this point represents the point where the water remaining in the soil is that which entered the micelle during the crystalline phase of expansion. Because that water is not easily removed when the soil is being dried in an oven at 105°C during measurements of water content, it appears as if there is a greater decrease in water content with an increase in suction. In fact, it represents a point where the drying process is not able to remove more water and the measurement of water content decreases more rapidly. Thus, there still is some water of hydration in the soil but it is not removed by air drying or even drying at 105°C.

The curves shown in Figure 4.7 indicate that the value of suction at zero water content is equal to 1,000 MPa (7 pF) (Fredlund and Xing 1994). McKeen (1992) observed the soil suction at zero water content to be near 174 MPa (6.25 pF). Chao (2007) measured the total suction of five oven-dried claystone samples using the filter paper method to be approximately 245 MPa (6.40 pF). Cumbers (2007) data that are depicted by the dotted line in Figure 4.8 intersect the zero suction axis at a total suction value much less than 1,000 MPa (7 pF). Thus, it appears that the soil suction at oven-dry water content lies in the range of about 175 MPa to 250 MPa (6.25 pF to 6.40 pF). It may be expected to differ from one soil to another.

A simplified explanation for the hysteresis between the wetting and drying SWCC was presented earlier based on the "inkwell" analogy. In clay soils, however, the difference between the two curves most likely reflects differences in the wetting mechanism and hydration of the micelles, as previously discussed. For applications relating to expansive soils, it is important to take into account whether the wetting or drying curve should apply. In applications where the water content is increasing with time, such as irrigation and elimination of evapotranspiration, the wetting curve would be most applicable. By contrast, for applications where wetting and drying cycles occur, such as between wet and dry seasons of the year, both curves would need to be used. Computer models would ideally take into account the fact that the stress state of the soil will have an effect on the SWCC. However, from a practical standpoint, some form of an idealized average SWCC is normally used for modeling. Fredlund, Rahardjo, and Fredlund

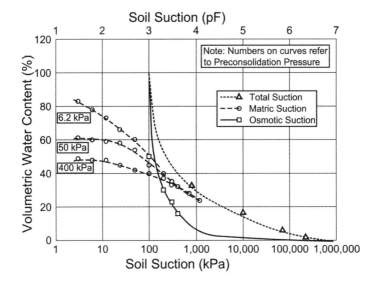

FIGURE 4.9. SWCC for Regina clay under various stress states (modified after Fredlund 2002).

(2012) present a detailed discussion of SWCCs and the various factors that will influence them.

4.2.3 Influence of Stress State on Soil Water Characteristic Relationships

Figure 4.9 shows the SWCCs for Regina clay that were determined under conditions of different applied stress (Fredlund 2002). The SWCCs are influenced by the overburden stress when the soil suction is below approximately 1,000 kPa (10 bars or 4 pF). At higher values of suction the overburden stress made less difference.

4.2.4 Effect of Suction on Groundwater Profiles

The *vadose zone* is the zone of soil above the groundwater table. In that zone, the pore water pressure is negative and the soil suction and water content must be in equilibrium. For a hydrostatic condition with continuous water and no evapotranspiration from the surface, the pore water pressure, u_w, at any point above the water table will be equal to

$$u_w = -z_{wt}\gamma_w \qquad (4\text{-}6)$$

where:

z_{wt} = height above the water table, and
γ_w = unit weight of water.

This hydrostatic pore water pressure is plotted as line A in Figure 4.10a. In an arid climate, such as usually exists in areas where expansive soil is of concern, the evapotranspiration from the soil exceeds the infiltration from precipitation events. As a result, there exists a moisture deficit to some depth below the ground surface. The decrease in water content causes an increase in the soil suction (i.e., the pore water pressure becomes more negative). The resulting pore water pressure profile for that condition is shown as line B in Figure 4.10a. After a site has been developed, the construction of structures and pavements generally causes a reduction in evapotranspiration. Irrigation of landscaping introduces infiltration of water. In that case, a downward moving wetting front will be developed in the soil. Above the wetting front, the soil may be unsaturated, in which case the pore water pressure would be negative but less negative than hydrostatic conditions. If the soil above the wetting front is saturated, the pore water pressure would be zero. The pore water pressure profile for that case is shown as line C in Figure 4.10a.

During a site investigation, it is more common to measure the water content profile and not the pore water pressure, especially when the pore water pressure is negative. Thus, the suction profile can be inferred from the water content profile. Figure 4.10b shows the water content profiles that correspond to the pore water pressure profiles shown in Figure 4.10a. For hydrostatic conditions, the water content profile will follow the SWCC and should look similar to the curve shown in Figure 4.2c. This is shown as line A in Figure 4.10b. For the case where there is a moisture deficit, the water content would be below that for the hydrostatic condition, as shown by line B in Figure 4.10b. For the case where excess water is being introduced at the surface, the water content profile will be as shown by line C in Figure 4.10b. In addition to effects of evapotranspiration, the water content profiles will change with temperature. Water will flow from a warm zone in the soil to a cool zone just due to thermal energy. Thus, the water content profiles will fluctuate between cool seasons and warm seasons.

4.3 MEASUREMENT OF MATRIC SUCTION

The most commonly used methods for measuring soil suction include (1) tensiometers, (2) the axis translation technique, (3) filter paper,

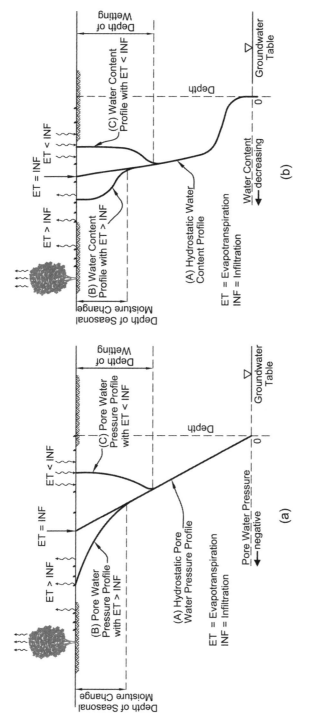

FIGURE 4.10. Soil suction and water content profiles: (a) idealized pore water pressure profiles; (b) idealized water content profiles.

91

(4) thermal conductivity sensors, (5) electrical resistance sensors, (6) the pore fluid extraction technique, and (7) psychrometers. This section will discuss the measurement of matric suction, and following sections will discuss measurement of osmotic and total suction.

A review of methods for measuring soil suction is presented in a number of publications. These include, for example, Fredlund and Rahardjo (1988), Nelson and Miller (1992), Lee and Wray (1992), Rahardjo and Leong (2006), Chao (2007), Bulut and Leong (2008), and Fredlund, Rahardjo, and Fredlund (2012).

4.3.1   Tensiometers

Tensiometers consist of a high air-entry pressure porous stone that is connected to a pressure measuring device such as a manometer or an electronic transducer. The simplest form of a tensiometer is shown in Figure 4.11a. The high air-entry pressure stone is saturated and the tube is filled with deaired water. The stone remains saturated when placed in contact with soil provided that its displacement pressure is greater than the soil suction being measured. It is important to ensure good contact between the stone and the soil. Upon contact, water will migrate between the tensiometer and the soil until equilibrium is reached. At this point, the water in the tensiometer has the same negative pressure as the pore water in the soil. If it is assumed that $u_a = 0$, the measured negative pore water pressure is equal to the matric suction of the soil. In Figure 4.11a, the height h_m is equal to the matric suction head.

The porous tip allows the migration of salts through the stone, and the salt concentration in the tensiometer tube will be the same as that in the water in the tube. Therefore, tensiometers measure only the matric component of suction.

Several types of tensiometers are commercially available. Soilmoisture Equipment Corporation in Santa Barbara, California, markets the *Quick Draw* tensiometer. It has been used with reasonable success for measurement of matric suction in the field and in the laboratory. The *Quick Draw* tensiometer is shown in Figure 4.11b. When using this equipment, the probe is inserted into an access hole made with a coring tool. The vacuum dial gauge is connected to the high air-entry pressure porous tip through a small bore capillary tube and is used to measure the negative pore water pressure.

Tensiometers are limited to use in measuring soil suction less than one atmosphere (100 kPa). Lambe and Whitman (1969) and Fredlund, Rahardjo, and Fredlund (2012) have shown that water in a small chamber can sustain tensions greater than 1 atmosphere (100 kPa) for a short

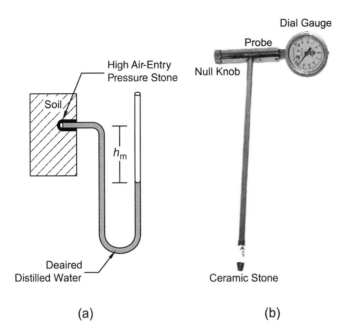

FIGURE 4.11. Tensiometer: (a) simplified diagram; b) Quick Draw tensiometer (photo courtesy of Soilmoisture Equipment Corp.).

period of time. Figure 4.12 shows the tensile stress measured in water. It is seen that the water could sustain the tensile stress for a period of time up to about 2.5 seconds, after which the tensile pressure reduced to less than 1 atmosphere. At that point, cavitation occurred, a process in which the water goes into a vapor phase as a result of the low pressure and forms bubbles in the water line. To reduce the potential for cavitation to occur, the water should be deaired as much as possible and small "nick points" in the apparatus that can serve as nucleation sites must be eliminated. This applies for almost all methods of measuring negative water pressure.

Because of the potential for cavitation to occur, the matric suction that a tensiometer can measure is limited to somewhat less than 1 atmosphere (100 kPa). At suction values greater than this, cavitation of the water in the tensiometer will occur. Figure 4.13 shows data obtained by Sweeney (1982) on residual soils in Hong Kong. The maximum readings obtained were about 90 kPa or 0.9 atmospheres. At suction values above that, it appears that cavitation occurred in the instrument.

Some high-suction-range tensiometers have been developed to measure the in situ matric suction for highly plastic expansive soils (Guan

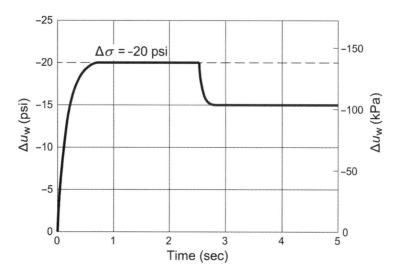

FIGURE 4.12. Tensile stress measured in pore water (Lambe and Whitman 1969).

1996; Guan and Fredlund 1997; Meilani et al. 2002). Guan and Fredlund (1997) indicated that they were able to measure matric suction up to about 12 atmosphere (1,250 kPa) with a high-suction-range tensiometer that they developed.

4.3.2 Axis Translation Technique

The axis translation technique allows for measurement and control of suction over one atmosphere (100 kPa) without cavitation taking place. It was developed by Hilf (1956) to measure matric suction of samples taken from the field. The principle on which the axis translation technique operates is illustrated in Figure 4.14. This figure depicts a soil sample consisting of soil particles, water, and continuous air voids. A soil sample is placed over a high air-entry ceramic stone, and air pressure is applied to the sample. As the air pressure in the soil is increased, water can flow through the porous stone and the pore water retreats to smaller void spaces in the soil. Air will not pass through the stone as long as the displacement pressure of the stone is greater than the applied air pressure, u_a. The water in the soil will remain connected through the stone with the water-measuring device depicted by the gauge marked u_w in Figure 4.14. In this way, the water pressure can be controlled throughout the system at a designated positive value. Matric suction is the difference between the pore air pressure and pore water pressure. The water pressure can be maintained at a positive value and the air

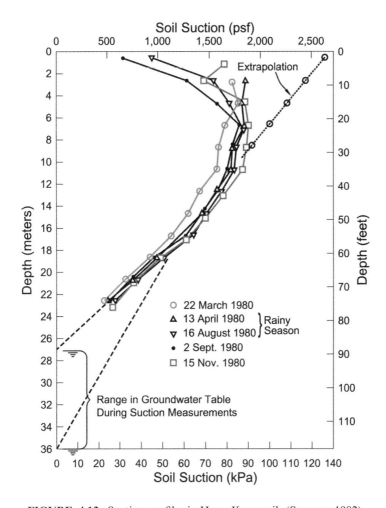

FIGURE 4.13. Suction profiles in Hong Kong soils (Sweeney 1982).

pressure can be increased such that even high values of matric suction can be controlled or measured. This provides a means of controlling the matric suction when conducting tests on unsaturated soils.

To illustrate this, assume that a soil sample having a matric suction of 30 psi (210 kPa) is placed over the high air-entry pressure stone in Figure 4.14. If no water was allowed to flow into or out of the sample and the sample were allowed to remain open to the atmosphere (i.e., u_a = 0), the water pressure gauge would read −30 psi (−210 kPa), which is the negative pore water pressure in the soil. Thus, the matric suction, ψ_m, would be,

$$(u_a - u_w) = 0 - (-30) = 30 \text{ psi} \qquad (4\text{-}7)$$

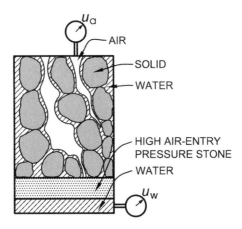

FIGURE 4.14. Axis translation technique.

In this case, the porous stone would be acting as a tensiometer. As discussed, this would not be possible because cavitation would increase in the gauge.

If the same sample is placed over the stone in Figure 4.14, and the water pressure is maintained at 20 psi (140 kPa) while the air pressure is increased to 50 psi (350 kPa), the matric suction is equal to

$$(u_a - u_w) = 50 - 20 = 30 \text{ psi} \tag{4-8}$$

Thus, the matric suction is the same for these two examples, but in the second case, equation (4-8), the water pressure is maintained at a positive value. The axis of reference against which the water pressure was measured was "translated" from 0 to 50 psi. Thus, the technique is called the axis translation technique. By translating the reference axis, both u_a and u_w can be maintained at positive values that can be measured without the potential for cavitation.

This technique is particularly useful for controlling matric suction in laboratory testing programs on unsaturated soils. In triaxial tests or closed-cell consolidation tests, the bottom stone in the test device can be replaced with a high air-entry pressure stone and the air pressure can be controlled through the top cap. In this way, it is possible to control the matric suction, $(u_a - u_w)$, during testing.

Hilf (1956) proposed the use of this technique to measure the suction of soil samples taken from the field. The apparatus used by Hilf is shown in Figure 4.15. A saturated high air-entry probe was inserted into the soil. The soil suction would tend to draw water into the soil.

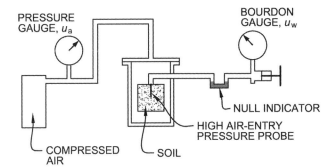

FIGURE 4.15. Hilf's apparatus for measuring soil suction.

A null-indicator system was used to sense the tendency for the flow of water into or out of the sample. The air pressure on the sample was increased until no further flow existed. When equilibration was reached, the value of $(u_a - u_w)$ was equal to the matric suction in the soil.

4.3.2.1   Pressure Plate Apparatus

The pressure plate apparatus is a commonly used device for measuring matric suction by the axis translation technique. In this test, a chamber is divided by a high air-entry pressure plate, as shown in Figure 4.16. In this test, the sample is placed on top of the saturated high air-entry pressure plate and the pore water in the sample comes into contact with the water below the plate. The water pressure can be maintained at any controlled value. For purposes of determining a drainage SWCC, the air pressure is increased in the cell. The water drains from the sample through the porous plate into the water below the high air-entry pressure plate. Air does not pass through the plate as long as the air-entry pressure of the porous plate is not exceeded. To measure a wetting SWCC, water is allowed to be imbibed by the sample through the porous stone from the water reservoir as the air pressure is reduced.

When the test has reached equilibrium (i.e., drainage or wetting has ceased), the matric suction is equal to the difference between the air and water pressure, $(u_a - u_w)$. The water content of the soil is determined at each equilibrium point. The results from this test provide data to plot the SWCC for the soil. The pressure plate apparatus is commercially available from a number of companies.

4.3.2.2   Fredlund SWCC Device

The Fredlund SWCC device was developed to directly measure matric suction using the axis translation technique on samples subjected to

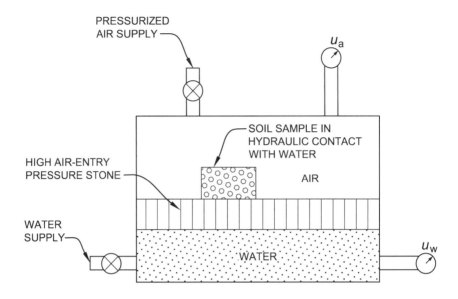

FIGURE 4.16. Pressure plate apparatus.

specific states of stress. This device has the ability to control the matric suction while applying total stress and measuring volume change. It can be used to control the matric suction over a range from near zero up to 1,500 kPa (15 bars or 4.2 pF).

The basic components of the device are shown in Figure 4.17. The device allows the use of a single soil specimen to obtain any number of data points on the SWCC within the range of the high air-entry pressure stone. This overcomes the problem of attempting to obtain several soil samples having identical soil structure. The device also allows for the applied stress to be controlled. Thus, stress-strain tests can be conducted while controlling both stress and matric suction.

4.3.3 Filter Paper Method for Matric Suction

The use of filter paper as a sensor for measuring soil suction has been used routinely by the Water Resources Division of the US Geological Survey for many years (McQueen and Miller 1968). The filter paper method has found widespread use for engineering applications (e.g., McKeen and Nielson 1978; McKeen 1981 and 1985; McKeen and Hamberg 1981; Ching and Fredlund 1984; Houston, Houston, and Wagner 1994; Bulut, Lytton, and Wray 2001; Leong, He, and Rahardjo 2002; Bulut and Wray 2005; Oliveira and Fernando 2006).

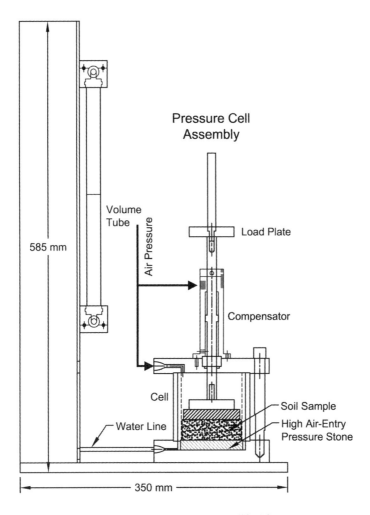

FIGURE 4.17. Fredlund SWCC device (modified from GCTS 2004).

An advantage of the filter paper method is the wide range of soil suction over which it can be used and its simplicity. The only special equipment that is required is an analytic balance capable of weighing to the nearest 0.0001 g.

The procedure for measuring soil suction by the filter paper method is relatively simple and is standardized in ASTM Standard D5298. A soil sample is placed in a small, sealed container such as a glass jar along with a piece of calibrated filter paper. The filter paper will either take up water or dry until the suction of filter paper has come to equilibrium with that of the soil specimen. The filter paper has its own unique relationship between water content and suction. Thus, by measuring the water content of the filter paper, its suction value can be determined, and hence, the suction of the soil as well.

The filter paper method can be used to measure either total suction or matric suction. On the one hand, if the filter paper is kept isolated from the soil, the relative humidity inside the container is controlled by the total suction of the soil, and hence, the suction value that is measured is the total suction. This is termed the *noncontact method*. On the other hand, if it is desired to measure only the matric suction of the soil, the filter paper can be placed in contact with the soil. This is termed the *contact method*. In this case, water transfer will take place by liquid transport and the salt concentration in the filter paper will be the same as that in the soil. Thus, this contact method measures the matric suction. Figure 4.18 shows the placement of the filter papers for measurement of total and matric suction in a container.

4.3.3.1   Principle of Measurement
In the contact method, a soil sample along with calibrated filter papers is placed in a closed container. The filter papers are placed in contact with the soil sample, as shown in Figure 4.18. The jar is sealed carefully and the samples are kept at a constant temperature for a period of time. After equilibrium is reached, usually about 7 days, the filter papers are removed and weighed to the nearest 0.0001 g before and after oven drying. The water content of the filter paper allows the soil suction of the sample to be calculated from the calibration curve of the filter paper. If

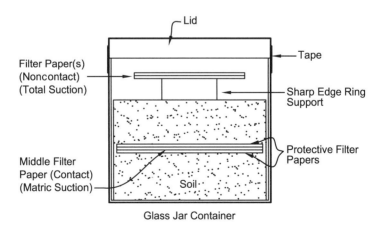

FIGURE 4.18. Placement of filter paper for measurement of total and matric suction by the contact and noncontact methods.

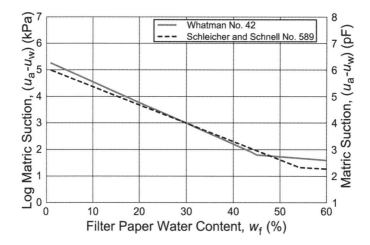

FIGURE 4.19. Filter paper calibration curves. (Reprinted with permission from ASTM Standard D5298-10, Standard Test Method for Measurement of Soil Potential (Suction) Using Filter Paper, copyright ASTM International 100 Barr Harbor Drive, West Conshohocken, PA 19428. A copy of the complete standard may be obtained from ASTM International, www.astm.org.)

calibration of the filter paper is done using the pressure plate apparatus its calibration curve will be for matric suction. The noncontact method for measuring total suction is discussed in Section 4.5.2.

4.3.3.2   Calibration Curves
Calibration curves must be developed for each specific filter paper being used. Variations even between different lots of the same type of filter paper can influence the calibration curves. Figure 4.19 shows calibration curves determined for two commonly used filter papers as presented in ASTM D5298. These curves were determined for wetting (imbibition) conditions. On a semilogarithmic scale, filter paper has a bilinear relationship between suction and water content.

The calibration curves shown in Figure 4.19 were determined using pressure plate and vacuum desiccator methods. The calibration curves shown in ASTM D5298 should be used only when measuring matric suction. Calibration curves for the matric suction measurement were also determined by other researchers (Fawcett and Collis-George 1967; Hamblin 1981; Chandler and Gutierrez 1986; Houston, Houston, and Wagner 1994; Leong, He, and Rahardjo 2002). They showed that there can be significant differences in the calibration curves, even for the same brand of filter paper. The differences in the calibration curves may result from (1) the quality of filter paper, (2) the suction source used in the calibration process, (3) hysteresis of filter paper, (4) equilibration time, and (5) simply variations between the different lots (Leong, He, and Rahardjo 2002; Walker, Gallipoli, and Toll 2005; Rahardjo and Leong 2006). Therefore, prior to use, samples from each lot of filter paper should be calibrated.

4.3.3.3   Filter Paper Hysteresis

Fawcett and Collis-George (1967) have shown experimental evidence of hysteresis for filter paper between desorption and absorption (drying and wetting). Al-Khafaf and Hanks (1974) suggested that since filter paper is always being wetted up during suction measurement, the filter paper should also be calibrated in the same manner to avoid the problem of hysteresis. On the other hand, Leong, He, and Rahardjo (2002) indicated that inadequate equilibration time will lead to larger effects of hysteresis. They concluded that hysteretic effects appear to be minor when equilibration time is sufficient. Nevertheless, it is reasonable that the calibration procedure for the filter paper should take into account whether the filter paper will be wetted or dried in the suction measurements for which it will be used. The ASTM method of test specifies that the filter paper be dried prior to use, and therefore, the calibration curves for wetting should be used, unless other factors dictate otherwise.

4.3.3.4   Time Required to Reach Equilibrium

During a filter paper test, it is important that the equilibration period be sufficiently long to allow enough time for the water content in the paper to reach equilibrium with the suction of the soil. The equilibration time for measuring matric suction using the filter paper contact method has been evaluated by a number of researchers, and the results have varied over a fairly large range depending on the soil type being tested. Table 4.3 lists the equilibration times reported by different investigators. The procedure outlined in ASTM D5298 recommends a minimum equilibration time of 7 days for the filter paper contact method. However, research has shown that the equilibration time is dependent on a large number of factors. These include the type of test (contact or noncontact), material type, initial water content of the soil specimen (suction level), the number of pieces of filter paper used, initial relative humidity of the air, and the soil mass and space in the container. The time required to reach equilibrium when measuring total suction using the filter paper noncontact method is discussed in Section 4.5.2.3 of this chapter.

TABLE 4.3   Summary of Equilibration Time for Matric Suction Measurement Using the Filter Paper Contact Method

Reference	Material Type	Time Required for Equilibrium (days)
Fawcett and Collis-George (1967)	—	6–7
Chandler and Gutierrez (1986)	London clay	5
Greacen, Walker, and Cook (1987)	Kapunda loam	7
Sibley and Williams (1990)	Cellulose, Millipore	3
Lee and Wray (1992)	Sandy clay (SC)	14
Houston, Houston, and Wagner (1994)	Sand, silt, clay	7
Harisson and Blight (1998)	Clayey sand (SC), clayey silt (ML/MH), and clay (CH)	10 days for wetting curve 20 days for drying curve
Burger and Shackelford (2001)	Processed diatomaceous earth (CG1 and CG2)	8 days for $(u_a - u_w) \leq 100$ kPa 14 days for $(u_a - u_w) > 100$ kPa
Leong, He, and Rahardjo (2002)	Clay (CL)	2–5
ASTM D5298 (2010)	—	7

4.3.4   Thermal Conductivity Sensors

Thermal conductivity sensors provide an indirect measurement of matric suction. They consist of a porous ceramic stone containing a heating element and a temperature sensor. Figure 4.20 shows the configuration of a commercially available thermal conductivity sensor developed by Geotechnical Consulting and Testing Systems, Inc. (GCTS). The sensor shown can be used to measure matric suction up to about 1,000 kPa (10 bar or 4 pF). Other instrumentation has also been developed that allows suction measurements up to 1,500 kPa (15 bars or 4.2 pF).

The basic principle on which these sensors work is similar to that for the filter paper method. When they are inserted into unsaturated soil, the sensor will take up or give off water until the suction in the porous stone is in equilibrium with that in the soil. The thermal conductivity of the material making up the porous stone is directly proportional to its water content, and the water content of the ceramic stone can be determined by measuring the rate of heat dissipation from the porous stone. The SWCC of the porous stone can be determined using conventional pressure plate equipment. Thus, the matric suction in the stone, and hence, the soil, can be deduced from the heat dissipation, which is a measure of the water content.

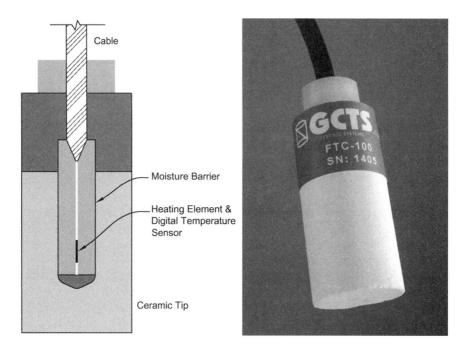

FIGURE 4.20. Thermal conductivity sensor: (a) schematic diagram; (b) FTC-100 sensor (courtesy of GCTS).

In a typical application, the sensor is inserted into a predrilled hole in the soil. After equilibrium is reached between the sensor and the soil, a controlled heat pulse is applied in the center of the porous ceramic stone. The change in temperature is measured over a fixed period of time, and is inversely proportional to the water content of the porous stone. The factory calibration of the sensor allows the user to calculate the matric suction in the soil. Fredlund, Rahardjo, and Fredlund (2012) suggested that effects of ambient temperature and hysteresis of the sensor should be taken into account when measuring matric suction in the soil using the thermal conductivity sensors. Puppala et al. (2012) noted that a good contact between the thermal conductivity sensor and the surrounding soil is essential to obtain reliable suction measurements. They also indicated that the thermal conductivity sensor may be suitable to provide measurements of high matric suction in the field over a sustained period of time.

4.3.5 Electrical Resistance Sensors

Electrical resistance sensors operate on the principle that the electrical conductivity of the sensor material is a function of its water content.

They operate in a manner similar to the thermal conductivity sensors discussed previously. The element of the sensor is placed in the soil and the suction in the sensor comes into equilibrium with that of the soil. The electrical conductivity is measured and correlated with the water content of the sensor. From that, the suction value can be determined.

4.4   MEASUREMENT OF OSMOTIC SUCTION

One method of measuring osmotic suction is a direct measurement using an osmotic tensiometer having a chamber containing a pre-scribed salt solution. Another is an indirect measurement that involves measuring the electrical conductivity of pore water that has been extracted from the soil.

Calibration of suction sensors is conducted by placing them in a closed chamber over salt solutions of various concentrations. The osmotic suction in the air space over a salt solution is a function of salt concentration as given by the following equation (Bulut, Lytton, and Wray 2001):

$$\psi_o = vRTm\phi \tag{4-9}$$

where:

ψ_o = osmotic suction,
v = number of ions from one molecule of salt (ex. $v = 2$ for NaCl, KCl, NH_4Cl, and $v = 3$ for Na_2SO_4, $CaCl_2$, etc.),
R = universal gas constant,
T = absolute temperature,
m = molality or molal concentration of the solution, and
ϕ = osmotic coefficient.

Values of the osmotic coefficient, ϕ, for various salt solutions are listed in Table 4.4. Values of the osmotic suction for the salt solutions shown in Table 4.4 as calculated using equation (4-9) are shown in Figure 4.21. In equation (4-9) the temperature is expressed in terms absolute temperature. Therefore, calculated values of osmotic suction using that equation are not sensitive to the temperature within the range of temperatures encountered in most engineering applications.

4.4.1   Osmotic Tensiometers

An osmotic tensiometer was developed at the University of Saskatchewan to provide a method of direct measurement of ten-sion in the pore water of soils (Bocking and Fredlund 1979). A high

TABLE 4.4   Summary of Osmotic Coefficient of Various Salt
Solutions (Hamer and Wu 1972)

Molality (m)	NaCl	KCl	NH$_4$Cl
0.001	0.988	0.988	0.988
0.002	0.984	0.984	0.984
0.005	0.976	0.976	0.976
0.010	0.968	0.967	0.967
0.020	0.959	0.957	0.957
0.050	0.944	0.940	0.941
0.100	0.933	0.927	0.927
0.200	0.924	0.913	0.913
0.300	0.921	0.906	0.906
0.400	0.920	0.902	0.902
0.500	0.921	0.900	0.900
0.600	0.923	0.899	0.898
0.700	0.926	0.898	0.897
0.800	0.929	0.898	0.897
0.900	0.932	0.898	0.897
1.000	0.936	0.898	0.897
1.200	0.944	0.900	0.898
1.400	0.953	0.902	0.900
1.600	0.962	0.905	0.902
1.800	0.973	0.908	0.905
2.000	0.984	0.912	0.908
2.500	1.013	0.923	0.917
3.000	1.045	0.936	0.926
3.500	1.080	0.950	0.935
4.000	1.116	0.965	0.944
4.500	1.153	0.981	0.953
5.000	1.191	0.997	0.960
5.500	1.231	—	0.966
6.000	1.270	—	0.970

air-entry disk in the tensiometer was placed in contact with the soil. The water in the ceramic disk was separated by a semipermeable membrane from a chamber containing a salt solution. The imbalance of the salt concentration in the two fluids created a situation similar to that shown in Figure 4.4. A pressure gauge measured the pressure in the salt solution. The use of this type of tensiometer is limited primarily to research purposes.

4.4.2   Pore Fluid Extraction Technique

The osmotic suction of a soil has been estimated directly from measuring the concentration of salts in the soil water. Water was extracted from soil by applying sufficient pressure to squeeze the pore fluid from

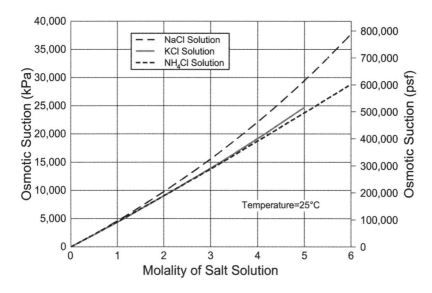

FIGURE 4.21. Osmotic suction versus molality of salt solution.

the soil (Manheim 1966). The osmotic suction of the soil was estimated from the measured salt concentration by means of equation (4-9). The salt concentration can be correlated to the electrical conductivity of the solution (USDA 1950).

Krahn and Fredlund (1972) compared the osmotic suction values obtained from the pore fluid extraction technique to those calculated by taking the difference between the total suction measurements obtained from a psychrometer and matric suction measured in a modified oedometer. They concluded that the extraction technique appears to be a satisfactory way of measuring osmotic suction. Peroni and Tarantino (2003) came to similar conclusions, but they suggested that the results appear to be affected by the magnitude of the extraction pressure applied and the initial water content of the soil sample.

4.5 MEASUREMENT OF TOTAL SUCTION

The measurement of total suction is based on the principle that the energy state of air is a function of its relative humidity, and the energy state of the soil water is the total suction. When a soil sample is placed in a closed volume the humidity of the air will come into equilibrium with the total suction. The relative humidity of the air can be measured by various devices and the total suction of the soil can be calculated using Kelvin's equation. Kelvin's equation is derived from the ideal gas law and is given as

$$\psi = -\frac{RT}{V} \ln \frac{P}{P_0} \qquad (4\text{-}10)$$

where:

> ψ = total suction,
> R = universal gas constant,
> T = absolute temperature,
> V = molar volume of the solution,
> P/P_0 = relative humidity,
> P = partial pressure of pore water vapor, and
> P_0 = saturation pressure of water vapor over a flat surface of pure water at the same temperature.

The total suction as a function of relative humidity calculated for a reference temperature of 25°C is plotted in Figure 4.22. The total suction calculated using equation (4-10) is not sensitive to the temperature within the range of temperatures normally encountered in engineering applications. At very low values of relative humidity, the total suction in expansive soil will be influenced by factors other than those considered in Kelvin's equation, including thermodynamic effects and hydration energy of the cations. Thus, this curve would not be expected to be valid at very high values of suction in expansive soil.

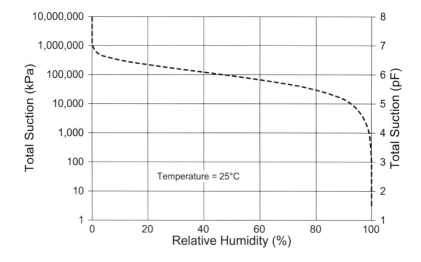

FIGURE 4.22. Total suction as a function of relative humidity as calculated by Kelvin's equation.

Devices used for measuring the humidity use noncontact methods such as psychrometers and noncontact filter paper. Calibration of these devices to measure the energy state of the air is usually done by placing them in a sealed container over a solution of sodium chloride, NaCl, or potassium chloride, KCl, or other suitable solute. In that procedure the osmotic suction of the salt solution is given by equation (4-9). The measurement techniques presented in the next sections include psychrometers and the filter paper noncontact method.

4.5.1 Psychrometers

A psychrometer is basically an instrument for measuring the humidity of the air in the suction measuring device. The thermocouple psychrometer and the chilled mirror psychrometer are discussed below.

4.5.1.1 *Thermocouple Psychrometers*

In basic terms, a thermocouple psychrometer (TCP) consists of a thermocouple in a porous element that is inserted into the soil. The thermocouple is used to determine the relative humidity of the air that is then correlated to the total suction of the soil.

A thermocouple consists of a welded junction joining two dissimilar metals, commonly either copper and constantan or chromel and constantan. Other metals could also be used. When two dissimilar metals are brought into contact, a voltage is produced at the junction, the magnitude of which is proportional to the temperature. This is known as the Seebeck effect. On the other hand, the temperature of the junction can be controlled by adjusting a voltage applied to the junction. The absorption or release of heat by a change in voltage applied across the junction is called the Peltier effect.

In order to measure the temperature with a thermocouple, the junction voltage must be in reference to some base. The reference is often provided by a second thermocouple that is placed in the circuit to provide an opposing voltage. A schematic diagram of the thermocouple circuit in a thermocouple psychrometer is shown in Figure 4.23. The reference junction is kept at a controlled temperature. The voltage E correlates to the difference between the temperature at the reference junction and that at the thermocouple junction. In modern thermocouple psychrometers the reference junction is provided electronically.

The technique used in a thermocouple psychrometer is to apply a voltage, E, to control the temperature at the thermocouple junction. If that junction is in a controlled space, it can be cooled until a drop of water condenses. The temperature at which the droplet condenses

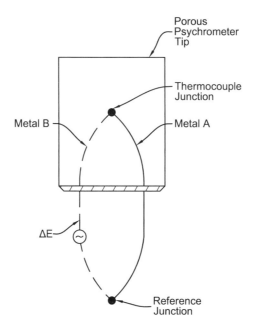

FIGURE 4.23. Peltier-type thermocouple psychrometer.

indicates the dewpoint of the air in that space. From the dewpoint, the humidity, and hence, the soil suction, can be determined.

4.5.1.2 Chilled Mirror Psychrometer

The chilled mirror psychrometer uses the same principle as the thermocouple psychrometer, except in this case the soil sample is placed in the psychrometer, whereas the thermocouple psychrometer is inserted into the soil. The soil sample is placed in a sealed chamber containing a small mirror. The temperature of the mirror is controlled by thermocouple circuitry as was discussed above. The mirror can be cooled to the dewpoint at which time condensation appears on the mirror. From the temperature of the mirror at this point, the vapor pressure of the air in the chamber, and hence, the total suction of the soil can be determined.

4.5.2 Filter Paper Method for Total Suction

The procedure for filter paper measurement of total soil suction is basically the same as that for matric suction, except that the filter paper is not allowed to come into contact with the soil. The placement of the filter paper for both methods was shown in Figure 4.18. The following sections present the principle of measurement and important factors that must be considered when using this technique.

4.5.2.1 Principle of Measurement

The principle of measurement of total suction by the filter paper method is basically the same as that by which psychrometers work. The filter paper is placed in a closed volume over the soil and allowed to come into equilibrium with the humidity in the air, which, in turn, is in equilibrium with the total suction in the soil. In contrast to the method for measuring matric suction, the filter paper is not allowed to come into contact with the soil in this procedure. Because the transfer of water between the soil and the filter paper takes place only by vapor transport, transfer of the solute does not occur. Thus, the water content of the filter paper is in equilibrium with the total suction of the soil. The total suction of the soil sample is determined from the calibration curve of the filter paper.

4.5.2.2 Calibration Curves

It was noted in Section 4.3.3.2 that the calibration curve included in ASTM D5298 should be used only in the measurement of matric suction. Leong, He, and Rahardjo (2002) developed wetting calibration curves for both total and matric suction. The calibration curves for total suction measurements were developed over several reservoirs of salt solutions each having a different concentration. Figure 4.24 shows the calibration curves that were determined for two different makes of filter paper. For suction values greater than about 1,000 kPa (10 bar or 4 pF), the curves for matric suction and total suction follow close together. For suction values below 1,000 kPa (10 bar or 4 pF) the two curves diverge significantly.

Agreement does not exist between researchers regarding the use of these curves. Walker, Gallipoli, and Toll (2005) and Bulut and Wray (2005) suggest that a unique single calibration curve (i.e., the matric suction calibration curve) can be used for both total and matric suction measurements. Harisson and Blight (1998) and Leong, He, and Rahardjo (2002) attributed the differences between the total and matric suction calibration curves to the initial condition of the filter paper. That is, if the calibration curves are for an initially wet filter paper condition (drying curve), then it may be possible that both calibration curves would be similar. However, if the calibration curves are for an initially dry filter paper condition (wetting curve), then the total and matric suction calibration curves would be different, as shown in Figure 4.24. For use in practical applications, the filter paper is usually put into the

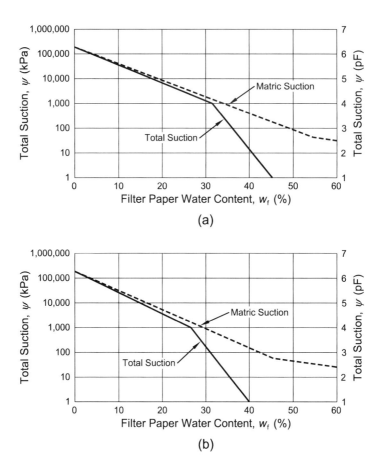

FIGURE 4.24. Filter paper calibration curves for total and matric suction measurements under wetting conditions: (a) Schleicher and Schuell No. 589 filter paper; (b) Whatman No. 42 filter paper (Leong, He, and Rahardjo 2002) (Reprinted with permission from Geotechnical Testing Journal, copyright ASTM International 100 Barr Harbor Drive, West Conshohocken, PA 19428).

container in a condition drier than the soil. Thus, the filter paper would be following a wetting curve, which is the condition under which Leong, He, and Rahardjo (2002) performed their measurements. Thus, it is reasonable to use those types of curves. It is recommended that prior to use of any filter paper, it should be calibrated, and the calibration conditions should follow the actual wetting or drying conditions that will be measured.

4.5.2.3   Time Required to Reach Equilibrium

Various researchers have investigated the time required for the filter paper to reach equilibrium with the total suction in the soil. Table 4.5 shows the time required to reach equilibrium for the filter paper

TABLE 4.5   Summary of Equilibration Time for Total Suction Measurement Using the Filter Paper Noncontact Method

Reference	Material Type	Time Required to Reach Equilibrium (days)
McQueen and Miller (1968)	$Na_2S_2O_3$, Na_2SO_4, and $CaSO_4$ solutions	7
Sibley and Williams (1990)	Cellulose, millipore	10
Lee and Wray (1992)	Sandy clay (SC)	14
Houston, Houston, and Wagner (1994)	Sand, silt, and clay	7
Marinho (1994)	NaCl solution	>30 days ($\psi = 0 - 100$ kPa)
		30 days ($\psi = 100 - 250$ kPa)
		15 days ($\psi = 250 - 1,000$ kPa)
		7 days ($\psi = 1,000 - 30,000$ kPa)
Bulut, Lytton, and Wray (2001)	Fine clay, sandy silt, and pure sand	7
Leong, He, and Rahardjo (2002)	Salt solution	6
Bulut and Wray (2005)	Salt solution	14
Chao (2007)	Remolded claystone	Up to 10 days, depending on water content of the soil sample (refer to Figure 4.25)
ASTM D5298 (2010)	-	7

noncontact method as determined by the various researchers. ASTM D5298 recommends a minimum equilibration time of 7 days for both the contact and noncontact methods. However, Table 4.5 indicates that an equilibration period up to 30 days for the full range of the filter paper calibration over salt solutions might be necessary. For soil samples, an equilibration period of up to 14 days may be required.

Chao (2007) measured the time required for equilibration of filter paper over claystone of the Pierre Shale. Several filter paper discs were stacked over the claystone samples in a plastic specimen container. Individual filter papers were removed after different time periods and their water content was measured.

The equilibration times determined for the claystone samples having various water contents, are depicted in Figure 4.25. The time required to reach equilibrium increased as the water content of the sample increased. If the volumetric water content of the sample is higher than 20 percent, the equilibration time can be longer than the 7 days recommended by ASTM D5298.

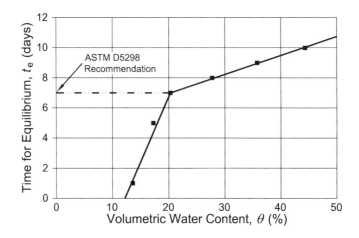

FIGURE 4.25. Variation of equilibration time as a function of soil water content (Chao 2007).

References

Al-Khafaf, S., and R. J. Hanks. 1974. "Evaluation of the Filter Paper Method for Estimating Soil Water Potential." *Soil Science* 117(4): 194–199.

ASTM D5298-10. 2010. "Standard Test Method for Measurement of Soil Potential (Suction) Using Filter Paper." ASTM International: West Conshohocken, PA.

Bocking, K. A., and D. G. Fredlund. 1979. "Use of the Osmotic Tensiometer to Measure Negative Pore Water Pressure." *Geotechnical Testing Journal*, ASTM 2(1): 3–10.

Bolt, G. H. 1976. "Soil Physics Terminology." *Bulletin of the International Society of Soil Science* 49, 26–36.

Brooks, R. H., and A. T. Corey. 1964. "Hydraulic Properties of Porous Media." Hydrology Paper No. 3, Civil Engineering Department, Colorado State University, Fort Collins, CO.

Bulut, R., and E. C. Leong. 2008. "Indirect Measurement of Suction." *Geotechnical and Geological Engineering* 26(6): 633–644.

Bulut, R., and W. K. Wray. 2005. "Free Energy of Water-Suction-in Filter Papers." *Geotechnical Testing Journal*, ASTM 28(4): 1–10.

Bulut, R., R. L. Lytton, and W. K. Wray. 2001. "Soil Suction Measurements by Filter Paper." *Proceedings of the Geo-Institute Shallow Foundation and Soil Properties Committee Sessions at the ASCE 2001 Civil Engineering Conference*, Houston, TX, 243–261.

Burger, C. A., and C. D. Shackelford. 2001. "Evaluating Dual Porosity of Pelletized Diatomaceous Earth Using Bimodal Soil-Water Characteristic Curve Functions." *Canadian Geotechnical Journal* 38, 53–66.

Chandler, R. J., and C. I. Gutierrez. 1986. "The Filter Paper Method of Suction Measurements." *Geotechnique* 36, 265–268.

Chao, K. C. 1995. "Hydraulic Properties and Heave Prediction for Expansive Soil." Master's thesis, Colorado State University, Fort Collins, CO.

———. 2007. *Design Principles for Foundations on Expansive Soils*. PhD dissertation, Colorado State University, Fort Collins, CO.

Chao, K. C., D. B. Durkee., D. J. Miller., and J. D. Nelson. 1998. "Soil Water Characteristic Curve for Expansive Soil." *Proceedings of the 13th Southeast Asian Geotechnical Conference*, Taipei, Taiwan, 1, 35–40.

Chao, K. C., J. D. Nelson, D. E. Overton, and J. M. Cumbers. 2008. "Soil Water Retention Curves for Remolded Expansive Soils." *Proceedings of the 1st European Conference on Unsaturated Soils*, Durham, UK, 243–248.

Ching, R. K. H., and D. G. Fredlund. 1984. "A Small Saskatchewan Town Copes with Swelling Clay Problems." *Proceedings of the 5th International Conference on Expansive Soils*, Adelaide, South Australia, 306–310.

Corey, A. T. 1994. *Mechanics of Immiscible Fluids in Porous Media*. Highlands Ranch, CO: Water Resources Publications.

Cumbers, J. M. 2007. "Soil Suction for Clay Soils at Oven-Dry Water Contents and the End of Swelling Conditions." Master's thesis, Colorado State University, Fort Collins, CO.

Durkee, D. B. 2000. *Active Zone Depth and Edge Moisture Variation Distance in Expansive Soils*. PhD dissertation, Colorado State University, Fort Collins, CO.

Fawcett, R. G., and N. Collis-George. 1967. "A Filter Paper Method for Determining the Moisture Characteristics of Soil." *Australian Journal of Experimental Agriculture and Animal Husbandry* 7, 162–167.

Fredlund, D. G. 2000. "The Implementation of Unsaturated Soil Mechanics into Geotechnical Engineering." *Canadian Geotechnical Journal*, 37, 963–986.

Fredlund, D. G. 2002. "Use of Soil-Water Characteristic Curve in the Implementation of Unsaturated Soil Mechanics." *Proceedings of the 3rd International Conference on Unsaturated Soils*, Recife, Brazil, 887–902.

Fredlund, D. G., and H. Q. Pham. 2006. "A Volume-Mass Constitutive Model for Unsaturated Soils in Terms of Two Independent Stress State Variables." *Proceedings of the 4th International Conference on Unsaturated Soils*, Carefree, AZ, 105–134.

Fredlund, D. G., and H. Rahardjo. 1988. "State-of-Development in the Measurement of Soil Suction." *Proceedings of the International Conference on Engineering Problems on Regional Soils*, Beijing, China, 582–588.

Fredlund, D. G., and A. Xing. 1994. "Equations for the Soil-Water Characteristic Curve." *Canadian Geotechnical Journal* 31(3): 521–532.

Fredlund, D. G., H. Rahardjo, and M. D. Fredlund. 2012. *Unsaturated Soil Mechanics in Engineering Practice*. Hoboken, NJ: John Wiley and Sons.

Gardner, W. R. 1958. "Some Steady State Solutions of the Unsaturated Moisture Flow Equation with Application of Evaporation from a Water Table." *Soil Science* 85(4): 228–232.

Geotechnical Consulting and Testing Systems, Inc. (GCTS). 2004. "*Fredlund SWCC Device Operating Instructions.*" Tempe, AZ.

Greacen, E. L., G. R. Walker, and P. G. Cook. 1987. "Evaluation of the Filter Paper Method for Measuring Soil Water Suction." *Proceedings of the International Conference on Measurement of Soil and Plant Water Status*, Logan, UT, 137–143.

Guan, Y. 1996. *The Measurement of Soil Suction*. PhD dissertation, University of Saskatchewan, Saskatoon.

Guan, Y., and D. G. Fredlund. 1997. "Use of the Tensile Strength of Water for the Direct Measurement of High Soil Suction." *Canadian Geotechnical Journal* 34(4): 604–614.

Hamblin, A. P. 1981. "Filter Paper Method for Routine Measurement of Field Water Potential." *Journal of Hydrology* 53(3/4): 355–360.

Hamer, W. J., and Yung-Chi Wu. 1972. "Osmotic Coefficients Mean Activity Coefficients of Uni-Univalent Electrolytes in Water at 25°C." *Journal of Physical and Chemical Reference Data* 1(4): 1047–1099.

Harisson, B. A., and G. E. Blight. 1998. "The Effect of Filter Paper and Psychrometer Calibration Techniques on Soil Suction Measurements." *Proceedings of the 2nd International Conference on Unsaturated Soils*, Beijing, China, 1, 362–367.

Hilf, J. W. 1956. *An Investigation of Pore-Water Pressure in Compacted Cohesive Soils*. PhD dissertation, Tech. Memo No. 654, US Department of the Interior, Bureau of Reclamation, Design and Construction Division, Denver, CO.

Houston, S. L., W. N. Houston, and A. M. Wagner. 1994. "Laboratory Filter Paper Suction Measurements." *Geotechnical Testing Journal*, ASTM, 17(2): 185–194.

Jury, W. A., W. R. Gardner., and W. H. Gardner. 1991. *Soil Physics* (5th ed.) New York: John Wiley and Sons.

Krahn, J., and D. G. Fredlund. 1972. "On Total Matric and Osmotic Suction." *Soil Science* 114(5): 339–348.

Lambe, T. W., and R. V. Whitman. 1969. *Soil Mechanics*. New York: John Wiley and Sons.

Lee, H. C., and W. K. Wray. 1992. "Evaluation of Soil Suction Instruments." *Proceedings of the 7th International Conference on Expansive Soils*, Dallas, TX, 1, 307–312.

Leong, E. C., and H. Rahardjo. 1997. "Review of Soil-Water Characteristic Curve Equations." *Journal of Geotechnical and Geoenvironmental Engineering*, ASTM 123(12): 1106–1117.

Leong, E. C., L. He, and H. Rahardjo. 2002. "Factors Affecting the Filter Paper Method for Total and Matric Suction Measurements." *Geotechnical Testing Journal*, ASTM 25(3): 322–333.

Manheim, F. T. 1966. "A Hydraulic Squeezer for Obtaining Interstitial Water from Consolidated and Unconsolidated Sediment." Professional Paper 550-C, US Geological Survey, 256–261.

Marinho, F. A. M. 1994. "Shrinkage Behavior of Some Plastic Soils." PhD dissertation, University of London, Imperial College of Science, Technology and Medicine, London, UK.

McKee, C. R., and A. C. Bumb. 1984. "The Importance of Unsaturated Flow Parameters in Designing a Monitoring System for Hazardous Wastes and Environmental Emergencies." *Proceedings of Hazardous Materials Control Research Institute National Conference*, Houston, TX, 50–58.

McKeen, R. G. 1981. "Design of Airport Pavements for Expansive Soils." US Department of Transportation, Federal Aviation Administration, Report No. DOT/FAA/RD-81/25.

———. 1985. "Validation of Procedures for Pavement Design on Expansive Soils." US Department of Transportation, Federal Aviation Administration.

———. 1992. "A Model for Predicting Expansive Soil Behavior." *Proceedings of the 7th International Conference on Expansive Soils*, Dallas, TX, 1, 1–6.

McKeen, R. G., and D. J. Hamberg. 1981. "Characterization of Expansive Soils." *Transportation Research Record* 790, Transportation Research Board, 73–78.

McKeen, R. G., and J. P. Nielson. 1978. "Characterization of Expansive Soils for Airport Pavement Design." US Department of Transportation, Federal Aviation Administration, Report No. FAA-120-78-59.

McQueen, I. S., and R. F. Miller. 1968. "Calibration and Evaluation of Wide Range Gravimetric Method for Measuring Soil Moisture Stress." *Soil Science* 10, 521–527.

McWhorter, D. B., and D. K. Sunada. 1977. *Ground-Water Hydrology and Hydraulics*. Highlands Ranch, CO: Water Resources Publications.

Meilani, I., H. Rahardjo, E. C. Lcong, and D. G. Fredlund. 2002. "Mini Suction Probe for Matric Suction Measurement." *Canadian Geotechnical Journal* 39(6): 1427–1432.

Miao, L. C., F. Jing, and S. L. Houston. 2006. "Soil-Water Characteristic Curve of Remolded Expansive Soils." *Proceedings of the 4th International Conference on Unsaturated Soils*, Carefree, AZ, 997–1004.

Miller, D. J. 1996. *Osmotic Suction as a Valid Stress State Variable in Unsaturated Soils*. PhD dissertation, Colorado State University, Fort Collins, CO.

Miller, D. J., and J. D. Nelson. 2006. "Osmotic Suction in Unsaturated Soil Mechanics." *Proceedings of the 4th International Conference on Unsaturated Soils*, Carefree, AZ, 1382–1393.

Nelson, J. D., and D. J. Miller. 1992. *Expansive Soils: Problems and Practice in Foundation and Pavement Engineering*. New York: John Wiley and Sons.

Oliveira, O. M., and F. A. M. Fernando. 2006. "Evaluation of Filter Paper Calibration." *Proceedings of the 4th International Conference on Unsaturated Soils*, Carefree, AZ, 1845–1851.

Peroni, N., and A. Tarantino. 2003. "Measurement of Osmotic Suction Using the Squeezing Technique." *Proceedings of the International Conference "from Experimental Evidence towards Numerical Modeling of Unsaturated Soils,"* Weimar, Germany, Springer Proceedings in Physics 93(2): 159–168.

Pham, H. Q., and D. G. Fredlund. 2008. "Equations for the Entire Soil-Water Characteristic Curve for a Volume Change Soil." *Canadian Geotechnical Journal* 45, 443–453.

Puppala, A. J., K. Punthutaecha, S. K. Vanapalli. 2006. "Soil-Water Characteristic Curves of Stabilized Expansive Soils." *Journal of Geotechnical and Geoenvironmental Engineering, ASCE* 132(6): 736–751.

Puppala, A. J., T. Manosuthkij, S. Nazarian, and L. R. Hoyos. 2012. "In Situ Matric Suction and Moisture Content Measurements in Expansive Clay During Seasonal Fluctuations." *Geotechnical Testing Journal*, ASTM 35(1): 1–9.

Rahardjo, H., and E. C. Leong. 2006. "Suction Measurements." *Proceedings of the 4th International Conference on Unsaturated Soils*, Carefree, AZ, 81–104.

Sibley, J. W., and D. J. Williams. 1990. "A New Filter Material for Measuring Soil Suction." *Geotechnical Testing Journal*, ASTM 13(4): 381–84.

Sweeney, D. J. 1982. "Some in Situ Soil Suction Measurements in Hong Kong's Residual Soil Slopes." *Proceedings of the 7th Southeast Asian Geotechnical Conference*, Hong Kong, 1, 91–106.

Taylor, D. W. 1948. *Fundamentals of Soil Mechanics*. New York: John Wiley and Sons.

US Department of Agriculture (USDA). 1950. "Diagnosis and Improvement of Saline and Alkali Soils." Agricultural Handbook No. 60. USDA, Washington, DC.

Van Genuchten, M. T. 1980. "A Closed-Form Equation for Prediction the Hydraulic Conductivity of Unsaturated Soils." *Soil Science Society of America Journal* 44(5): 892–898.

Walker, S. C., D. Gallipoli, and D. G. Toll. 2005. "The Effect of Structure on the Water Retention of Soil Tested Using Different Methods of Suction Measurement." *International Symposium on Advanced Experimental Unsaturated Soil Mechanics*, Trento, Italy, 33–39.

Williams, J., R. E. Prebble, W. T. Williams, and C. T. Hignett. 1983. "The Influence of Texture, Structure, and Clay Mineralogy on the Soil Moisture Characteristics." *Australian Journal of Soil Research* 21, 15–32.

5

State of Stress and Constitutive Relationships

The behavior of a foundation on expansive soil is related directly to the response of the soil to changes in the stress state of the soil. Expansive soils are a subset of the more global context of unsaturated soil, and the state of stress is defined in a manner somewhat different than for ordinary saturated soils. Of particular relevance to the state of stress in expansive soils is the fact that one of the stress state variables relates to changes in water content. Section 5.1 presents the appropriate concept of stress state for an expansive soil.

The response of soil to changes in the state of stress is characterized by relationships between the stress state variables and the physical properties of the soil. These relationships are termed *constitutive relationships*. In the case of expansive soil, the soil response of primary concern is volume change. Volume change is related to both change in applied stress and change in water content. Therefore, the relationship between water content and the stress state variables is also of concern. Sections 5.2 and 5.3 will discuss the constitutive relationships for volume change and water content.

5.1 STATE OF STRESS AND STRESS STATE VARIABLES

At any particular point within a soil mass, the state of stress is defined by a system of stresses acting in different directions. For a single phase material such as steel or aluminum, the stress is expressed as the force applied per unit area. The state of stress is defined by specifying the normal and shear stresses acting on each of three orthogonal planes at a point. This is illustrated in Figure 5.1. The normal stresses are designated by the symbol σ and the shear stresses by the symbol τ. The first subscript for each stress designates the direction of the outer normal vector[1] of the face on which the stress is acting, and the second

---

[1] The outer normal vector is a vector perpendicular to the face of the element acting in the direction away from the element.

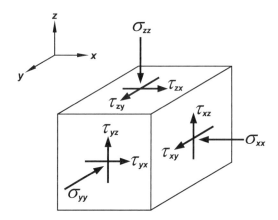

FIGURE 5.1. State of stress at a point in a solid material.

subscript designates the direction in which the stress is acting. Thus, for example, the stress σ_{zz} is acting on the face whose outer normal is in the z direction, and the stress is also acting in the z direction. Thus, a stress for which the two subscripts are the same is a normal stress. The stress τ_{zx} is also acting on the same face but is acting in the x direction. Thus, this stress is a shear stress. The sign convention used herein is defined in Figure 5.1. Compressive stresses are positive. Shear stresses are positive if the direction in which they are acting has the same sign (positive or negative) as that of the outer normal of the face on which they are acting. In Figure 5.1 the shear stresses are shown as being positive.

The state of stress shown in Figure 5.1 can be written as a tensor in the form shown in equation (5-1). The properties of the tensor are such that if the values for the stresses in the brackets are known, the stresses at any other orientation such as σ_{ij} can be determined. Thus, an equation of the form given by equation (5-1) defines the stress at any orientation at that point. For a complete discussion of a stress tensor, the reader is referred to any book on continuum mechanics or theory of elasticity.

$$\sigma_{ij} = \begin{bmatrix} \sigma_{xx} & \tau_{xy} & \tau_{xz} \\ \tau_{yx} & \sigma_{yy} & \tau_{yz} \\ \tau_{zx} & \tau_{zy} & \sigma_{zz} \end{bmatrix} \quad (5\text{-}1)$$

At any point there exists a particular orientation such that the shear stresses on all three orthogonal planes are zero. These planes are called the principal planes and the stresses are the principal stresses. Convention dictates that the normal stresses on those planes are designated as

σ_1, σ_2, and σ_3 where:

$$\sigma_1 \geq \sigma_2 \geq \sigma_3 \tag{5-2}$$

The largest stress, σ_1, is called the major principal stress, with σ_2 and σ_3 being called the intermediate and minor principal stresses, respectively. In terms of the principal stresses, the stress tensor can be written as

$$\sigma_{ij} = \begin{bmatrix} \sigma_1 & 0 & 0 \\ 0 & \sigma_2 & 0 \\ 0 & 0 & \sigma_3 \end{bmatrix} \tag{5-3}$$

In contrast to materials such as metals and plastics, which compose a single solid material, soils are made up of three phases: solid, liquid, and gas. Saturated soils contain no gas and comprise two phases, the solid phase consists of a mineral and the liquid phase consists of water. When a stress is applied to a saturated soil, some of the stress is carried by the mineral and some is carried by the water. The sum of the two stresses is the total stress at that point. Thus, for the normal stress,

$$\sigma = \sigma' + u_w \tag{5-4}$$

where:

σ = total stress,
σ' = component carried by the mineral component, termed the *effective stress*, and
u_w = pore water pressure.

Equation (5-4) can be rewritten in the form shown in equation (5-5), which describes the principle of effective stress.

$$\sigma' = \sigma - u_w \tag{5-5}$$

It should be noted that, because the pore fluid has no shear strength, equation (5-5) does not apply to shear stresses.

The state of stress shown in Figure 5.1 is defined in terms of the variables σ_{xx}, τ_{xy}, and so on. Because the stress variables define the state of stress, they are termed *stress state variables*. For a saturated soil, the effective stress is the stress state variable that governs the behavior of the soil. It is the controlling variable for shear strength of a soil, and it controls the volume change of the soil when subjected to stress.

In an unsaturated soil, there are three phases: solid, liquid, and gas (i.e., mineral, water, and air). This gives rise to the need for another stress state variable in order to consider the contribution of each phase in carrying the applied stress. The stress, or pressure, in each of the mineral, water, and air are designated σ, u_w, and u_a, respectively. Analogous to equation (5-5), there are three possible stress state variables that can be defined by these parameters. Those are shown in equations (5-6a, b, and c):

$$\sigma' = \sigma - u_w \tag{5-6a}$$

$$\sigma'' = \sigma - u_a \tag{5-6b}$$

$$\psi_m = u_a - u_w \tag{5-6c}$$

These stress state variables are generally referred to as the effective stress, σ', the net normal stress, σ'', and the matric suction, ψ_m. If any two of equations (5-6) are known, the third one can be determined. Thus, only two of the three stress state variables can be considered as being independent.

Matyas and Radhakrishna (1968) and Barden, Madedor, and Sides (1969) considered that two of those independent stress state variables can describe the behavior of soil. They presented a relationship for volume changes of unsaturated soil in terms of $(\sigma - u_a)$ and $(u_a - u_w)$. Fredlund and Morgenstern (1977) showed experimentally that these are valid stress state variables for unsaturated soil. They showed that each variable acts independently in governing the behavior of the soil. A detailed and very comprehensive discussion of the state of stress in unsaturated soil can be found in Fredlund and Rahardjo (1993) and Fredlund, Rahardjo, and Fredlund (2012).

For unsaturated soil, and particularly expansive soil, the net normal stress, σ'' or $(\sigma - u_a)$, and the matric suction, $(u_a - u_w)$, are the two stress state variables that are most commonly used to define the state of stress. In general terms, the state of stress in an unsaturated soil is defined by the two stress tensors shown in Equations (5-7a and b).

$$\sigma'' = \begin{bmatrix} (\sigma_x - u_a) & \tau_{xy} & \tau_{xz} \\ \tau_{yx} & (\sigma_y - u_a) & \tau_{yz} \\ \tau_{zx} & \tau_{zy} & (\sigma_z - u_a) \end{bmatrix} \tag{5-7a}$$

$$\psi_m = \begin{bmatrix} (u_a - u_w) & 0 & 0 \\ 0 & (u_a - u_w) & 0 \\ 0 & 0 & (u_a - u_w) \end{bmatrix} \tag{5-7b}$$

In terms of principal stresses, the state of stress would be

$$\sigma'' = \begin{bmatrix} (\sigma_1 - u_a) & 0 & 0 \\ 0 & (\sigma_2 - u_a) & 0 \\ 0 & 0 & (\sigma_3 - u_a) \end{bmatrix} \tag{5-8}$$

and the suction tensor would be the same as that in equation (5-7b).

It should be noted that in the stress tensor for $(u_a - u_w)$, the off-diagonal terms are always zero. This reflects the fact that, since water and air are both fluids, they cannot resist shear stress, and the pressure is the same in all directions. Thus, this is an isotropic tensor and is the same for both the principal stress and the general stress state.

Several early proposals have been presented for the definition of effective stress in unsaturated soils. Most of these proposals attempted to use a single stress state variable. One equation that has found widespread use, and is still in use by some investigators, was an equation proposed by Bishop (1959) having the form

$$\sigma' = (\sigma - u_a) + \chi(u_a - u_w) \tag{5-9}$$

where χ is the chi parameter.

If that equation is used in constitutive relationships and compared with equations (5-7) or (5-8), it is seen that the value of χ will be different when the equation is being used to describe shear strength than when it is being used to describe volume change. It also differs for different soils. Thus, such a parameter would be a material property. A stress state variable that would satisfy equations of equilibrium cannot depend on material properties. Another argument against the use of the χ parameter rests in the fact that χ has been shown to vary with water content. Since water content is a function of the matric suction, $(u_a - u_w)$, as was discussed in chapter 4, χ would also be a function of $(u_a - u_w)$. Thus, the χ parameter cannot legitimately define the state of stress in a soil. A comprehensive discussion of the χ parameter is presented in Morgenstern (1979) and Fredlund, Rahardjo, and Fredlund (2012).

The constitutive relationships presented next use the net normal stress and the matric suction as the independent stress state variables that control the soil behavior. This is consistent with the comprehensive and detailed treatment of the mechanics of unsaturated soils presented by Fredlund, Rahardjo, and Fredlund (2012).

5.2 STRESS–VOLUME RELATIONSHIPS

For a saturated soil under confined one-dimensional loading, the most commonly used constitutive relationship is defined by an equation of the form shown in equation (5-10):

$$\Delta e = -C_c \Delta \log(\sigma - u_w) \tag{5-10}$$

where:

e = void ratio,
σ = normal stress, and
u_w = pore water pressure.

The constitutive parameter C_c is termed the compression index.

For an unsaturated soil, two independent stress state variables must be used to define the state of stress. Most commonly the stress state variables $(\sigma - u_a)$ and $(u_a - u_w)$ are used. When the second stress state variable is introduced, equation (5-10) takes the form

$$\Delta e = -[C_t \Delta \log(\sigma - u_a) + C_m \Delta \log(u_a - u_w)] \tag{5-11}$$

where:

C_t = compression index with respect to the net normal stress, and
C_m = compression index with respect to the matric suction.

Equation (5-11) is written in terms of changes in void ratio, but it can also be written in terms of strain, or percent strain. For an element of soil with initial void ratio e_0 that undergoes a volume change of Δe, the volumetric strain is as given by equation (5-12a). For an expansive soil that undergoes swell, the percent swell would be as given in equation (5-12b).

$$\varepsilon_{vol} = \frac{\Delta e}{1 + e_0} \tag{5-12a}$$

$$\varepsilon_{s\%} = \varepsilon_{vol} \times 100 = \frac{\Delta e}{1 + e_0} \times 100 \tag{5-12b}$$

Equation (5-11) could be written in terms of either equation (5-13) or (5-14).

$$\varepsilon_{vol} = \frac{\Delta e}{1 + e_0} = -\left[\frac{C_t}{1 + e_0} \Delta \log(\sigma - u_a) + \frac{C_m}{1 + e_0} \Delta \log(u_a - u_w)\right] \tag{5-13}$$

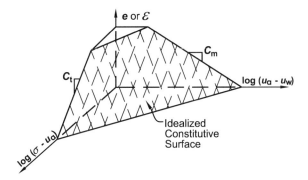

FIGURE 5.2. Idealized three-dimensional constitutive surface for unsaturated soils.

and

$$\varepsilon_{s\%} = - \left[\frac{C_t}{1 + e_0} \Delta \log(\sigma - u_a) + \frac{C_m}{1 + e_0} \Delta \log(u_a - u_w) \right] \times 100 \quad (5\text{-}14)$$

The constitutive surface for equations (5-11), (5-13), or (5-14) can be represented as a planar surface, as shown in Figure 5.2. This constitutive surface will be used in later sections to explain stress paths during various loading conditions.

5.3 STRESS–WATER RELATIONSHIPS

In a fashion similar to the stress–volume relationships, a constitutive relationship relating volumetric water content, θ, to the stress state variables can be expressed as

$$\theta = \frac{V_w}{V} = F_1(u_a - u_w) + F_2(\sigma - u_a) + F_3(\psi_o) \quad (5\text{-}15)$$

where:

$$V_w = \text{volume of water in an element of soil,}$$
$$V = \text{total volume of soil,}$$
$$F_1, F_2, \text{ and } F_3 = \text{functions of the various stress state variables, and}$$
$$\psi_o = \text{osmotic suction.}$$

For practical engineering applications, the osmotic suction is not generally used as a stress state variable. Also, in general use the constitutive relationship relating water content to matric suction, $(u_a - u_w)$, is generally determined for a constant applied normal stress. Thus, the matric suction is normally the only stress state variable used to define the soil–water constitutive relationship. The constitutive equation then takes the form

$$\theta = F_1(u_a - u_w) \tag{5-16}$$

The function F_1 is the soil water characteristic relationship (SWCR), and the graphical form is referred to as the soil water characteristic curve (SWCC) that was discussed in chapter 4. The effect of applied net normal stress (i.e., the function F_2) was also discussed in chapter 4.

References

Barden, L., A. O. Madedor, and G. R. Sides. 1969. "Volume Change Characteristics of Unsaturated Clay." *Journal of the Soil Mechanics and Foundation Division*, ASCE 95(SM1): 33–52.

Bishop, A. W. 1959. "The Principle of Effective Stress." *Teknisk Ukeblad*, Norwegian Geotechnical Institute, 106(39): 859–863.

Fredlund, D. G., and N. R. Morgenstern. 1977. "Stress State Variables for Unsaturated Soils." *Journal of the Geotechnical Engineering Division*, ASCE 103(5): 447–466.

Fredlund, D. G., and H. Rahardjo. 1993. *Soil Mechanics for Unsaturated Soils*. New York: John Wiley and Sons.

Fredlund, D. G., H. Rahardjo, and M. D. Fredlund. 2012. *Unsaturated Soil Mechanics in Engineering Practice*. Hoboken, NJ: John Wiley and Sons.

Matyas, E. L., and H. S. Radhakrishna. 1968. "Volume Change Characteristics of Partially Saturated Soils." *Geotechnique* 18(4): 432–448.

Morgenstern, N. R. 1979. "Properties of Compacted Soils." *Proceedings of the 6th Pan American Conference on Soil Mechanics and Foundation Engineering*, Lima, Peru, 3, 349–354.

6

Oedometer Testing

When Terzaghi first set forth his concept of effective stress, his hypothesis was verified on the basis of experimental data obtained using a piece of equipment termed the *oedometer* (Terzaghi 1925 and 1943). This apparatus is also termed a *consolidometer*. This test has found widespread use for measurement of soil properties used in computing expected settlement of foundations. The test entails one-dimensional loading of a laterally confined soil sample to replicate geostatic conditions. Thus, it was only natural for this type of testing to be used for determining the requisite parameters for computing expected heave, or collapse, of unsaturated soils (Chen 1988; Nelson and Miller 1992; Fredlund, Rahardjo, and Fredlund 2012).

The use of the oedometer test to measure swelling has a distinct advantage over other tests because the testing equipment is commonly available, and most geotechnical engineers are familiar with the testing methods. A typical oedometer is shown in Figure 6.1.

For expansive soils two basic types of oedometer tests are commonly performed. The most common type is the consolidation-swell (CS) test, in which the sample is initially subjected to a prescribed vertical stress in the oedometer, and inundated under that vertical stress. The vertical strain that occurs due to wetting, termed the percent swell, $\varepsilon_{s\%}$, is measured. After the swelling has been completed the sample may or may not be subjected to additional vertical load. The stress that would be required to restore the sample to its original height is termed the "consolidation-swell swelling pressure," σ''_{cs}.[1]

The constant volume (CV) test is another common method of test. In this test the sample is initially subjected to a prescribed vertical stress, but during inundation the sample is confined from swelling and the stress that is required to prevent swell is measured. That stress is termed

---

[1] The term *pressure* is normally used to denote an isotropic stress such as hydrostatic pressure. The swelling pressure is one component of an anisotropic state of stress. Nevertheless, in the interest of maintaining consistency with common usage, the term *pressure* will be used when referring to swelling pressure.

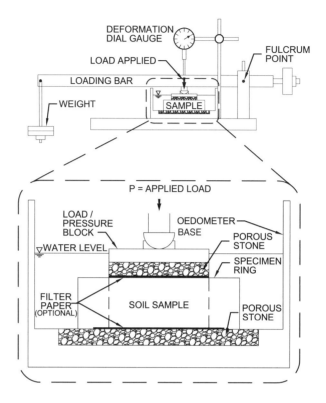

FIGURE 6.1. Oedometer apparatus.

the *constant volume swelling pressure, σ''_{cv}.* These tests will be described and discussed in detail below.

Standards for the performance of oedometer tests to measure expansion potential are set forth in ASTM D4546. In contrast to ASTM D2435/D2435M, which relates to compressibility of nonswelling soils, these test methods were developed specifically for expansive soils.

Oedometer test results are normally plotted in the form of vertical strain as a function of the applied stress, which is plotted on a logarithmic scale. The typical forms of each test are shown in Figure 6.2. Although the test results are plotted in two-dimensional form, it must be recognized that the percent swell takes place as a result of change in suction due to inundation. Thus, Figure 6.2 actually represents the projection of a three-dimensional plot onto a plane. This will be discussed in more detail at a later point with reference to Figure 6.3.

Oedometer testing is a rigorous test method that can provide reliable results. Testing procedures and correction factors that must be taken

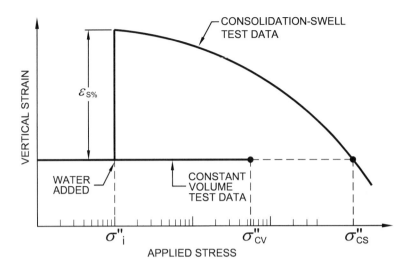

FIGURE 6.2. Oedometer test results.

into account to ensure accuracy are discussed in the following sections. The final stress conditions in these tests assume that the matric suction has been reduced to zero and, thus, represents the maximum swelling that can occur. The engineer using these results must recognize the difference between field and test conditions, and take into account the actual stress state conditions that are represented.

6.1 CONSOLIDATION-SWELL AND CONSTANT VOLUME TESTS

In the CS test, a soil specimen is placed in a consolidation ring and subjected to a prescribed vertical stress, termed the *inundation stress, σ_i''*. After loading under the inundation stress for a period of time, usually about 24 hours, the specimen is inundated and allowed to swell while still being loaded at the inundation stress. The inundation stress may represent the overburden stress, overburden stress plus the applied load from the structure, or some other arbitrary value. An inundation stress of 500 psf (24 kPa) or 1,000 psf (48 kPa) is commonly used for foundations. Inundation stresses of 200 psf (10 kPa) are commonly used for pavement designs. After swelling, the specimen is subjected to additional load in increments, and may be unloaded in decrements.

Figure 6.2 showed the two-dimensional depiction of oedometer test results. In consideration of the fact that the constitutive relationship is three-dimensional in nature as shown in Figure 6.3, the actual stress

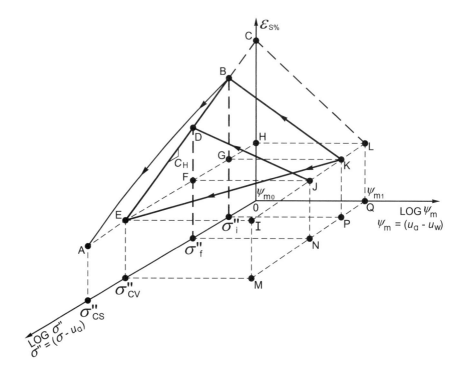

FIGURE 6.3. Stress path for soil expansion (data from Reichler 1997).

paths are three-dimensional in nature as well. The stress paths are shown in Figure 6.3. The initial state of the soil in an oedometer test is represented by the point labeled K in Figure 6.3. At that point the soil suction is equal to a value designated as ψ_{m1}. The net normal stress is that at which the sample will be inundated, i.e., the inundation stress, σ_i''. When the sample is inundated, the suction is reduced to ψ_{m0} and the soil swells along the path KB. The projection of that stress path on the plane defined by the axes for $\varepsilon_{s\%}$ and log σ'' is the line GB. The sample is then loaded back to its original height along the path BA. The value of stress corresponding to point A is the consolidation-swell swelling pressure, σ_{cs}''.

In a conventional constant volume (CV) oedometer test, the sample begins at point K, but because it is constrained from swelling it develops a confining stress as the suction decreases to ψ_{m0} and the stress path would be along a line such as KE. The value of stress corresponding to point E is the constant volume swelling pressure, σ_{cv}''. Due to hysteretic effects, the value of σ_{cv}'' is generally less than that of σ_{cs}''. The reason for this is somewhat intuitive in that it should be easier to prevent water molecules from entering into the soil lattice than to force the water out once it has entered into the soil. All of the factors contributing to the hysteresis are not known. However, one important reason for this hysteresis is because soil expansion takes place in two distinct ways. As was discussed in chapter 2, the initial expansion is due to hydration of adsorbed cations on the soil particles. This is *crystalline* expansion. After that, expansion is a result of *osmotic* expansion in which the soil is probably developing a diffuse double layer (Norrish 1954).

For general purposes, one could argue that a sample inundated at σ''_{cv} would exhibit no swell. Although nonlinearity and secondary effects may indicate that this is not exactly true, this assumption is accurate enough for purposes of computing heave (Justo, Delgado, and Ruiz 1984; Reichler 1997; Nelson, Reichler, and Cumbers 2006; Fredlund, Rahardjo, and Fredlund 2012).

We can consider next a sample inundated at a stress condition for which the initial stress conditions correspond to those existing at some point in an actual soil stratum. Hypothetical conditions are labeled as point J in Figure 6.3. When inundated that sample would swell along a stress path such as JD. Point D will fall between points B and E. It has been shown that the line BDE is close to being a straight line (Justo, Delgado, and Ruiz 1984; Reichler 1997; Nelson, Reichler, and Cumbers 2006). Thus, the slope of the line BDE defines the relationship between the percent swell, $\varepsilon_{s\%}$, that a soil will exhibit when wetted and the stress to which it is subjected when wetted. The slope of that line is a constitutive parameter, C_H, which is termed the *heave index*.

It is important to note that as shown in Figure 6.3, the line BDE, which defines C_H, depicts the expansion due to suction changes. Thus, it is a constitutive relationship that recognizes both of the independent stress state variables, σ'' and $(u_a - u_w)$.

For practical purposes, it is not necessary to plot the entire three-dimensional stress paths in order to determine C_H. Figure 6.4 shows the projection of the stress paths from Figure 6.3 onto the $\varepsilon_{s\%}$ and log σ'' plane. The results of both the consolidation-swell test and the constant volume test are shown as the paths GBA and GFE, respectively.

The parameter C_H, is the slope of the line BDE in Figure 6.4 and is given by equation (6-1):

$$C_H = \frac{\varepsilon_{s\%}}{100 \times \log\left[\frac{\sigma''_{cv}}{\sigma''_i}\right]} \qquad (6\text{-}1)$$

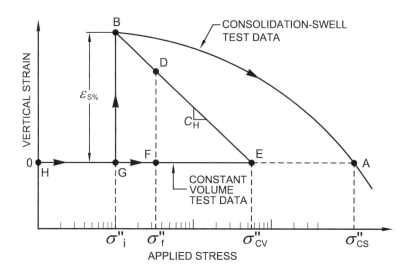

FIGURE 6.4. Determination of heave index, C_H.

where:

$\varepsilon_{s\%}$ = percent swell corresponding to the particular value of σ''_i
expressed as a percent, and

σ''_i = vertical stress at which the sample is inundated.

The value of C_H can be determined, therefore, from the results of a CS test and a CV test using identical samples of the same soil. In practice, it is virtually impossible to obtain two identical samples from the field. Therefore, it is convenient to develop a relationship between σ''_{cs} and σ''_{cv} such that C_H can be determined from a single CS test. The relationship between σ''_{cs} and σ''_{cv} will be discussed in Section 6.3.

6.2   CORRECTION OF OEDOMETER TEST DATA

Oedometer test data requires corrections to ensure reliable results (Fredlund 1969; Chen 1988; Nelson and Miller 1992; Fredlund, Rahardjo, and Fredlund 2012). Corrected values of the swelling pressure and swelling index may differ significantly from uncorrected values due to (1) compressibility of the oedometer apparatus and (2) specimen disturbance during sampling. Fredlund, Rahardjo, and Fredlund (2012) recommended that a correction first be applied for the compressibility of the oedometer apparatus itself, and then a second correction be applied for sampling disturbance. The effects of these corrections on the soil expansion properties are discussed next.

6.2.1 Correction for Oedometer Compressibility

Desiccated expansive soils are generally not very compressible. They may have been subjected to high preconsolidation pressures, and high swelling pressures may be developed during tests. At the stresses that are encountered, the compressibility of the oedometer apparatus may be significant. The ASTM D2435/D2435M standard recommends that the oedometer test data be corrected for oedometer compressibility whenever the calibration correction exceeds 0.1 percent of the initial specimen height and in all tests where filter paper is used. The deflection of the apparatus can be measured by substituting a smooth copper, brass, or hard steel disk for the soil specimen, and performing a loading and unloading cycle. The measured oedometer data may then be corrected for each load increment and decrement. Examples 6.1 and 6.2 demonstrate the calculation for both types of tests and show the results.

EXAMPLE 6.1

Given:

Uncorrected CS test data and oedometer compressibility readings are shown in columns (1), (2), and (3) of Table E6-1. The dial gauge for the oedometer test readings was initially set at a reading of 1.000 in. An increase in reading indicates downward movement of the top of the soil sample in the oedometer. The initial thickness of the soil sample was 1.0 in.

Find:

1. Plot uncorrected and corrected oedometer test results.
2. Determine the percent swell, $\varepsilon_{s\%}$, and the CS swelling pressure, σ''_{cs} for both conditions.

Solution:

Part 1:
The calculation procedure to correct the CS oedometer test data is shown in Table E6-1. Figure E6-1 plots the uncorrected and corrected CV oedometer test data.

Part 2:
Because column (3) is the same before and after inundation, the value of $\varepsilon_{s\%}$ is the same for the uncorrected and corrected data (i.e., $8.8 - (-0.3) = 9.1 - 0.0$). The values of CS swelling pressure, σ''_{cs}, for the uncorrected and corrected data can be read off Figure E6-1b or interpolated in Table E6-1. (In the interpolation, it must be noted that the stress is plotted on a log scale.) The uncorrected and corrected values are 8,833 and 10,822, respectively. This represents a difference of 23 percent.

TABLE E6-1   Correction of Consolidation-Swell Test Data for Oedometer Compressibility

(1)	(2)	(3)	(4)	(5)	(6)
Applied Stress (psf)	Uncorrected CS Test Dial Reading (in.)	Oedometer Compressibility Reading (in.)	Corrected CS Test Reading (in.)	Uncorrected Vertical Strain (%)	Corrected Vertical Strain (%)
			$(2)-(3)$	$\frac{1.000\,\text{in.}-(2)}{1.0\,\text{in.}} \times 100$	$\frac{1.000\,\text{in.}-(4)}{1.0\,\text{in.}} \times 100$
100 (seating)	1.0000	0.0000	1.0000	0.0	0.0
1,000	1.0030	0.0027	1.0003	−0.3	0.0
1,000 (inundated)	0.9120	0.0027	0.9093	8.8	9.1
2,000	0.9340	0.0050	0.9290	6.6	7.1
4,000	0.9620	0.0079	0.9541	3.8	4.6
8,000	0.9940	0.0110	0.9830	0.6	1.7
(8,833) (interpolated)	—	—	—	0.0	—
(10,822) (interpolated)	—	—	—	—	0.0
16,000	1.0360	0.0142	1.0218	−3.6	−2.2
4,000 (unload)	1.0080	0.0131	0.9949	−0.8	0.5

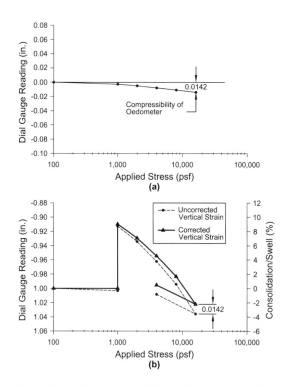

FIGURE E6-1. Corrections for compressibility of oedometer: (a) compressibility of oedometer; (b) correction applied to CS test; (c) correction applied to CV test.

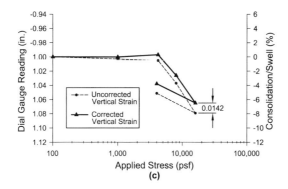

FIGURE E6-1. (*continued*)

EXAMPLE 6.2

Given:

Uncorrected CV test data and oedometer compressibility readings are listed in columns (1), (2), and (3) in Table E6-2. The dial gauge for the oedometer test readings was initially set at a reading of 1.000 in. The initial thickness of the soil sample was 1.0 in.

Find:

Plot uncorrected and corrected oedometer test results.

Solution:

The calculation procedure to correct the CV oedometer test data is shown in Table E6-2. The uncorrected and corrected CV oedometer test data are plotted in Figure E6-1.

TABLE E6-2   Correction of Constant Volume Test Data for Oedometer Compressibility

(1)	(2)	(3)	(4)	(5)	(6)
Applied Stress (psf)	Uncorrected CV Test Dial Reading (in.)	Oedometer Compressibility Reading (in.)	Corrected CV Test Reading (in.)	Uncorrected Vertical Strain (%)	Corrected Vertical Strain (%)
			(2)–(3)	$\frac{1.000\,\text{in.}-(2)}{1.0\,\text{in.}} \times 100$	$\frac{1.000\,\text{in.}-(4)}{1.0\,\text{in.}} \times 100$
100 (seating)	1.0000	0.0000	1.0000	0.0	0.0
1,000 (inundation)	1.0030	0.0027	1.0003	−0.3	0.0
4,200*	1.0051	0.0080	0.9971	−0.5	0.3
8,000	1.0371	0.0110	1.0261	−3.7	−2.6
16,000	1.0791	0.0142	1.0649	−7.9	−6.5
4,000 (unload)	1.0511	0.0131	1.0380	−5.1	−3.8

*4,200 psf represents applied vertical stress where dial gauge begins to shown sample compression.

ASTM D2435/D2435M recommends that porous stones be used at the top and bottom of the specimen in the oedometer test so that the sample can drain or imbibe water. ASTM D4546 states that the stones shall be smooth ground and fine enough to minimize intrusion of soil into the stone if filter paper is not used. It should also reduce false displacements caused by seating of the soil specimen against the surface of porous stones. A suitable pore size of the stone is 10 μm if filter paper is not used. The porous stones may contribute a significant part of the oedometer compressibility. Therefore, it is necessary to use the actual porous stones that will be used in the tests when measuring the compressibility of the oedometer apparatus.

The ASTM D4546 standard recommends against the use of filter paper when measuring the swell of stiff natural clays and compacted soils because of its high compressibility. If filter paper is used, the ASTM recommends the use of dry filter paper, but significant errors can be introduced by using dry filter paper. Figure 6.5 shows the results of oedometer tests conducted on a sample of filter paper in which a steel block was used in place of the soil. Significant deformation of the filter paper took place on the first loading after inundation. Subsequent unloading and reloading cycles show much less deformation. After about the third cycle of loading, the hysteresis loop became smaller and the loading–unloading cycles of filter paper were nearly the same.

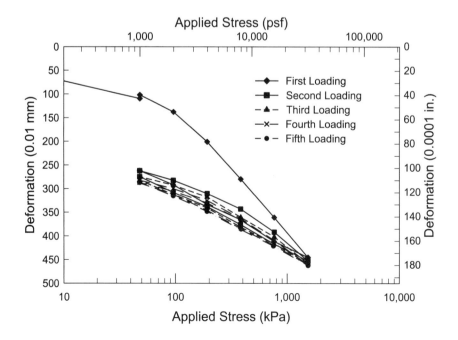

FIGURE 6.5. Compressibility of filter paper.

Therefore, it is recommended that if filter paper is used, each individual filter paper being used for the test should be calibrated. The filter paper should be inundated and it should be loaded and unloaded at least three times in the calibration apparatus.

6.2.2 Correction for Specimen Disturbance in the CV Test

Disturbance to the soil structure during sampling can cause a reduction in matric suction, and that can cause the measured value of the CV swelling pressure to be underestimated. The effect of specimen disturbance on the stress paths for the CV test is depicted in Figure 6.6. In Figure 6.6, the initial test conditions are represented by point 0. During the initial load application, the specimen follows an undefined path between points 0 and 2. At point 2 the specimen is inundated and the effective stress increases so as to prevent volume change. If the specimen were totally undisturbed, it would follow path 2–A during inundation and A–5 during subsequent loading. However, due to specimen disturbance, the specimen follows the path 2–3 during inundation and 3–4–5 during subsequent loading. The swelling pressure at point A for the undisturbed specimen is significantly larger than that at point 3 for the

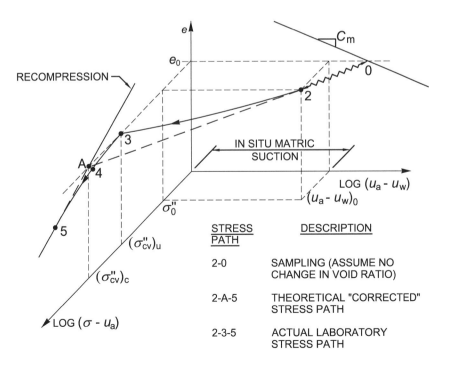

FIGURE 6.6. Three-dimensional constant volume test data (modified from Nelson and Miller 1992).

disturbed specimen. Thus, it is necessary to correct the laboratory curve to determine the corrected swelling pressure.

Based on the stress paths shown in Figure 6.6, Nelson and Miller (1992) proposed the following method for correction for sample disturbance. Figure 6.7a shows the part of the curve that is observed in the laboratory tests. The points in Figure 6.7a are labeled to be consistent with corresponding points in Figure 6.6:

1. Plot the laboratory curve adjusted for oedometer compressibility (2′–3′–5′-N).
2. Draw a line tangent to the curve along the segment just below point 5′. Extend this line to where it meets the extension of line 2′–3′.
3. The corrected swelling pressure, σ_{cv}'', is the intersection between the tangent to the curve and the horizontal line (point A).

This method is similar to the method proposed by Schmertmann (1955) to determine the maximum past pressure for an overconsolidated clay in the oedometer test.

Fredlund, Rahardjo, and Fredlund (2012) proposed an alternate method for correction of swelling pressure that is similar to Casagrande's (1936) method to determine the maximum past pressure of overconsolidated clay. Their suggested procedure is shown in Figure 6.7b and is described as follows:

1. Determine the point of maximum curvature, point O, where the void ratio versus pressure curve bends downward onto the recompression branch.
2. Draw a horizontal line OA and a tangent line OB at the point of maximum curvature.
3. Draw a line OC that bisects the angle AOB.
4. Draw a line DE tangent to the curve that is parallel to the slope of the rebound curve.
5. The corrected swelling pressure σ_{cv}'' is designated as the intersection of lines OC and DE.

6.2.3   Effect of the Corrections on Expansion Properties

Fredlund (1969) stated that percentage errors without the corrections can be in excess of 100 percent for the swelling pressure and up to 50 percent for the swelling index. Fredlund, Rahardjo, and Fredlund (2012) compared corrected and uncorrected swelling pressure data for constant volume tests. They indicated that the corrected swelling pressure could be 300 percent more than the uncorrected swelling pressure.

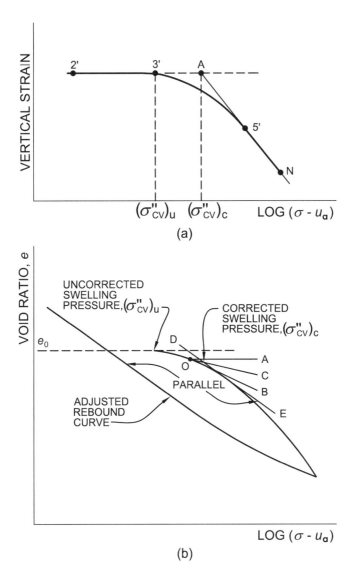

FIGURE 6.7. Correction of constant volume test data for specimen disturbance: (a) procedure proposed by Nelson and Miller (1992); (b) procedure proposed by Fredlund, Rahardjo, and Fredlund (2012).

The authors' experience has been that for highly overconsolidated claystones, the difference can be up to 100 percent or more. The effect of the correction required is greater at the higher values of swelling pressure. As shown in Example 6.1, the percent swell, $\varepsilon_{s\%}$, is not affected by the correction, but the swelling pressure can be influenced greatly.

6.3   RELATIONSHIP BETWEEN CS AND CV SWELLING PRESSURES (THE m METHOD)

As shown in Figure 6.2, the swelling pressure, σ''_{cs}, determined in the CS test is significantly higher than the swelling pressure, σ''_{cv}, measured in the CV test (Edil and Alanazy 1992; Reichler 1997; Nelson, Overton, and Chao 2003; Nelson, Reichler, and Cumbers 2006). Reasons for the value of σ''_{cs} to be greater than σ''_{cv} have been discussed previously. They relate to the concept of crystalline and osmotic swell, as was discussed in chapter 2. Simply put, it is easier to keep the sample from swelling than it is to recompress it.

As was shown in Figure 6.4, both the percent swell at a particular value of σ''_i and the CV swelling pressure are needed to determine the constitutive parameter C_H. In normal geotechnical engineering practice, generally only the CS test is conducted, and only the CS swelling pressure is measured. Therefore, it is convenient to have a relationship between the swelling pressure, σ''_{cs}, as measured in the CS test, and σ''_{cv}, so that the value of C_H can be determined from only a single CS test.

A number of investigators have proposed relationships between σ''_{cv} and σ''_{cs}. These include Edil and Alanazy (1992), Reichler (1997), Bonner (1998), Thompson, Perko, and Rethamel (2006), Nelson, Reichler, and Cumbers (2006), and Nelson et al. (2012). Those investigators have generally proposed some form of equation that uses a simple ratio between σ''_{cv} and σ''_{cs}. Another approach can be derived by observations of oedometer test results from a series of tests performed on identical samples (Nelson and Chao 2014). For ease of reference this method will be referred to as the "m method." Oedometer tests performed on the same soil at different values of σ''_i have shown that as σ''_i increases, the swelling pressure obtained in the CS test decreases (Gilchrist 1963; Reichler 1997). An idealized form of their data is shown in Figure 6.8. If the soil does not swell when it is inundated (i.e., a constant volume oedometer test), the inundation stress would correspond to the swelling pressure. Thus, if σ''_{cs} is plotted against σ''_i, the value of σ''_{cs} will converge to σ''_{cv} at the point where σ''_i is equal to σ''_{cv}. This is shown in Figure 6.9. In that figure, the values of σ''_{cs1}, σ''_{cs2}, and σ''_{cs3} are plotted against the corresponding values of σ''_{i1}, σ''_{i2}, and σ''_{i3}. Point M in Figure 6.8 corresponds to point M in Figure 6.9 and represents the point where σ''_i equals σ''_{cv}. This point plots on a line with a slope of 1:1 in Figure 6.9. It represents the point where the soil was wetted at an inundation stress that is equal to the constant volume swelling pressure. Data for remolded soils taken from Gilchrist (1963) and Reichler (1997) are shown in Figure 6.10, which confirm the idealized form shown in Figure 6.9.

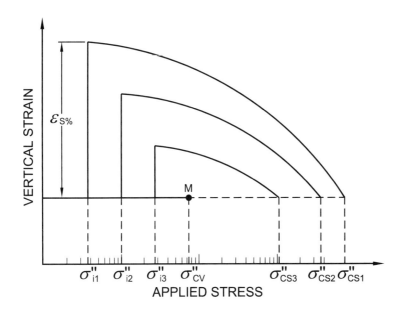

FIGURE 6.8. Idealized oedometer test results for different values of σ_i''.

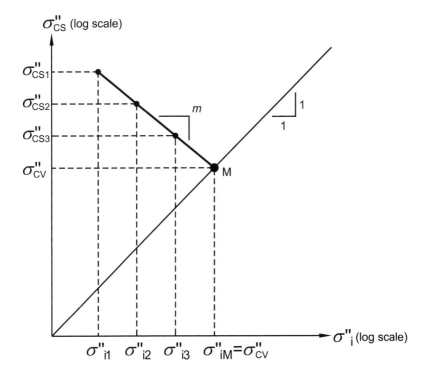

FIGURE 6.9. Convergence of σ_{cs}'' and σ_i'' to σ_{cv}''.

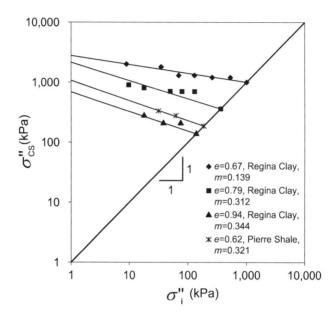

FIGURE 6.10. Relationship between σ_{cs}'' and σ_i''. (Pierre shale data from Reichler [1997]; Regina clay data from Gilchrist (1963) as presented in Fredlund, Rahardjo, and Fredlund [2012].)

The equation of the line in Figure 6.9 can be written for any particular value of σ_{cs}'' and its corresponding value of σ_i'' as Equation (6-2).

$$\frac{\log \sigma_{cs}'' - \log \sigma_{cv}''}{\log \sigma_{cv}'' - \log \sigma_i''} = m \tag{6-2}$$

where:

  m = slope of the line.

It is noted that although the slope of the line in Figure 6.9 is negative, the order of the terms in equation (6-2) is written such that the value of m is positive. Thus, the value of m is actually the absolute value of the slope of the line in Figure 6.9.

Equation (6-2) can be rewritten to obtain the relationship between σ_{cs}'' and σ_{cv}'' as

$$\log \sigma_{cv}'' = \frac{\log \sigma_{cs}'' + m \times \log \sigma_i''}{1 + m} \tag{6-3}$$

The parameter, m, depends on the particular soil, its expansive nature, and other properties of the soil.

The authors have compiled a database of corresponding values of σ_{cs}'' and σ_{cv}'' based on various sources including their own data, review of many soils reports, and values published in the literature (Gilchrist 1963; Porter 1977; Reichler 1997; Bonner 1998; Feng, Gan, and Fredlund 1998; Fredlund 2004; Al-Mhaidib 2006; Thompson, Perko, and Rethamel 2006). The types of the soils that were analyzed included claystone, weathered claystone, clay, clay fill, and sand-bentonite.

Figure 6.11 shows values of m for each sample sorted according to the value of m. The data indicate that all values of m fall below 2. In Figure 6.11, there appear to be two groups of data. The values for the clay tend to be grouped at lower values than those of the claystone or shale. This is believed to be due to physicochemical differences and the existence of diagenetic bonds in the highly overconsolidated claystone and shale. Histograms for those two groups of data are shown in Figure 6.12.

Equation (6-3) was developed on the basis of observed behavior of expansive soil in oedometer tests. Thus, it has a rational engineering basis. The values of m computed for the different soils do not range widely. When considering the wide variety of soil types to which these values of m correspond, this is a remarkably small range of m values.

Computations of heave for hypothetical strata for the remolded Regina clay, the undisturbed Pierre shale, and the Texas clay showed that within the ranges of values appropriate for a particular soil type, the computed heave is not overly sensitive to the value of m (Nelson and

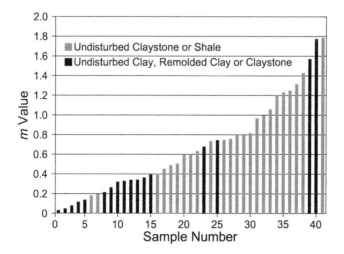

FIGURE 6.11. Calculated values of m sorted by value.

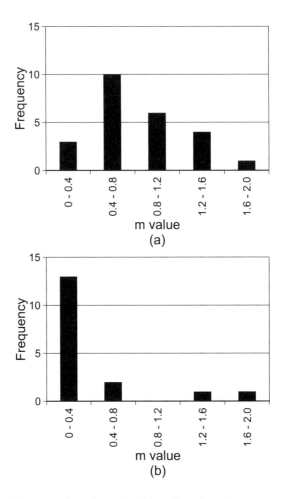

FIGURE 6.12. Histogram for values of *m*: (a) undisturbed claystone or shale; (b) undisturbed clay, remolded clay, or claystone.

Chao 2014). It was observed that if the value of *m* is varied outside of the appropriate range the predicted heave will be more sensitive to the value of *m*. However, the ranges of values for particular soils are sufficiently broad so that the values for a particular soil type or geological area can be determined by careful testing of a number of samples of the same soil.

6.4   FACTORS INFLUENCING OEDOMETER TEST RESULTS

Standards have been written by ASTM for the conduct of oedometer tests. However, there are some factors that will influence the results of the CS test that are different for very dense highly overconsolidated claystones and clayshales than for the soils that are addressed in ASTM D4546. Also, some parts of ASTM D4546 are inadequately followed by some commercial testing laboratories. Because of differences in the nature of these soils and bedrocks, local practices have evolved that deviate from ASTM D4546. One such revised test is used in the Front Range area and is commonly referred to as the *Denver consolidation-swell test*.

The factors affecting CS test data and a review of the current testing procedures based on observations and research performed by the authors will be discussed in this section. The commentaries are not intended as criticism of the ASTM standards. Rather, they are observations by the authors, and present consideration for practitioners concerned with oedometer testing of expansive soil.

6.4.1   Initial Stress State Conditions

In an oedometer test, the percent swell and the CS swelling pressure, σ_{cs}'', decrease as the inundation stress increases up to the point where the inundation stress is equal to the CV swelling pressure, σ_{cv}''. Figure 6.13

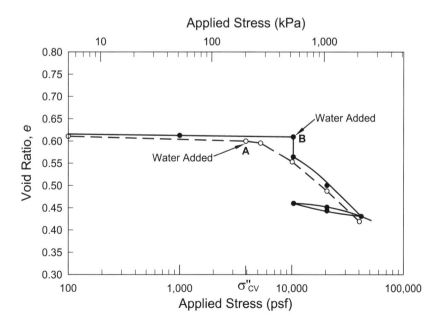

FIGURE 6.13. Oedometer test data for samples inundated at stresses equal to and greater than the constant volume swelling pressure, σ_{cv}'' (data from Reichler 1997).

shows the results of oedometer tests that were inundated at stresses equal to and greater than σ''_{cv} as shown by points A and B, respectively. The sample that was inundated at a stress equal to σ''_{cv} exhibited no volume change. The sample that was inundated at a stress greater than σ''_{cv} underwent a decrease in volume (i.e., hydrocollapse). These data are consistent with the double oedometer test proposed by Jennings and Knight (1957). They showed that depending on the value of the inundation stress, the samples may swell or collapse.

6.4.2   Soil Fatigue

The phenomenon of soil fatigue is another factor affecting the swelling pressure and percent swell of an expansive soil (Chen 1965; Chu and Mou 1973; Popescu 1980). Chen (1965) observed that expansive soil showed decreasing values of percent swell after each cycle of drying and wetting. It was observed that pavements founded on expansive clays that have undergone seasonal movement due to wetting and drying have a tendency to reach a point of stabilization after a number of years (Chen 1988). This phenomenon is likely related to the different mechanisms of swell and shrinkage, as discussed in chapter 2.

6.4.3   Initial Consolidation of Sample

In the Denver consolidation-swell test, the specimen is initially loaded to the inundation stress and allowed to remain under that stress for a period of time until compression of the sample stops. Chen (1988) noted that in the Denver area, the standard procedure for loading the sample was to place the undisturbed sample in a consolidometer under a surcharge load of 1,000 psf (48 kPa) for at least 24 hours prior to saturating the sample. The ASTM standard does not require that the inundation stress be applied for sufficient consolidation to be completed prior to adding water. Experience has shown that placing the sample under the surcharge load for at least 24 hours is necessary to allow consolidation of the sample under the applied load. This also provides time to allow for "healing" of sampling disturbance and closure of micro-cracks caused by stress release during sampling. By allowing the sample to consolidate for about 24 hours or until consolidation has stopped the sample will regain some of its natural undisturbed condition.

During this initial loading, ASTM D4546 specifies that the sample should be covered with plastic wrap, aluminum foil, or moist filter paper so as to avoid water content loss. The authors conducted a series of experiments in which some samples were covered using a loosely fitting plastic wrap during initial consolidation and others were not. The uncovered samples were observed to lose as much as 3 to 4 percent water content during a period of 4 to 6 hours. The samples that were covered with plastic wrap had negligible water content loss during the same period. The importance of this procedure is evident.

6.4.4   Time and Method of Inundation

The ASTM standard specifies that the sample be inundated "until primary swell or collapse volume change is completed and changes in deformation reading for secondary swell/collapse phase is small." It is stated that this inundation period is "typically 24 to 72 hours." The typical inundation time used by many commercial laboratories is 24 hours, but this may not be sufficiently long to saturate the soil sample and dissolve or displace the entrapped air in the pores. If the soil is unable to reach saturation, the maximum swell potential will not be realized.

The manner in which the samples are inundated can also influence the results. Comparative CS tests were conducted on a shale sample to determine the effect of inundation time and method of wetting. The shale had a liquid limit of 85 percent and plasticity index of 67 percent. An inundation stress of 1,000 psf (48 kPa) was applied for all tests. The first test followed the ASTM D4546 procedure for inundation. The oedometer base was filled with water to a depth above the height of the porous stone and the initial load was applied for 30 hours prior to increasing the load. For the second test, the sample was inundated in the same manner, but the initial load was applied for 40 days before consolidation loads were applied. In the third test, the sample was inundated by filling the oedometer just to the top of the sample but not allowing the water to cover the sample. After 31 days, the sample was covered with water for 9 days prior to applying the consolidation loads. The purpose of wetting the sample using the method described for the third test was to provide a pathway for the air in the sample to escape. In the first two samples, the water covering the samples restricted the escape of air out of the sample.

The swelling pressure and percent swell of the three samples are summarized in Table 6.1. These results show that increasing the time of application of the initial load increased the percent swell significantly. Also, by allowing a pathway for the air to escape during inundation caused more complete wetting of the sample and an even greater increase in the percent swell. It also increased the swelling pressure.

TABLE 6.1   **Effect of Inundation Procedure**

Inundation Procedure	Swelling Pressure (psf / kPa)	Percent Swell (%)
30-hour inundation	8,540 / 409	3.9
40-day inundation	8,250 / 395	4.3
31-day immersion/9-day inundation	9,250 / 443	5.2

These procedures for wetting the samples may not be practical for use in commercial soil testing. It is important to realize that soil samples may not be fully saturated within the inundation time recommended by the ASTM standard. Good laboratory practice would be to record the swell over the inundation time, and if the sample continues to exhibit significant swell after 24 hours, the sample should be allowed to swell until the rate of swell becomes insignificant. If this procedure is not followed, the test results may underpredict heave.

6.4.5   Storage of Samples

The ASTM standard states: "storage in sampling tubes is not recommended for swelling soils." It goes on to state that "containers for storage of extra samples may be either cardboard or metal and should be approximately 25 millimeters greater in diameter and 40 to 50 millimeters greater in length than the sample to be encased." However, these procedures can lead to significant disturbance of the sample due to the release of confinement by extruding the samples. A lined split-barrel sampler with brass ring liners is a common sampling technique for expansive soils. The use of brass liners eliminates concerns with the influence of rust, and the samples remain confined. The caps on the ends of the brass liners should be waxed as soon as possible after sampling to minimize moisture loss.

In chapter 3, it was recommended that continuous core sampling be conducted, along with the driven samples. The entire core should be retained and taken to the laboratory for review of the soil profiles. The core may be kept in cardboard or metal storage containers. The majority of such cores are usually too disturbed to be used for testing. However, they are invaluable for identification and classification, as well as to identify features of the soil profile that would otherwise be unidentified.

6.4.6   Competency of Laboratory Personnel

Note 2 in ASTM D4546 states: "The precision of this test method is dependent on the competence of the personnel performing the test and the suitability of the equipment and facilities used." Although this is a relatively obvious comment, lapses in the competency of the personnel frequently occur when communication is not well established between the engineering staff and the laboratory staff. Depending on the nature of the testing and the purpose to which the data are to be put, variations in the testing procedures may be required. It is important for the engineering staff to be knowledgeable in oedometer testing, and to be fully involved in the laboratory testing that is being conducted. It is also important for the laboratory staff to be educated in the basic engineering principles underlying the testing and their results.

References

Al-Mhaidib, A. I. 2006. "Swelling Behavior of Expansive Shale, a Case Study from Saudi Arabia." In *Expansive Soils—Recent Advances in Characterization and Treatment*, edited by A. A. Al-Rawas and M. F. A. Goosen. London: Taylor & Francis Group.

ASTM D2435/D2435M-11. 2011. "Standard Test Methods for One-Dimensional Consolidation Properties of Soils using Incremental Loading." ASTM International: West Conshohocken, PA.

ASTM D4546-08. 2008. "Standard Test Methods for One-Dimensional Swell or Collapse of Cohesive Soils." ASTM International: West Conshohocken, PA.

Bonner, J. P. 1998. "Comparison of Predicted Heave Using Oedometer Test Data to Actual Heave." Master's thesis, Colorado State University, Fort Collins, CO.

Casagrande, A. 1936. "The Determination of the Pre-Consolidation Load and Its Practical Significance." Discussion D-34, vol. 3, *Proceedings of the 1st International Conference on Soil Mechanics and Foundation Engineering* (June 22-26, Cambridge, MA), 60–64.

Chen, F. H. 1965. "The Use of Piers to Prevent the Uplifting of Lightly Loaded Structures Founded on Expansive Soils." *Engineering Effects of Moisture Changes in Soils, Concluding Proceedings International Research and Engineering Conference on Expansive Clay Soils*. College Station: Texas A & M Press.

———. 1988. *Foundations on Expansive Soils*. New York: Elsevier Science.

Chu, T., and C. H. Mou. 1973. "Volume Change Characteristics of Expansive Soils Determined by Controlled Suction Tests." *Proceedings of the 3rd International Conference on Expansive Soils*, Haifa, Israel, 177–185.

Edil, T. B., and A. S. Alanazy. 1992. "Lateral Swelling Pressures." *Proceedings of the 7th International Conference on Expansive Soils*, Dallas, TX, 227–232.

Feng, M., K-M Gan, and D. G. Fredlund. 1998. "A Laboratory Study of Swelling Pressure Using Various Test Methods." *Proceedings of the International Conference on Unsaturated Soils*, Beijing, China, International Academic Publishers, VI, 350–355.

Fredlund, D. G. 1969. "Consolidometer Test Procedural Factors Affecting Swell Properties." *Proceedings of the 2nd International Conference on Expansive Soils*, College Station, TX, 435–456.

———. 2004. "Analysis of Decorah Clay-Shale at the Sandstone Ridge Housing Development in Cannon Falls, Minnesota." Project No. 98823-B.

Fredlund, D. G., H. Rahardjo, and M. D. Fredlund. 2012. *Unsaturated Soil Mechanics in Engineering Practice*. Hoboken, NJ: John Wiley and Sons.

Gilchrist, H. G. 1963. "A Study of Volume Change in Highly Plastic Clay." Master's Thesis, University of Saskatchewan, Saskatoon, SK (as presented in Fredlund, Rahardjo, and Fredlund 2012).

Jennings, J. E. B., and K. Knight. 1957. "The Prediction of Total Heave from the Double Oedometer Test." *Proceedings of Symposium on Expansive Clays, South African Institute of Civil Engineers*, Johannesburg, 7(9): 13–19.

Justo, J. L., A. Delgado, and J. Ruiz. 1984. "The Influence of Stress-Path in the Collapse-Swelling of Soils at the Laboratory." *Proceedings of 5th International Conference on Expansive Soils*, Adelaide, Australia.

Nelson, J. D., and K. C. Chao. 2014. "Relationship between Swelling Pressures Determined by Constant Volume and Consolidation-Swell Oedometer Tests." *Proceedings of the UNSAT2014 Conference on Unsaturated Soils*, Sydney, Australia.

Nelson, J. D., K. C. Chao, D. D. Overton, and R. W. Schaut. 2012. "Calculation of Heave of Deep Pier Foundations." *Geotechnical Engineering Journal of the Southeast Asian Geotechnical Society and Association of Geotechnical Societies in Southeast Asia* 43(1): 12–25.

Nelson, J. D., and D. J. Miller. 1992. *Expansive Soils: Problems and Practice in Foundation and Pavement Engineering*. New York: John Wiley and Sons.

Nelson, J. D., D. O. Overton, and K. C. Chao. 2003. "Design of Foundations for Light Structures on Expansive Soils." *Proceedings of the California Geotechnical Engineers Association Annual Conference*, Carmel, CA.

Nelson, J. D., D. K. Reichler, and J. M. Cumbers. 2006. "Parameters for Heave Prediction by Oedometer Tests." *Proceedings of the 4th International Conference on Unsaturated Soils*, Carefree, AZ, 951–961.

Norrish, K. 1954. "The Swelling of Montmorillonite." *Discussions of the Faraday Society* 18, 120–134.

Popescu, M. 1980. "Behavior of Expansive Soils with a Crumb Structure." *Proceedings of the 4th International Conference on Expansive Soils*, Denver, CO, 158–171.

Porter, A. A. 1977. "The Mechanics of Swelling in Expansive Clays." Master's thesis, Colorado State University, Fort Collins, CO.

Reichler, D. K. 1997. "Investigation of Variation in Swelling Pressure Values for an Expansive Soil." Master's thesis, Colorado State University, Fort Collins, CO.

Schmertmann, J. H. 1955. "The Undisturbed Consolidation Behavior of Clay." *Transactions ASCE* 120(1): 1201–1227.

Terzaghi, K. 1925. *Erdbaumechanik auf Bodenphysikalischer Grundlage*. Leipzig u. Wien, F. Deuticke.

Terzaghi, K. 1943. *Theoretical Soil Mechanics*. New York: John Wiley and Sons.

Thompson, R. W., H. A. Perko, and W. D. Rethamel. 2006. "Comparison of Constant Volume Swell Pressure and Oedometer Load-Back Pressure." *Proceedings of the 4th International Conference on Unsaturated Soils*, Carefree, AZ, 1787–1798.

7

Water Migration in Expansive Soils

The development of sites in arid climates generally results in an increase in water content in the soil profile due primarily to two factors. One factor is irrigation at the surface that introduces water into the soil. A second factor is the construction of impermeable surfaces such as roadways and structures that stop or reduce the evapotranspiration of water from the surface. Additional sources of water include broken pipes, leaky sewers, off-site sources, storage of snow, and others. The soil may become saturated or remain unsaturated during the wetting process, depending on the rate of infiltration, site conditions, and soil properties.

The depth of potential heave was defined in chapter 1 as the greatest depth to which the overburden vertical stress equals or exceeds the swelling pressure of the soil. This represents the maximum depth of the active zone that could occur. This entire zone may or may not become fully wetted within the design life of a structure. The analysis of the depth to which wetting is expected to occur is an important part of the design of foundations on expansive soils. At some sites the depth of potential heave is well within the depth that could be expected to be wetted. In other cases, the depth of potential heave may be much greater than the depth to which water would be expected to migrate.

In this chapter, the terms *full wetting* and *partial wetting* will be introduced. Full wetting is the degree of wetting needed for a soil to achieve the maximum amount of swelling of which it is capable. Partial wetting is some degree of wetting less than that.

Overton, Chao, and Nelson (2010) presented a simplified procedure to hand-calculate the final water content of the soil based on the work by McWhorter and Nelson (1979). That approach only estimates the final water content if the entire depth of potential heave is wetted. However, for sites with highly expansive soils, the depth of potential heave may be so great that the depth of wetting may not extend through its entire depth during the design life of the structure. In that case, it may be prudent to perform a water migration analysis.

152

The depth of wetting and corresponding degree of saturation can be calculated using commercially available computer software such as Vadose/W, SVFlux, Hydrus 2-D, Modflow-Surfact, etc. Numerical modeling provides an opportunity to allow a wide variety of factors to be considered when analyzing the depth and degree of wetting. If the results of these analyses indicate that, by the end of the design life, full wetting is not expected to occur, the calculation of heave can be conducted for the partial wetting conditions. Such analyses result in a more economical design of the foundation.

A general overview of the fundamental theory regarding water flow in unsaturated soils is introduced in this chapter. The depth and degree of wetting of the soil that is contributing to heave is discussed, and determination of final water content profiles for various conditions and methods of analysis are presented. At the end of this chapter, challenges in water migration modeling, particularly for expansive soils, are discussed.

7.1   WATER FLOW IN UNSATURATED SOILS

7.1.1   Darcy's Law for Unsaturated Soils

Darcy (1856) postulated that the flow rate through porous media is proportional to the head loss and inversely proportional to the length of the flow path. In general, a one-dimensional form of Darcy's law may be written as follows. The negative sign in equation (7-1) indicates that water flows in the direction of a decreasing hydraulic head.

$$q = -K_s \frac{\Delta H_t}{L} = -K_s i \qquad (7\text{-}1)$$

where:

q = flow rate of water,
K_s = coefficient of permeability, assumed to be constant in a saturated soil,
ΔH_t = change in hydraulic (total) head,
L = length of the flow path over which ΔH_t occurs, and
$i = \Delta H_t/L$ = hydraulic head gradient.

Buckingham (1907) proposed a modification of Darcy's law to describe water flow through unsaturated soil. The modification made by Buckingham was based primarily on two assumptions:

1. The driving force that causes water to flow in isothermal, rigid, unsaturated soil is the sum of the matric and gravitational potentials.
2. The hydraulic conductivity of unsaturated soil is a function of the water content or matric potential of the unsaturated soil.

The general one-dimensional form of Buckingham–Darcy flux law for vertical flow may be expressed as follows:

$$q = -K(h)\frac{\partial H_t}{\partial z} = -K(h)\frac{\partial}{\partial z}(h+z) = -K(h)\left(\frac{\partial h}{\partial z} + 1\right) \qquad (7\text{-}2)$$

where:

H_t = total hydraulic head,
h = pressure head,
z = elevation head, and
$K(h)$ = coefficient of unsaturated hydraulic conductivity.

Equation (7-2) is similar to equation (7-1), except that under conditions of unsaturated flow, the coefficient of unsaturated hydraulic conductivity varies with changes in water content. Thus, it indirectly varies with changes in matric suction. Therefore, even though Darcy's law was originally derived for a saturated soil, it has been shown that it can also be applied to the flow of water through an unsaturated soil (Richards 1931; Childs and Collis-George 1950; McWhorter and Sunada 1977).

7.1.2   Water Mass Balance Equation

As water flows through soil, the matric suction and water content vary as functions of time and space. Such transient flows are time dependent and can be mathematically described by the *water mass balance equation,* also called the continuity equation or water conservation equation. The water mass balance equation is related to water flux, storage changes, and sources or sinks of water. The equation can be formulated by calculating the mass balance for a one-dimensional system as follows (Jury Gardner, and Gardner 1991).

$$\frac{\partial q}{\partial z} + \frac{\partial \theta}{\partial t} + r_w = 0 \qquad (7\text{-}3)$$

where:

q = flow rate of water,
z = distance in z direction (elevation),
θ = volumetric water content of soil,
t = time, and
r_w = sources or sinks of water.

If water is allowed to flow in all three directions (x-, y-, and z-directions), the water conservation equation would be written as follows:

$$\frac{\partial q_x}{\partial x} + \frac{\partial q_y}{\partial y} + \frac{\partial q_z}{\partial z} + \frac{\partial \theta}{\partial t} + r_w = 0 \qquad (7\text{-}4)$$

where q_x, q_y, and q_z are the components of the water flux vector.

The Richards equation (Richards 1931) in the form shown in equation (7-5) is derived by combining equations (7-2) and (7-3) and assuming $r_w = 0$. The Richards equation states that the rate of change of water content is equal to the rate of change of flow in a soil system:

$$\frac{\partial \theta}{\partial t} = \frac{\partial}{\partial z}\left[K(h)\left(\frac{\partial h}{\partial z} + 1 \right) \right] \qquad (7\text{-}5)$$

Equation (7-5) contains two unknowns, θ and h. Solution of the equation, therefore, requires another equation, normally the soil water characteristic function, $h(\theta)$. The soil water characteristic function was discussed in chapter 4.

7.1.3   Vertical Seepage in Unsaturated Soil

McWhorter and Nelson (1979, 1980) presented a method for analyzing water movement in an unsaturated soil that lends itself easily to hand calculation or programming on a computer spreadsheet. Figure 7.1 shows a *wetting front* moving downward with time due to infiltration from the surface. The analysis is complicated somewhat by the fact that across the wetting front the water content, and hence the soil suction, is varying. In Figure 7.1, it is shown that the wetting front transitions over some distance from the higher water content above the wetting front to that of the native soil below. There is not a distinct singular point that defines the actual depth of wetting. In clays and claystones with low permeability, the transition zone may extend over a relatively long distance, and the heave in the transition zone may be significant. Computation of heave in that zone must consider the

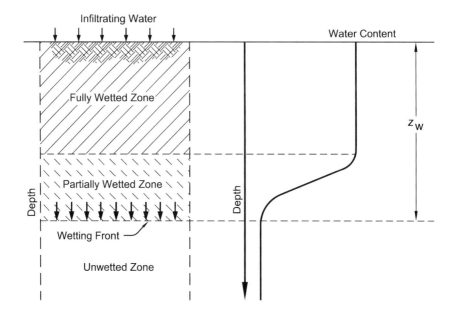

FIGURE 7.1. Progression of wetting in unsaturated soil.

percentage of ultimate heave caused by the actual degree of partial wetting. Therefore, in Figure 7.1 the depth of wetting is shown as the depth to the bottom of the transition zone.

McWhorter and Nelson (1979, 1980) described the migration of the wetting front underneath a constant source of water in different stages. Figure 7.2 shows the water content profiles for Stages I and II. In Figure 7.2a, the wetting front is shown as the transition zone. During Stage I, the wetting front advances downward and the soil above the wetting front may or may not be saturated. The water content above the wetting front, shown as θ_f in Figure 7.2a, depends on the rate at which water can be supplied to the soil from the ground surface. Stage I ends when the wetting front contacts either an impermeable stratum or a groundwater table.

During Stage II a groundwater mound develops and the phreatic surface moves upward, as shown in Figure 7.2b. The soil below the rising groundwater table is saturated. The rate of rise of the groundwater table depends on the amount of storage available above the wetting front (i.e., the difference between soil porosity, n, and θ_f and the ability of the groundwater mound to expand laterally).

The equation for water content above the wetting front was derived by McWhorter and Nelson (1980) as equation (7-6). It can be used to

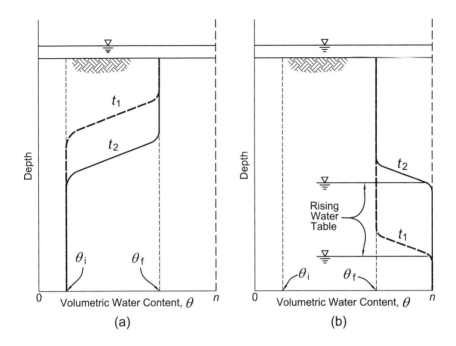

FIGURE 7.2. Distribution of water content profiles for Stages I and II: (a) Stage I; (b) Stage II (modified from McWhorter and Nelson 1979).

perform simple hand calculations of the degree of wetting that will exist above a continuously downward moving wetting front:

$$\theta_f = \theta_r + (n - \theta_r)\left(\frac{q_m}{K_s}\right)^{\frac{\lambda}{2+3\lambda}} \tag{7-6}$$

where:

θ_f = volumetric water content above the wetting front,
θ_r = residual water content as discussed in chapter 4,
n = porosity of the soil,
K_s = saturated hydraulic conductivity of the soil above the
        wetting front,
λ = pore size distribution index, and
q_m = mean rate of infiltration at the ground surface.

If q_m is known, equation (7-6) can be used to estimate the degree of wetting that would ultimately develop in a deep uniform soil profile. This method is demonstrated in Examples 7.1 and 7.2.

7.1.4 Flow through Fractured Rocks and Bedding Planes

Water movement in bedrock is governed primarily by discontinuities in the rock mass. The term *discontinuity* is a collective term commonly used to include bedding planes, fractures, joints, rock cleavage, foliation, shear zone, faults, and other contacts (Singhal and Gupta 2010). The most common discontinuities in rocks are bedding planes and fractures (Bell 2000). Bedding planes are a characteristic of practically all sedimentary rocks such as sandstones, siltstones, and claystones. Fractures are planes in rock along which stress has caused some form of local failure. They include joints, faults, and fissures (Singhal and Gupta 2010).

Evidence of water migration through cracks and fissures and along bedding planes was seen in core samples that were shown in Figure 3.1. Iron-stained features that are an indication of such water transport were observed throughout entire sections of the core samples.

Because the bedrock is less permeable than the overlying soil, water that infiltrates from the ground surface generally forms a perched water table above bedrock. Water continues to permeate from the perched water table into bedrock along bedding planes and fractures. Figure 7.3 depicts the movement of groundwater through bedding planes and fractures. The water moving in the fractures and bedding planes migrates outward into the blocks of expansive material by suction, and results in heave. The presence of steeply dipping bedding

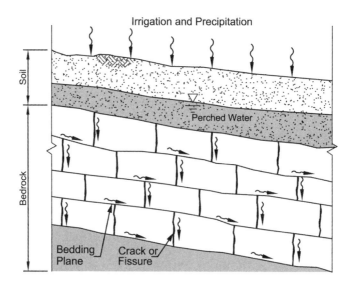

FIGURE 7.3. Movement of groundwater through bedding planes and fractures.

FIGURE 7.4. Layers of steeply dipping claystone exposed in a roadcut near Denver, Colorado.

planes in expansive bedrock can exacerbate the differential movement associated with heave.

In the Front Range area of Colorado, the uplift of the Rocky Mountains caused the beds to be steeply dipping as was shown in Figure 2.22. Figure 7.4 shows layers of steeply dipping claystone exposed in a roadcut near Denver, Colorado. In some of the older (lower) members of the Pierre Shale, there are beds of sandy and silty clay interbedded with beds of highly expansive claystone. The relatively high hydraulic conductivity of the sandy silty beds allows water to infiltrate deeply. Water migrates laterally into the highly expansive beds as well. The difference in the expansion potential between the sandy, silty beds and the highly expansive beds causes linear heave features that extend along the strike of the beds, as shown in Figure 7.5. Figure 7.6 shows the linear heave ridges along a street in a western area of Denver, Colorado (Noe 2007). Noe has documented the presence of longitudinal surface features along the direction of regional bedrock strike in areas where the bedrock is steeply dipping and expansive. He has also documented evidence of water movement in dipping clayshale bedrock through fractures to depths up to 100 ft (30 m). The migration of water into the expansive beds from the sandy, silty beds due to soil suction can be a very slow process due to the relatively low permeability of claystone bedrock. For this reason, heave can continue over a very long period of time.

The movement of water through fractures may cause an uneven distribution of water content in expansive bedrock. Consequently,

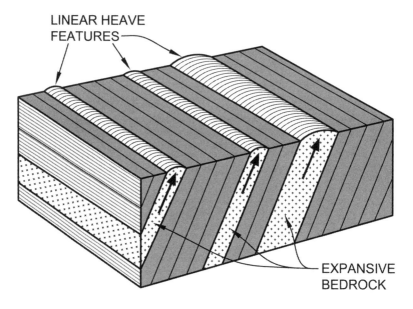

FIGURE 7.5. Differential heave in steeply dipping expansive bedrock showing linear heave features (modified from Noe 2007).

FIGURE 7.6. Linear heave ridges in a steeply dipping bedrock area near Denver, Colorado (Noe 2007).

when taking samples to measure the water content erratic results may be obtained. Chen (1988) noted that "Changes of environment can … alter the entire picture. Construction operations such as pier drilling can break through the system of interlaced seams and fissures in the claystone structure, allowing water to saturate [an] otherwise dry area, thus allowing further swelling of otherwise stable material."

The rate of water migration will be mainly controlled by the effective hydraulic conductivity of the fractured rock. The bedding planes and fractures provide a secondary soil structure that allows water to move through the fractured rock at a faster rate. Table 7.1 shows typical ranges of hydraulic conductivity for various types of geologic materials. As shown in Table 7.1, the hydraulic conductivity of natural materials has wide variation. Fractures increase the hydraulic conductivity by

TABLE 7.1   Range of Hydraulic Conductivity for Various Types of Geologic Materials (Modified from Singhal and Gupta 2010)

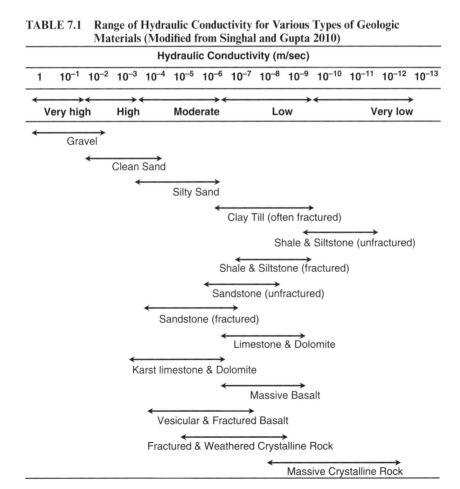

several orders of magnitude over that of unfractured rock mass. For example, the hydraulic conductivity of unfractured shale is usually below 1×10^{-10} m/sec. It is evident that even under very high hydraulic gradients, groundwater in unfractured shale would not move at rates greater than a few centimeters per century. However, the hydraulic conductivity of fractured shale can be up to 5×10^{-7} m/sec or more. The secondary soil structure that fractures create in shale, even if hairline fractures exist at relatively wide spacing, can produce secondary hydraulic conductivity of magnitudes that exceed the primary hydraulic conductivity of the rock matrix (Freeze and Cherry 1979).

7.2 DEPTH AND DEGREE OF WETTING

Engineers have attempted to describe the zone of soil that is contributing to heave using different terms, each of which considers a particular emphasis. The term *active zone* has been in common usage among engineers in the field of expansive soils. However, the usage of that term has taken different meanings at different times and in different places. Five distinct zones were defined in chapter 1. They include:

- The active zone, z_A,
- The zone of seasonal moisture fluctuation, z_s,
- The zone (depth) of wetting, z_w,
- The depth of potential heave, z_p, and
- The depth of design active zone, z_{AD}.

The amount of heave that occurs in the active zone depends mainly on three factors: (1) the depth to which water content in the soil has increased (i.e., depth of wetting), (2) the amount of water content changes within the zone of soil that is being wetted (i.e., degree of wetting), and (3) the expansion potential of the various soil strata. As water migrates through a soil profile, different strata become wetted, some of which may have more swell potential than others. Consequently, the active zone and the extent of heave varies with time.

7.2.1 Depth of Wetting

As noted above, the extent of the active zone varies with time. The depth to which the soil is expected to contribute to heave within the design life of the structure (i.e., the depth of heave for which the foundation will be designed), is termed the *design active zone*. The estimation of the design active zone is one of the most important factors in the design of foundations. In some areas, it is common practice to just assume a particular depth of wetting and use that for general design (CAGE 1996; Walsh et al. 2009 and 2011). However, to make general assumptions without regard to specific site conditions is not sound practice (Nelson, Chao, and Overton 2011). If the depth of potential heave is computed to be of a depth to which wetting can be expected to occur, it would be prudent to use the depth of potential heave as the design active zone. However, many factors such as a nonuniform soil profile, zones of deep wetting, perched water conditions, and others complicate the wetting process in the soil. The following sections discuss the various factors that must be considered in the determination of a design active zone.

7.2.2   Degree of Wetting

As water migrates into the soil, it may or may not become fully wetted. As noted in chapter 1, the term *fully wetted* refers to the amount of wetting, or degree of saturation, required to cause the soil to expand to its maximum amount. That may correspond to a degree of saturation less than 100 percent. The term *degree of wetting* refers to the amount by which the water content has increased relative to the fully wetted condition.

It must be noted that fully wetted conditions do not necessarily correspond to a particular degree of saturation. For example, in collapsible soils, the full collapse potential may be achieved at a degree of saturation as low as 60 percent (Houston et al. 2001). Thus, the fully wetted condition for such soils corresponds to a degree of saturation well below 100 percent. On the other hand, an expansive soil will generally not achieve fully wetted conditions until the soil is nearly saturated. Cumbers (2007) measured the degree of saturation at fully wetted conditions in oedometer tests to be between 95 and 100 percent. However, if a soil sample is very highly stressed such that all air is forced out of the void spaces, and the adsorbed cations have not been fully hydrated or if some osmotic swell is still possible, it would theoretically be possible for a saturated soil to continue to take up more water and expand. In that case the soil is only partially wetted even though the degree of saturation is 100 percent. In the authors' experience there have been occasions where soil samples with a very high degree of initial saturation have exhibited a high percent swell when inundated.

For a downward progressing wetting front, the water content below the depth of the wetting front is the same as that which existed prior to introduction of the water source, as shown in Figure 7.2a. Above the wetting front the soil may or may not be fully wetted. Full wetting of the soil profile would be expected when rising groundwater or perched water conditions develop, as shown in Figure 7.2b, or where the infiltration rate exceeds the hydraulic conductivity of the soil. Where a rising groundwater table is anticipated, the full wetting conditions should be used to make calculations (Houston et al. 2001). Thus, if a water table is within the depth of the design active zone and the groundwater is anticipated to rise to the surface, it would be reasonable to assume that fully wetted conditions could develop in the design active zone within the design life of the structure.

For sites subject to surface water sources, and where an aquitard does not exist at depth, the soil may not become fully wetted. The depth of wetting may or may not extend to the depth of potential heave. Table 7.2 presents observations of the degree of wetting from several cases of downward infiltration into moisture sensitive soil deposits (Houston and Nelson 2012). At expansive soil sites where the soil has a low hydraulic conductivity and relatively high soil suction, the degree of saturation of the soil may increase to amounts higher than 90 percent. However, for soils with higher values of hydraulic conductivity the degree of wetting is much less.

7.2.3　Perched Water Tables in Layered Strata

Typically, an expansive soil profile will comprise various strata of claystone or other sedimentary rock, each having a different value of hydraulic conductivity. Water will accumulate on top of the low permeability lenses, forming perched water tables. Further downward migration into a lower stratum is often neglected. However,

TABLE 7.2　Examples of Partial Wetting for Various Sites (Houston and Nelson 2012)

Site Location	Partial Wetting Conditions[1]	Reference
New Mexico Site	$S_i \sim 15\%$; $S_f \sim 35\%$. Irrigated lawn.	Walsh, Houston, and Houston (1993)
Groundwater Recharge Sites in Arizona	$S_i \sim 15\%$; $S_f \sim 60\%$ in upper 3 ft and essentially no change below that. Ponding for extended period.	Houston, Duryea, and Hong (1999)
China Loess	Reports of settlement of about 10% of full collapse potential. Therefore, partial wetting.	Houston (1995)
Arizona Collapsible Soils Study Sites	$S_i \sim 15$ to 20%; $S_f \sim 50$ to 70%. Ponding for extended periods.	El-Ehwany and Houston (1990)
Denver, CO site	$S_i \sim 68\%$; $S_f \sim 96\%$ in the upper 25 ft for 4 years. Irrigated lawn.	Chao, Overton, and Nelson (2006)

[1] S_i = initial degree of saturation; and S_f = final degree of saturation.

as was shown in Figure 7.3, water will continue to move downward through bedding planes and fractures. Freeze and Cherry (1979) stated that: "The existence of a low-permeability clay layer...can lead to the formation of a discontinuous saturated lense, with unsaturated conditions both above and below....Saturated zones of this type dissipate with time under the influence of downward percolation and evaporation from the surface." Corey (1994) also discussed flow through stratified media and stated that: "...the water is interconnected (continuous) throughout all layers, including at the boundaries between the layers, otherwise there could be no flow." Edgar, Nelson, and McWhorter (1989) presented a theory for modeling flow through several unsaturated soil layers when heave or compression of the layers and temperature effects are considered. In general, it must be assumed that the migration of a wetting front will continue to move downward with time. In foundation design, therefore, it cannot be assumed that the existence of a shallow perched water table will define a boundary for downward migration of water.

Reed (1985) noted, "Design errors consist of under-prediction of soil movement and pressure, [and] inadequate prediction of the total zone of soil movement." Taylor and Wassenaar (1989) stated, "The presence of a shallow water table or the absence of a water table does not assure the engineer of the absence of deep movement problems resulting from unstable bedrock." Additionally, Taylor and Wassenaar stated that "On a site with underlying bedrock consisting of lenses of claystone-sandstone or claystone-siltstone, the entire pier length and the bedrock strata below the bottom of a pier have a potential to become saturated, or as a minimum, be subjected to an increase in moisture content."

In summary, it must be recognized that flow will occur through layered soil strata and perched water zones. Serious design errors can result from inadequate analysis of the depth of the design active zone. It is important to determine an appropriate depth of design active zone for the specific site conditions, particularly for a site with multilayered bedrock and perched water zones.

7.2.4 Wetting Profiles

Figure 7.7 shows measured water content profiles before construction and approximately 4 years after construction for a residential building located in Denver, Colorado. The structure was measured to be out of level by up to 3 in. (76 mm) at about 3 years after construction. The presence of a wetting front that is progressing downward is evident.

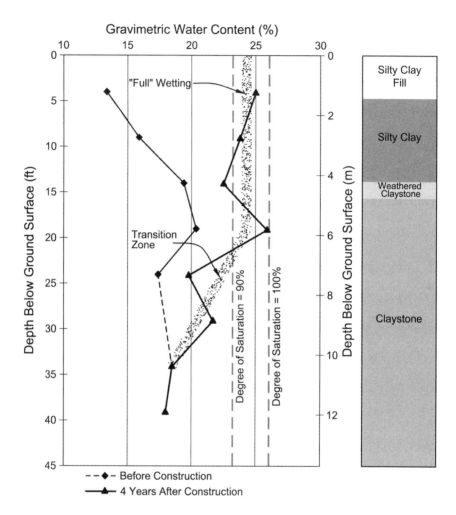

FIGURE 7.7. Water content profiles for a site located on flat-lying bedrock.

The depth of wetting is clearly shown to extend to a depth of about 35 ft (11 m) with full wetting to a depth of about 20 ft (6 m) and the transition zone between 20 ft and 35 ft (6 m and 11 m). Wetting of the soil at the site was attributed largely to poor surface grading and drainage around the structure. The silty clay soils above the claystone were wetted to a degree of saturation of about 92 percent, and one sample in the claystone was fully saturated. Even with the degree of wetting that is shown, the soils and bedrock within the zone of wetting exhibited a significant expansion potential with a swelling pressure up to 27,900 psf (1,340 kPa).

The steeply dipping bedrock along the foothills of the Rocky Mountains in the area west of Denver was depicted in Figure 2.22. At some locations the dip is almost vertical. The steeply dipping beds pose particular problems, as was discussed in Section 7.1.4. The frequency of distress to structures in this area is significantly higher than those in areas with more flat-lying beds (Thompson 1992). The area in which the beds dip at angles greater than 30 degrees is the Designated Dipping Bedrock Area. The *Designated Dipping Bedrock Area Guide* by Jefferson County, Colorado (2009) has noted:

> [T]he Designated Dipping Bedrock Area (DDBA) defines an area of Jefferson County where heaving bedrock is possible under certain geological and human-influenced conditions. The conditions warrant special consideration in all phases of development, including site exploration and evaluation, facilities, design, construction, and subsequent maintenance. In some areas, avoidance may be the best mitigation method.

When evaluating the depth and degree of wetting, it is important to take into account the time at which the observations were made. Diewald (2003) evaluated post-construction data from 133 investigations and determined that the depth of wetting for 7- to 10-year-old residences can be up to 40 ft (12 m). Diewald also indicated that there continues to be an increase in the depth of wetting over time. The authors' experience has shown that the depth of wetting can extend to the depth of potential heave depending on site conditions (Nelson, Overton, and Durkee 2001; Overton, Chao, and Nelson 2006; Chao, Overton, and Nelson 2006; Chao 2007). For many structures on expansive soils, distress has resulted from designs being based on underestimation of the depth of the design active zone. The following sections discuss means of evaluating the appropriate depth of the design active zone.

7.3 DETERMINATION OF FINAL WATER CONTENT PROFILES FOR DESIGN

Determination of the design active zone and the expected degree of wetting in that zone requires careful analysis. In the absence of more rigorous analyses, it is reasonable to assume that the soil profile to the depth of potential heave will become fully wetted within the design life of the structure. For sites where the depth of potential heave is large, such an assumption may lead to overly conservative and costly design.

In those cases, it may be cost-effective to conduct modeling of long-term water migration in the vadose zone for site-specific conditions. The predicted heave calculated on the basis of the results of numerical modeling would be more realistic than that calculated, assuming a full wetting condition to the entire depth of potential heave.

7.3.1   Hand Calculation of Final Water Contents for Design

Simple hand calculations can be performed to determine the final water content of the soils, as described in Section 7.1.3. The calculations will provide a final water content profile, but that profile is assumed to extend through the entire depth of potential heave. This type of analysis is useful if the depth of potential heave is within depths that can reasonably be expected to be wetted. However, the calculations do not readily lend themselves to determine the depth of wetting with time.

Examples 7.1 and 7.2 are presented using hand calculations to determine equilibrium water contents in the soil profile.

EXAMPLE 7.1

Given:

A sandy clay soil profile extending to large depth. Overirrigation has caused a mean infiltration rate of 0.08 in./day (7.5×10^{-9} ft/sec). The properties of the sandy clay are given in Table E7.1.

TABLE E7.1   Soil Properties for Sandy Clay

Property	Value
Initial volumetric water content, θ_i (%)	15
Porosity, n (%)	54
Pore size distribution index, λ	2
Saturated hydraulic conductivity, K_s (ft/sec)	6×10^{-7}
Displacement pressure head, h_d (ft)	2.6
Residual water content, θ_r (%)	2
Specific gravity of solids, G_s	2.64

Find:

The equilibrium volumetric water content and degree of saturation in the soil profile.

Solution:

The equilibrium volumetric water content, θ_f, will be estimated using equation (7-6). All soil parameters needed are given in Table E7.1. Although values for volumetric water content and porosity are listed in units of %, the decimal form must be used in equation (7-6). Thus,

$$\theta_f = \theta_r + (n - \theta_r)\left(\frac{q_m}{K_f}\right)^{\frac{\lambda}{2+3\lambda}}$$

$$= 0.02 + (0.54 - 0.02)\left(\frac{7.5 \times 10^{-9}}{6 \times 10^{-7}}\right)^{\frac{2}{2+3\times2}} = 0.32$$

The degree of saturation is related to the porosity and volumetric water content by the relationship shown below.

$$S = \frac{\theta}{n} = \frac{0.32}{0.54} = 0.59 = 59\%$$

EXAMPLE 7.2

Given:

An expansive soil profile consisting of homogeneous claystone to a large depth. Overirrigation has caused a mean infiltration rate of 0.08 in./day (7.5×10^{-9} ft/sec). The claystone properties are shown in Table E7.2.

TABLE E7.2   Soil Properties for Example 7.2

Soil Property	Value
Initial volumetric water content, θ_i (%)	21
Porosity, n (%)	37
Displacement pressure head, h_d (ft)	32
Pore size distribution index, λ	0.41
Saturated hydraulic conductivity, K_s (ft/sec)	8.3×10^{-9}
Specific gravity of solids, G_s	2.75

Find:

The equilibrium volumetric water content and degree of saturation in the soil profile.

Solution:

Equation (7-6) will be used to compute the equilibrium water content. For claystone, it was shown in chapter 4 that the residual water content approaches zero at a suction between 6 and 7 pF. Thus, the value for θ_r will be taken as 0.0. This has negligible influence on the result.

$$\theta_f = 0.0 + (0.37 - 0.0)\left(\frac{7.5 \times 10^{-9}}{8.3 \times 10^{-9}}\right)^{\frac{0.41}{2+3\times0.41}} = 0.36$$

From the relationship shown in Example 7.1,

$$S = \frac{\theta}{n} = \frac{0.36}{0.37} = 0.97 = 97\%$$

The computations presented in Examples 7.1 and 7.2 demonstrate the significant influence that the hydraulic conductivity of the soil has on the degree of wetting that will result. In Example 7.1, the resulting degree of saturation was much less than 100 percent. In Example 7.2, the hydraulic conductivity was of the same order of magnitude as the infiltration rate and the soil was nearly saturated. For the situation demonstrated by Example 7.1, if the wetting front were to encounter an aquitard of some material, or a water table, a rising water table would develop, as shown in Figure 7.2b. In that case, a fully wetted condition could develop through the entire design active zone if the aquitard were at a relatively shallow depth. Similarly, limited perched zones of fully wetted soil could develop over lenses of low permeability material.

7.3.2 Computer Modeling of Water Migration

Computer modeling of water migration in soils allows multiple factors to be considered. These are generally taken into account by simplifying assumptions. The water flow in expansive bedrock is generally modeled using one of the following three possible conceptualizations: (1) an equivalent porous medium model, (2) a discrete fracture network model, or (3) a dual-porosity medium model. General descriptions of the three approaches are presented as follows. Full discussion of the details of the approaches is beyond the scope of this overview.

The first of these approaches assumes that the medium is fractured to such an extent that it behaves hydraulically as a porous medium, in which the soil properties can be averaged over the fractures and the intact blocks of claystone forming the matrix. This is a commonly used approach. Many researchers have shown that flow in a large enough volume of fractured medium can be modeled with reasonable accuracy using the equivalent porous medium model (Singhal and Gupta 2010). This approach is discussed further in the following sections.

The second approach assumes a network of discrete fractures through which most or all of the groundwater moves. For simplicity, the matrix blocks are considered to be impermeable. This approach requires that the geometric character of each fracture (e.g., aperture, length, density, spacing, and orientation) is known, as well as the pattern of connection among fractures (National Research Council 1990).

It is extremely difficult to define the geometry of a complex fracture rock system with accuracy. Furthermore, this approach does not take into account the fact that the soil suction in the unsaturated rock matrix causes the migration of water into the matrix blocks. Although this approach may be meaningful for purposes of modeling overall directions and quantities of flow, it does not characterize the change in water content in the overall soil sufficiently for use in computing heave in expansive soil.

The third approach considers a dual-porosity medium in which the water in the matrix blocks is separate from the water in the fractures. The matrix blocks are considered to have high primary porosity but low hydraulic conductivity. The fractures are considered to have low storativity but high hydraulic conductivity. Therefore, both matrix blocks and fractures contribute to groundwater flow, but fractures are the main contributors. This model can explain why low water contents can be observed in discrete samples even under perched water tables or ponds that have existed for long periods of time.

The development of sophisticated software has facilitated modeling of complex systems, taking into account a complex set of factors including climatologic conditions, surface grading and drainage, introduction of surface water and sources of deep wetting, heterogeneous soil conditions, and many others. In computer applications care must be taken to assure that the computer model that is developed and subsequently applied is representative of the actual situation. The modeler must show that the computer code actually represents the physical system. This is done by calibrating the model. In the calibration process, the model is shown to be capable of producing results that can be compared with some actual measurements. This is usually accomplished by performing a series of computations and adjusting the material properties in the model until the computed water content profile satisfactorily matches a water content profile measured at the beginning of the project.

Subsequent to calibration, the model is *validated*. The validation process involves a series of computations performed forward in time from the calibration, which are compared to values measured at some time subsequent to the time of the calibration. For purposes of design, this may be problematical because at the time the foundation is being designed, validating data are seldom available. In forensic investigations, however, the initial design data provide calibration data, and measurements at the time of the distress provide validation data. Consequently, computer modeling of water migration can be valuable in identifying sources of water and in predicting future heave.

The case history presented here demonstrates the application of the above procedure at a site at Denver International Airport.

CASE STUDY—WATER MIGRATION MODELING AT DENVER INTERNATIONAL AIRPORT

Computer modeling of water migration was conducted at the Denver International Airport (DIA) site, where the deep benchmarks described in Section 3.2 were installed (CSU 2004; Chao 2007). The foundation soils at the site consist of silty clay, weathered claystone, and interlayered claystone and sandstone bedrock with several coal seams. The weathered claystone and claystone bedrock are highly expansive. One-dimensional analyses of water migration were conducted. The finite element program that was used can model two-dimensional flow under both saturated and unsaturated conditions. It is possible to take into account infiltration, climatic conditions, precipitation, surface water runoff and ponding, plant transpiration, evaporation, and heat flow.

Downhole nuclear gauge access tubes were installed at several locations on the site. Ground surface boundary conditions varied due to differences in landscaping or pavements. For purposes of illustration in this case study, the analyses and results at only one location will be presented. At that location, the ground surface was subjected to natural climate conditions with no irrigation and sparse vegetation. The time period for which the analysis was performed was based on the design life of the foundation, which was specified to be 50 years.

The input parameters for the soil are summarized in Table 7.3. Where available, measured values were obtained from laboratory test results, and other values were estimated using a commercial database.

TABLE 7.3   Summary of Soil Parameters Used in the Seepage Analyses

Soil Type	Saturated Hydraulic Conductivity (m/sec)	Saturated Volumetric Water Content (vol./vol.)	Residual Volumetric Water Content (vol./vol.)
Silty clay fill	1.6×10^{-8}*	0.40*	0.22*
Silty clay	1.0×10^{-8}**	0.40*	0.22*
Weathered claystone	3.6×10^{-9}***	0.46**	0.07**
Claystone	8.5×10^{-10}***	0.46**	0.07**
Coal	5.0×10^{-6}***	0.48*	0.04*
Sandstone	1.5×10^{-7}*	0.44*	0.13*

*Parameters estimated using SoilVision database
**Laboratory data
***Calibrated values determined during model calibration

The surface climate data that were used in the seepage models included the daily precipitation, the maximum and minimum daily temperature, the maximum and minimum daily relative humidity, and the average daily wind speed. The climate data for the years 1949 to 2005 were obtained from a National Oceanic and Atmospheric Administration database.

Whitney (2003) conducted a two-dimensional analysis across one cross-section of the site and showed that the primary source of water to the soil at this site was flow through the coal seams. The water level in the coal seams was measured to fluctuate from 0 ft to 19 ft (0 m to 5.8 m) below the ground surface.

The four steps in the water migration analyses were as follows:

1. The initial water content profiles were measured in May 2001 using the downhole nuclear gauge. Values of water content obtained from soil samples taken during installation of a deep benchmark were used to supplement the water content data for the deeper portions of the soil profile. Figure 7.8 shows measured values of volumetric water content and the initial water content profile that was specified in the model. This water content profile was used as the initial condition for the model calibration process.
2. The model was calibrated on the basis of the initial water content profile and changes that took place between the initial readings in 2001 and readings taken in 2004. The SWCC for the claystone was measured in the laboratory. The SWCC for the weathered claystone was assumed to be the same as that for the claystone. The SWCCs and hydraulic conductivity functions for the soils were initially estimated based on measured values and by means of a commercially available soil database program. The model was calibrated

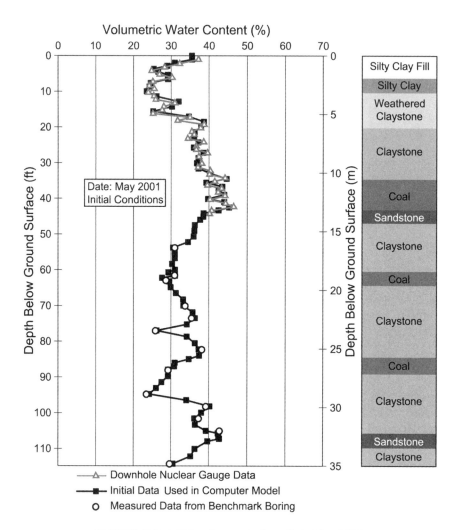

FIGURE 7.8. Initial water content profile in May 2001.

by varying the input parameters until the predicted water contents matched the observed water content data in June 2004. Figure 7.9 presents the predicted volumetric water content computed by the commercial program and the measured water content data for June 2004. Based on the close agreement between the predicted and observed values of water content, the model was considered to be calibrated.

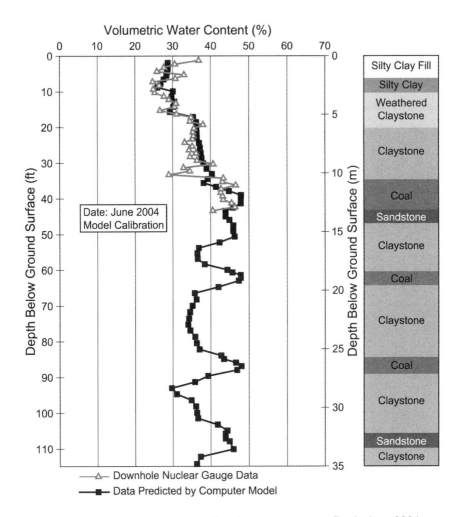

FIGURE 7.9. Measured and predicted water content profiles in June 2004.

3. The calibrated model was then validated by extending the computed results forward in time from June 2004. The validation compared the model output with downhole nuclear gauge data that was obtained in August 2006. Figure 7.10 shows the comparison between the predicted and measured volumetric water content profiles. The results of this process validated that the model was capable of predicting the water content distribution.

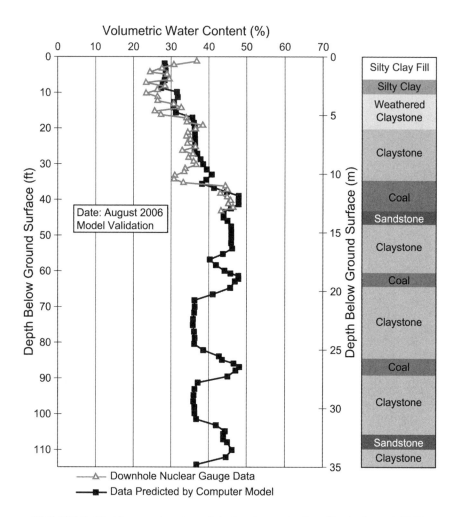

FIGURE 7.10. Measured and predicted water content profiles in August 2006.

4. The water content profile was then computed over the time period, extending to the end of the design life. This covered the time period from the years 2006 to 2040. Figure 7.11 shows the predicted water content profiles through the year 2040. This figure would define the design active zone. Of particular interest is the fact that the results show that water will migrate both upward and downward from the coal seams and sandstone layers. Figure 7.11 indicates that climate conditions will have a significant influence only in the top 20 ft below the ground surface. At deeper depths, the soils are influenced by the deep wetting. This upper 20-ft zone represents the zone of seasonal moisture fluctuation.

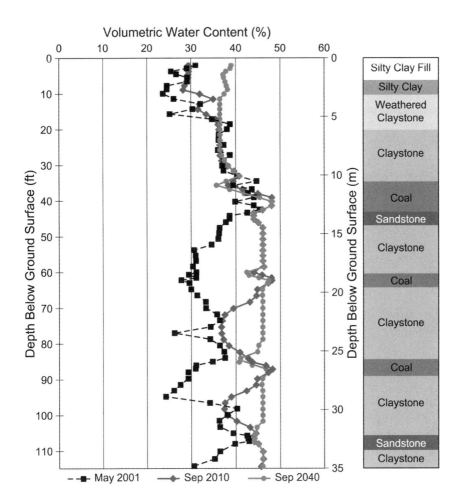

FIGURE 7.11. Predicted long-term water content profiles.

7.4 CHALLENGES IN WATER MIGRATION MODELING FOR EXPANSIVE SOILS

Numerical modeling of water migration in expansive soils is made complex by the large number of factors that influence water movement. These include, among others, volume change due to wetting of the soils, the state of stress and its effects on unsaturated soil properties, surface boundary conditions, and patterns of bedding planes and fractures in the bedrock (Chao, Kang, and Nelson 2014). One challenge to computer modeling relates largely to the hysteretic nature of the soil water characteristic curves of expansive soils under wetting and drying conditions and various stress state conditions. Also, in a natural process, the migration of water could be either into the ground due to precipitation and/or irrigation (i.e., the wetting process), or out of the ground due to evaporation or evapotranspiration (i.e., the drying process). In most practical applications, the water migration model involves both drying and wetting processes in the same model. Pham, Fredlund, and Barbour (2003) developed an elaborate model to include the hysteresis in the curves. However, most commonly used available commercial software does not consider both processes in the same model. Other challenges are posed by the complexity involved in including the heterogeneity of the system into the computer-based mathematical model. These challenges were discussed in more detail by Chao, Kang, and Nelson (2014).

Some of the challenges can be addressed by performing a sensitivity analysis of those factors previously described to evaluate the relative influence of each factor on the input and output of the numerical model. A sensitivity analysis provides a quantitative evaluation of the influence that each factor has on model output. Once the most influential parameters are identified, additional field investigation and/or laboratory testing may be performed to verify certain parameters. Field measurements and monitoring also provide greater confidence with respect to the results of the modeling.

The value of numerical modeling lies in its ability to allow a wide variety of factors to be considered when deciding on an appropriate depth of design active zone to be used for foundation design.

References

Bell, F. G. 2000. *Engineering Properties of Soils and Rocks* (4th ed.). Malden, MA: Blackwell Science.

Buckingham, E. 1907. "Studies of the Movement of Soil Moisture." USDA Bureau of Soils, Bulletin No. 38.

Chao, K. C. 2007. "Design Principles for Foundations on Expansive Soils." PhD dissertation, Colorado State University, Fort Collins, CO.

Chao, K. C., D. D. Overton, and J. D. Nelson. 2006. "The Effects of Site Conditions on the Predicted Time Rate of Heave." *Proceedings of the 4th International Conference on Unsaturated Soils*, Carefree, AZ, 2086–2097.

Chao, K. C., K. B. Kang, and J. D. Nelson. 2014. "Challenges in Water Migration Modeling for Expansive Soils." *Proceedings of Geoshanghai International Conference*, Shanghai, China.

Chen, F. H. 1988. *Foundations on Expansive Soils*. New York: Elsevier Science.

Childs, E. C., and N. Collis-George. "The Permeability of Porous Materials." *Proceedings of the Royal Society*, London, UK, Series A, 201a, 392–405.

Colorado Association of Geotechnical Engineers (CAGE). 1996. "Guideline for Slab Performance Risk Evaluation and Residential Basement Floor System Recommendations (Denver Metropolitan Area) Plus Guideline Commentary." CAGE Professional Practice Committee, Denver, CO.

Colorado State University (CSU). 2004. "Moisture Migration in Expansive Soils at the TRACON Building at Denver International Airport, Phase II. (January 2003–June 2004)." Fort Collins, CO.

Corey, A. T. 1994. "Mechanics of Immiscible Fluids in Porous Media." Highland Ranch, CO: Water Resources Publications.

Cumbers, J. M. 2007. "Soil Suction for Clay Soils at Oven-Dry Water Contents and the End of Swelling Conditions." Master's thesis, Colorado State University, Fort Collins, CO.

Darcy, H. 1856. *Les fontaines publiques de la ville de Dijon*. Paris: V. Dalmont.

Diewald, G. A. 2003. "A Modified Soil Suction Heave Prediction Protocol: With New Data from Denver Area Expansive Soil Sites." Master's thesis, University of Colorado at Denver, Denver, CO.

Edgar, T. V., J. D. Nelson, and D. B. McWhorter. 1989. "Nonisothermal Consolidation in Unsaturated Soil." *Journal of Geotechnical Engineering*, ASCE 115(10): 1351–1372.

El-Ehwany, M., and S. L. Houston. 1990. "Settlement and Moisture Movement in Collapsible Soils." *Journal of Geotechnical Engineering*, ASCE 116(10): 1521–1535.

Freeze, R. A., and J. A. Cherry. 1979. *Groundwater*. Englewood Cliffs, NJ: Prentice-Hall.

Houston, S. L. 1995. "State of the Art Report on Foundations on Collapsible Soils." *Proceedings of the 1st International Conference on Unsaturated Soils*, Paris, 3, 1421–1439.

Houston, S. L., P. D. Duryea, and R. Hong. 1999. "Infiltration Considerations for Groundwater Recharge with Waste Effluent." *Journal of Irrigation and Drainage Engineering*, ASCE 125(5): 264–272.

Houston, S. L., W. N. Houston, C. E. Zapata, and C. Lawrence. 2001. "Geotechnical Engineering Practice for Collapsible Soils." *Geotechnical and Geological Engineering* 19, 333–355.

Houston, W. N., and J. D. Nelson. 2012. "The State of the Practice in Foundation Engineering on Expansive and Collapsible Soils." *Geotechnical Engineering State of the Art and Practice: Keynote Lectures from the Geocongress 2012 Conference*, Oakland, CA, 608–642.

Jefferson County, Colorado. 2009. *The Designated Dipping Bedrock Area Guide*. Golden, CO.

Jury, W. A., W. R. Gardner, and W. H. Gardner. 1991. *Soil Physics* (5th ed.). New York: John Wiley and Sons.

McWhorter, D. B., and Nelson, J. D. 1979. "Unsaturated Flow Beneath Tailings Impoundments." *Journal of the Geotechnical Engineering Division*, ASCE 105(GT11): 1317–1334.

McWhorter, D. B., and J. D. Nelson. 1980. "Seepage in the Partially Saturated Zone beneath Tailings Impoundments." *Mining Engineering* 32(4): 432–439.

McWhorter, D. B., and D. K. Sunada. 1977. *Ground-Water Hydrology and Hydraulics*. Highlands Ranch, CO: Water Resources Publications.

National Research Council (Water Science and Technology Board, Committee on Ground Water Modeling Assessment, Commission on Physical Sciences, Mathematics and Resources). 1990. *Ground Water Models, Scientific and Regulatory Applications*. Washington, DC: National Academy Press.

Nelson, J. D., K. C. Chao, and D. D. Overton. 2011. "Discussion of 'Method for Evaluation of Depth of Wetting in Residential Areas' by Kenneth D. Walsh, Craig A. Colby, William N. Houston, and Sandra L. Houston." *Journal of Geotechnical and Geoenvironmental Engineering,* ASCE 137, 293–296.

Nelson, J. D., D. D. Overton, and D. B. Durkee. 2001. "Depth of Wetting and the Active Zone." *Expansive Clay Soils and Vegetative Influence on Shallow Foundations*, ASCE, Houston, TX, 95–109.

Noe, D. C. 2007. "A Guide to Swelling Soil for Colorado Homebuyers and Homeowners, (2nd ed.)." Colorado Geological Survey Special Publication 43.

Overton, D. D., K. C. Chao, and J. D. Nelson. 2006. "Time Rate of Heave Prediction in Expansive Soils." *Proceedings of the Geocongress 2006 Conference*, Atlanta, GA, 1–6.

———. 2010. "Water Content Profiles for Design of Foundations on Expansive Soils." *Proceedings of the 5th International Conference on Unsaturated Soils*, Barcelona, Spain, 1189–1194.

Pham, H. Q., D. G. Fredlund, and S. L. Barbour. 2003. "A Practical Hysteresis Model for the Soil-Water Characteristic Curve for Soils with Negligible Volume Change." *Geotechnique* 53(2): 293–298.

Reed, R. F. 1985. "Foundation Performance in an Expansive Clay." *Proceedings of the 38th Canadian Geotechnical Conference, Theory and Practice in Foundation Engineering*, Edmonton, Alberta, Canada, 305–313.

Richards, L. A. 1931. "Capillary Conduction of Liquids through Porous Media." *Physics* 1, 318–333.

Singhal, B. B. S., and R. P. Gupta. 2010. *Applied Hydrogeology of Fractured Rocks* (2nd ed.). New York: Springer.

Taylor, D. L., Jr., and A. G. Wassenaar. 1989. "Deep Bedrock Movement and Foundation Performance." *Colorado Engineering* 7(12): 5.

Thompson, R. W. 1992. "Performance of Foundations on Steeply Dipping Claystone." *Proceedings of the 7th International Conference on Expansive Soils*, Dallas, TX, 1, 438–442.

Walsh, K. D., W. N. Houston, and S. L. Houston. 1993. "Evaluation of In-Place Wetting Using Soil Suction Measurements." *Journal of the Geotechnical Engineering Division*, ASCE 119(15): 862–873.

Walsh, K. D., C. A. Colby, W. N. Houston, and S. L. Houston. 2009. "Method for Evaluation of Depth of Wetting in Residential Areas." *Journal of Geotechnical and Geoenvironmental Engineering*, ASCE 2, 169–176.

———. 2011. "Closure to Discussion of 'Method for Evaluation of Depth of Wetting in Residential Areas' by Kenneth D. Walsh, Craig A. Colby, William N. Houston, and Sandra L. Houston." *Journal of Geotechnical and Geoenvironmental Engineering*, ASCE, 137, 299–309.

Whitney, J. M. 2003. "Water Migration in Expansive Soils at the FAA TRACON Building at Denver International Airport." Master's thesis, Colorado State University, Fort Collins, CO.

8

Computation of Predicted Heave

The calculation of predicted heave should be the first step in the design of shallow and deep foundations or other ground supported structures, such as slab-on-grade floors and pavements. The amount of heave that can be expected at a site is the fundamental parameter in selecting a foundation type, designing a foundation, and assessing the risk of movement. The predicted heave should be included in the soil report for the site and used to quantify the risk of movement.

Several methods for calculating heave have been published in the technical literature and can easily be applied to predict heave. The various methods can be grouped into three main categories. These include oedometer methods, soil suction methods, and empirical methods. The oedometer and suction methods are so named because the calculations use results from oedometer tests and suction tests. Empirical methods correlate empirical heave data to various characteristics of the soil. Vanapalli and Lu (2012) present a fairly comprehensive review of various methods that have been proposed over the past several decades. The following sections discuss some of the more rigorous methods.

Factors that must be considered when predicting heave include the following:

- Expansion properties of the soil
- Depth of wetting
- Degree of wetting
- Initial and final effective stress state conditions
- Soil profile and thicknesses of the soil strata
- Groundwater conditions

It is important to be explicit with respect to what is meant by the nature of the heave being discussed. Chapter 7 discussed the active zone and other associated zones within the soil profile that affect the heave.

Different aspects of heave were defined in chapter 1. They include free-field heave, ultimate heave, design heave, and current heave.

8.1   OEDOMETER METHODS

Heave prediction methods were first developed in the late 1950s. They originated as an extension of methods used to estimate volume changes due to consolidation in saturated soils using results of one-dimensional oedometer (consolidation) tests. Those tests are the most commonly used to predict heave. The two main types of tests, namely, the "consolidation-swell" (CS) test and "constant volume" (CV) test, have been described in chapter 6. Prediction methods using oedometer test results have the distinct advantage of using conventional testing equipment with which most geotechnical engineers are familiar. Heave prediction methods have been refined continuously as understanding of unsaturated soil behavior has increased. Jennings and Knight (1957) first proposed the double oedometer method which used the results from two oedometer tests, one on a dry sample of soil and another on a wetted sample of identical soil. Salas and Serratosa (1957) presented an oedometer heave prediction model which incorporated the "swelling pressure" of a soil into the equation. The "swelling pressure" of a soil had been defined by Palit (1953) as the pressure in an oedometer test required to prevent a soil sample from swelling after being saturated. That would be what was termed the CV swelling pressure in chapter 6.

Fredlund, Hasan, and Filson (1980) set forth the theoretical framework to include soil suction in the prediction of heave. The US Department of the Army (1983) presented two approaches to predict heave. In 1983, Fredlund proposed the basis for the form of the heave prediction equation that will be developed below. The method presented in Nelson and Miller (1992) uses the same equation as Fredlund (1983). Fredlund and Rahardjo (1993) and Fredlund, Rahardjo, and Fredlund (2012) also published a heave prediction method based on Fredlund (1983). Chen (1988) referenced the method presented in Fredlund (1983) and presented an example of its use to predict heave. The primary difference between these methods and the one that will be presented below is the manner in which the heave index, C_H, is determined. The heave index was discussed in chapter 6 and is shown in Figures 6.3 and 6.4. Nelson, Durkee, and Bonner (1998) and Nelson, Reichler, and Cumbers (2006) introduced the heave index in the heave equation. That method will be presented below.

8.1.1 The Heave Equation

The equation for predicting heave is based on the fundamental definition of strain:

$$\varepsilon_s = \frac{\varepsilon_{s\%}}{100} = \frac{\Delta H}{H} \tag{8-1}$$

where:

$_s$ = strain in the layer that is heaving,
$\varepsilon_{s\%}$ = percent swell that will occur when that stratum becomes wetted (strain in percent),
H = thickness of a layer of soil, and
ΔH = change in thickness of that layer due to heave.

Equation (8-1) can be solved for ΔH to obtain the equation for heave of a soil layer. Thus,

$$\Delta H = \varepsilon_s H = \frac{\varepsilon_{s\%} H}{100} \tag{8-2}$$

The strain, ε_s or $\varepsilon_{s\%}$, in equation (8-2) is the strain that will result in the soil due to an increase in water content, or *wetting*. It is also a function of the vertical stress that exists in the stratum at the time it becomes wetted. Thus, it is necessary to develop a relationship between the amount of swell that a soil will experience when wetted, and the stress that is applied at the time of wetting. This relationship was developed in chapter 6 and Figure 6.4 and used the parameter, C_H, which is termed the heave index. The heave index can be determined from test results obtained in the CV and CS tests. It can also be determined from CS test results using the relationship between σ_{cv}'' and σ_{CS}'' that was presented in Section 6.3.

From equation (6-1), the strain in the layer, $\varepsilon_{s\%}$, can be related to the stress as,

$$\frac{\varepsilon_{s\%}}{100} = C_H \log \left[\frac{\sigma_{cv}''}{\sigma_i''} \right] \tag{8-3}$$

In the soil layer, the inundation stress is the vertical net normal stress acting at the midpoint of the layer. The final vertical net normal stress, σ_f'', that exists in the soil after construction has been completed is calculated from the weight of the soil above that point along with the increment of stress, $\Delta\sigma_v''$, due to the load applied by the structure. This value of final vertical stress is tantamount to the inundation stress that will exist in the soil profile as it becomes wetted. The value of σ_f'' increases with depth because the weight of the soil (overburden stress) above each point increases. The value of applied stress, $\Delta\sigma_v''$, due to loads produced by a structure decreases with depth. Thus, the value of σ_f'' at various points within the soil where increments of heave are computed is generally not the same as that at which the oedometer test was conducted. This is accounted for in the heave equation by the heave index, C_H.

In Figure 6.4 the heave index defines a linear relationship between percent swell and applied stress. Because the relationship is linear, the value of C_H will apply for any value of stress less than σ_{cv}''. At the time the soil stratum is inundated the average stress applied to the stratum will be equal to σ_f''. For any particular value of applied stress, σ_f'', equation (8-3) can be written as

$$\varepsilon_s = \frac{\varepsilon_{s\%}}{100} = C_H \log\left[\frac{\sigma_{cv}''}{\sigma_f''}\right] \tag{8-4}$$

Equation (8-4) can be substituted into equation (8-2) to give

$$\Delta H = C_H H \log\left[\frac{\sigma_{cv}''}{\sigma_f''}\right] \tag{8-5}$$

In actual application of equation (8-5), a soil profile will be divided into layers of thickness H, and the value of heave for each layer will be computed. The incremental values will be added to determine the total free-field heave, ρ. Thus, the general equation for predicting total free-field heave, ρ, of a soil profile is as shown in equation (8-6). The value of σ_f'' to be used in equation (8-6) is the stress at the midpoint of each layer.

$$\rho = \sum_{i=1}^{n} \Delta H_i = \sum_{i=1}^{n} \left\{ C_H H \log\left[\frac{\sigma_{cv}''}{\sigma_f''}\right] \right\}_i \tag{8-6}$$

where:
ρ = total free-field heave,
ΔH_i = heave of layer i,
C_H = heave index of layer i,
H_i = initial thickness of layer i,
σ_{cv}'' = CV swelling pressure of layer i, and
σ_f'' = final vertical net normal stress of layer i.

Some simplified methods for predicting heave use equation (8-2) directly and use the value for $\varepsilon_{s\%}$ as measured in an oedometer test inundated at a value of the overburden stress at the midpoint of the layer. That value is applied over the entire height of the soil layer. Such a method does not take into account the nonlinearity of the strain over the thickness of the layer, nor does it account appropriately for the applied stress.

8.1.2   Computation of Free-Field Heave

As noted, the free-field heave is the heave that takes place with no other loads applied to the soil such as by a foundation or an embankment. Free-field heave is computed using equation (8-6). The soil profile is divided into a number of layers and the heave of each layer is computed. Because there are no loads on the soil layers other than the weight of the overburden, the average stress σ_f'' over the thickness of the layer is equal to the overburden stress at the midpoint of the layer. Because the value of σ_f'' can vary nonlinearly over the depth of the layer, and because equation (8-6) is not linear across the thickness of the layer, it is desired to choose a layer thickness as small as practical in performing the computations. Usually, a thickness of 2 ft (0.5 m) or less is adequate.

The method of computation is presented in Examples 8.1 and 8.2. In these examples, the ultimate heave is computed, and therefore, the soil is assumed to have become fully wetted over the entire depth of potential heave.

EXAMPLE 8.1

Given:

The soil profile at a site containing a thick layer of expansive claystone. The dry density, γ_d, of the claystone is 115 pcf, with a gravimetric water content, w, of 12 percent and an initial degree of saturation of 64.7 percent. The specific gravity of the claystone solids is 2.79. The claystone was tested in a CS test and exhibited a percent swell, $\varepsilon_{s\%}$, of 4 percent when inundated at an inundation stress, σ_i'', of 1,000 psf. The results of the CS oedometer test are shown in Figure E8.1-1. The swelling pressure, σ_{cs}'', for the claystone was measured to be 9,500 psf. A CV test was also performed on the same soil and the CV swelling pressure was measured to be 4,390 psf.

Find:

Ultimate free-field heave under conditions of full wetting.

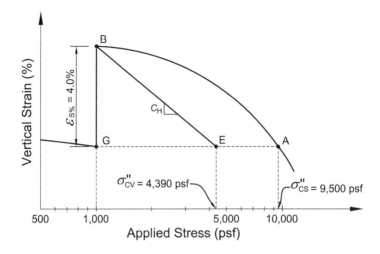

FIGURE E8.1-1. Consolidation-swell oedometer test.

Solution:

In a manner similar to that shown in Figure 6.4, the value for C_H can be determined from the slope of the line BE in Figure E8.1-1. It is computed by equation (6-1).

$$C_H = \frac{0.04}{\log\left[\dfrac{4,390}{1,000}\right]} = 0.0623$$

To compute ultimate free-field heave it is assumed that the soils over the entire depth of potential heave will become fully wetted. Therefore, the saturated unit weight will be used as the total unit weight to compute the overburden stress. If it is assumed that at full wetting $S = 100$ percent, or at least close to 100 percent, γ_{sat} can be computed from the block diagram shown in Figure E8.1-2 to be 136.22 pcf. The depth of potential heave, z_p, is computed by equating the overburden stress to the swelling pressure.

$$\gamma_{sat} \times z_p = \sigma''_{cv}$$

$$136.22 \text{ pcf} \times z_p = 4,390 \text{ psf}$$

$$z_p = 32.23 \text{ ft}$$

The soil throughout the depth of potential heave is divided into several layers and the heave of each layer is computed. For purposes of simplifying this example, the soil was divided into 10 equal layers. Therefore, each layer is 3.22 ft (32.23 ft/10) thick. In practice, it would be appropriate to use a larger number of layers to increase the accuracy of the calculations. To illustrate the computations of free-field heave, computation of heave of the top two layers will be demonstrated.

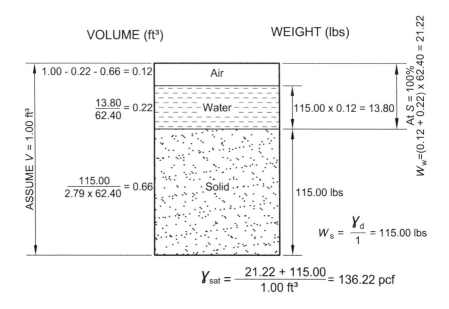

FIGURE E8.1-2. Block diagram.

Top layer: 0 to 3.22 ft

The midpoint of the first layer is 1.61 ft below the ground surface. The vertical net normal stress at that depth is,

$$\sigma''_{f1} = \gamma_{sat}z_1 = 136.22 \text{ pcf} \times 1.6114 \text{ ft} = 219.50 \text{ psf}$$

From equation (8-6), the heave of that layer is

$$\rho_1 = C_H H_1 \log\left[\frac{\sigma''_{cv}}{\sigma''_f}\right]$$

$$\rho_1 = 0.0623 \times 3.22 \text{ ft} \times \log\left[\frac{4,390 \text{ psf}}{219.50 \text{ psf}}\right] = 0.261 \text{ ft} = 3.13 \text{ in.}$$

Second layer: 3.22 to 6.45 ft

The midpoint of the second layer is 4.834 ft below the ground surface. The final vertical net normal stress at that depth is

$$\sigma''_{f2} = \gamma_{sat}z_2 = 136.22 \text{ pcf} \times 4.834 \text{ ft} = 658.50 \text{ psf}$$

and

$$\rho_2 = 0.0623 \times 3.22 \text{ ft} \times \log\left[\frac{4,390 \text{ psf}}{658.50 \text{ psf}}\right] = 0.165 \text{ ft} = 1.98 \text{ in.}$$

The calculations for each succeeding layer are similar and lend themselves well to solution by a spreadsheet. The results for the succeeding computations are shown in Table E8.1.

TABLE E8.1 Free-Field Heave Computations for Example 8.1

Depth to Bottom of Layer (ft)	Depth Below Excavation (ft)	Interval (ft)	Soil Type	Total Unit Weight (pcf)	Incremental Stress @ midlayer (psf)	Overburden Stress (psf)	Incremental Heave (in.)	Cumulative Heave (in.)	Cumulative Heave from z_p (in.)
0.00	0.00								10.10
3.22	3.22	3.223	Claystone	136.2	219.50	219.50	3.13	3.13	6.97
6.45	6.45	3.223	Claystone	136.2	219.50	658.50	1.98	5.11	4.99
9.67	9.67	3.223	Claystone	136.2	219.50	1,097.5	1.45	6.56	3.54
12.89	12.89	3.223	Claystone	136.2	219.50	1,536.5	1.10	7.66	2.44
16.11	16.11	3.223	Claystone	136.2	219.50	1,975.5	0.84	8.50	1.60
19.34	19.34	3.223	Claystone	136.2	219.50	2,414.5	0.63	9.13	0.97
22.56	22.56	3.223	Claystone	136.2	219.50	2,853.5	0.45	9.58	0.52
25.78	25.78	3.223	Claystone	136.2	219.50	3,292.5	0.30	9.88	0.22
29.01	29.01	3.223	Claystone	136.2	219.50	3,731.5	0.17	10.05	0.05
32.23	32.23	3.223	Claystone	136.2	219.50	4,170.5	0.05	10.10	

The ultimate free-field heave is the sum of the heave values calculated for all 10 layers. From Table E8.1 it is seen that the ultimate free-field heave, ρ_{total}, is 10.1 in.

In Example 8.1, the unit weight of the soil was corrected to reflect the unit weight after wetting. If the computations were performed using the initial total unit weight as measured during the site investigation the depth of potential heave would be calculated to be 34.08 ft and the calculated ultimate free-field heave would be 10.7 in. This represents an increase of less than 6 percent. Depending on the application of the analysis, this difference in computed value may or may not be considered important.

The data given for Example 8.1 included CV test results. If a CV test had not been performed, the value of σ_{cv}'' could be estimated using the m method discussed in Section 6.3. Those computations are shown in Example 8.2.

EXAMPLE 8.2

Given:

The same soil profile and data as for Example 8.1, except that a CV swell test was not performed.

Find:

Ultimate free-field heave under conditions of full wetting.

Solution:

The CV swelling pressure, σ_{cv}'', can be estimated from the relationship given by equation (6-3). From Figure 6.12, a reasonable value of m for claystone is estimated to be 0.8. Thus,

$$\log \sigma_{cv}'' = \frac{\log \sigma_{cs}'' + m \log \sigma_i''}{1 + m} = \frac{\log 9{,}500 + 0.8 \times \log 1{,}000}{1 + 0.8} = 3.5432$$

$$\sigma_{cv}'' = 3{,}493 \text{ psf}$$

The value for C_H for this value of σ_{cv}'' is then

$$C_H = \frac{0.04}{\log\left[\dfrac{3{,}493}{1{,}000}\right]} = 0.074$$

The depth of potential heave will be computed as in Example 8.1.

$$z_p = \frac{3{,}493 \text{ psf}}{136.22 \text{ pcf}} = 25.64 \text{ ft}$$

The remaining computations are conducted in the same manner as those for Example 8.1. The spreadsheet computations are presented in Table E8.2. The ultimate free-field heave was computed to be 9.5 in. This represents a difference of about 6 percent from the value computed in Example 8.1.

TABLE E8.2 Free-Field Heave Computations for Example 8.2

Depth to Bottom of Layer (ft)	Depth Below Excavation (ft)	Interval (ft)	Soil Type	Total Unit Weight (pcf)	Incremental Stress @ midlayer (psf)	Overburden Stress (psf)	Incremental Heave (in.)	Cumulative Heave (in.)	Cumulative Heave from z_p (in.)
0.00	0.00								9.51
2.56	2.56	2.56	Claystone	136.2	174.3	174.3	2.95	2.95	6.56
5.12	5.12	2.56	Claystone	136.2	174.3	523.0	1.87	4.82	4.69
7.68	7.68	2.56	Claystone	136.2	174.3	871.6	1.36	6.18	3.33
10.24	10.24	2.56	Claystone	136.2	174.3	1,220.3	1.03	7.21	2.30
12.80	12.80	2.56	Claystone	136.2	174.3	1,568.9	0.79	8.00	1.51
15.36	15.36	2.56	Claystone	136.2	174.3	1,917.6	0.59	8.59	0.92
17.92	17.92	2.56	Claystone	136.2	174.3	2,266.2	0.43	9.02	0.49
20.48	20.48	2.56	Claystone	136.2	174.3	2,614.9	0.28	9.30	0.21
23.04	23.04	2.56	Claystone	136.2	174.3	2,963.5	0.16	9.46	0.05
25.60	25.60	2.56	Claystone	136.2	174.3	3,312.2	0.05	9.51	

Examples 8.1 and 8.2 showed computations of heave for relatively uniform soil conditions. More often than not, the soil profile is more complex. In Example 8.3 the heave is computed for a soil profile that contains several different strata, some of which are highly expansive and one of which consists of nonexpansive sandstone. The incremental and cumulative heave is plotted over the depth of heave. This profile is modeled after an actual highly expansive soil site.

EXAMPLE 8.3

Given:

Nonuniform soil profile shown in Table E8.3-1.

TABLE E8.3-1   Soil Profile for Example 8.3

Depth (ft)	Soil Type	Water Content (%)	Dry Density (pcf)	CS % Swell, $\varepsilon_{s\%}$ @ 1,000 psf (%)	CS Swelling Pressure, σ_{cs}'' @ 1,000 psf (psf)
0–8	Weathered claystone	16	112	2	3,600
8–18	Gray claystone	12	121	4	11,200
18–22	Sandy claystone	10	118	1.1	2,160
22–27	Sandstone	8	115	0	0
27–40	Brown claystone	11	124	6	12,500

Find:

Ultimate free-field heave and plot incremental heave with depth.

Solution:

At this point, in order to facilitate computation of the depth of potential heave, it is convenient to compute the CV swelling pressure for each material in the soil profile using the m method. Those computations are shown in Table E8.3-2.

TABLE E8.3-2   Computation of σ_{cv}'' from σ_{cs}''

Material	σ_{cs}'' (psf)	m	σ_{cv}'' (psf)
Weathered claystone	3,600	0.4	2,497
Gray claystone	11,200	0.8	3,827
Sandy claystone	2,160	0.4	1,733
Sandstone	0	—	—
Brown claystone	12,500	0.8	4,068

Based on the data shown in Figure 6.15, the value of m to be used in equation (6-3) for the weathered claystone and the sandy claystone was assumed to be 0.4. For both the gray claystone and the brown claystone, m was assumed to be 0.8.

The depth of potential heave must be checked for each layer. It is assumed that it will not be in the weathered claystone. At the bottom of the gray claystone, the overburden stress is

$$(\sigma''_{vo})_{GC} = (112 \times 1.16 \times 8) + (121 \times 1.12 \times 10) = 2,395 \text{ psf}$$

This value is less than the swelling pressure of that stratum, and therefore, the depth of potential heave will be deeper than 18 ft. Although the computed value of overburden stress is greater than the value of σ''_{cv} for the sandy claystone, the swelling pressure in the brown claystone is higher. At the top of the brown claystone, the overburden stress is equal to

$$(\sigma''_{vo})_{BC} = 2,395 \text{ psf} + (118 \times 1.10 \times 4) + (115 \times 1.08 \times 5) = 3,535 \text{ psf}$$

The swelling pressure of the brown claystone is greater than the overburden at the top of that layer. Therefore, the depth of potential heave, z_p, will be in the brown claystone. The depth of z_p is calculated by equating the overburden stress to the value of σ''_{cv} for the brown claystone.

$$3,535 + (z_p)_{BC}(124 \times 1.11) = 4,068 \text{ psf}$$

$$(z_p)_{BC} = 3.87 \text{ ft}$$

Thus, the depth of potential heave is 3.9 ft below the top of the brown claystone, or 30.9 ft below the ground surface.

The soil profile from the ground surface to a depth of 30.9 ft is divided into 16 layers, each of which is 1.93 ft thick. The computations are similar to those in the preceding examples and are presented in Table E8.3-3. The predicted heave is calculated to be 7.0 in.

In Table E8.3-3 it is seen that the boundary between the sandstone and the brown claystone lies at the middle of a layer. To be exact, that layer should be divided into two layers, each being 1 ft thick. This would make the computations cumbersome and would not actually improve the accuracy of the heave prediction. To demonstrate that, computations were performed in which the soil profile was divided into 1 ft thick layers. The predicted heave was calculated to be 6.9 in., a difference of only 0.1 in. Thus, if sufficiently thin layers are used in the computations, the importance of having each layer coincide with a soil boundary is small.

The profile of incremental heave and total heave throughout the depth of potential heave is shown in Figure E8.3.

TABLE E8.3-3 Free-Field Heave Computations for Example 8.3

Depth to Bottom of Layer (ft)	Depth Below Excavation (ft)	Interval (ft)	Soil Type	Total Unit Weight (pcf)	Incremental Stress @ midlayer (psf)	Overburden Stress (psf)	Incremental Heave (in.)	Cumulative Heave (in.)	Cumulative Heave from z_p (in.)
0.00	0.00								7.00
1.93	1.93	1.93	Weathered claystone	129.9	125.5	125.5	1.52	1.52	5.48
3.86	3.86	1.93	Weathered claystone	129.9	125.5	376.4	0.96	2.48	4.52
5.79	5.79	1.93	Weathered claystone	129.9	125.5	627.3	0.70	3.18	3.82
7.72	7.72	1.93	Weathered claystone	129.9	125.5	878.2	0.53	3.71	3.29
9.66	9.66	1.93	Gray claystone	135.5	130.9	1,134.5	0.84	4.55	2.45
11.59	11.59	1.93	Gray claystone	135.5	130.9	1,396.2	0.70	5.25	1.75
13.52	13.52	1.93	Gray claystone	135.5	130.9	1,657.9	0.58	5.83	1.17
15.45	15.45	1.93	Gray claystone	135.5	130.9	1,919.7	0.48	6.31	0.69
17.38	17.38	1.93	Gray claystone	135.5	130.9	2,181.4	0.39	6.70	0.30
19.31	19.31	1.93	Sandy claystone	129.8	125.3	2,437.6	0.00	6.70	0.30
21.24	21.24	1.93	Sandy claystone	129.8	125.3	2,688.3	0.00	6.70	0.30
23.17	23.17	1.93	Sandstone	124.2	119.9	2,933.5	0.00	6.70	0.30
25.10	25.10	1.93	Sandstone	124.2	119.9	3,173.4	0.00	6.70	0.30
27.03	27.03	1.93	Brown claystone	137.6	132.9	3,426.2	0.17	6.87	0.13
28.96	28.96	1.93	Brown claystone	137.6	132.9	3,692.0	0.10	6.97	0.03
30.89	30.89	1.93	Brown claystone	137.6	132.9	3,957.9	0.03	7.00	

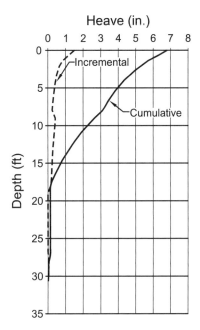

FIGURE E8.3. Heave profile throughout depth of potential heave.

8.1.3 Computation of Heave under an Applied Load

The computation of heave under an applied load such as a footing or embankment fill is similar to that for free-field heave except that the load applied to the soil must be included in the equation. A discussion of the calculations for a footing load is included in chapter 11.

8.1.4 Computation of Design Heave

In Example 8.1, the depth of potential heave was computed to be 32 ft. Wetting to that depth has been observed at various sites within periods of less than 10 years after construction. In a situation such as that, it is prudent to assume that the depth of wetting will extend to the depth of potential heave within the design life of the structure.

However, for a site at which the depth of potential heave is much deeper, or where the entire depth of potential heave is not expected to be fully wetted within the useful life of the structure, it would not be practical to design for such a case. In that situation, it would be worth the cost that would be required to perform a water migration analysis and design the foundation for the wetting conditions that are expected to occur during the design life of the structure.

Figure 7.1 showed the shape of the water content profile that would exist for a downward progressing wetting front. That figure shows that the water content varies throughout the transition zone and it was noted that heave in this partially wetted zone may be significant. In predicting heave for a design active zone that is not fully wetted throughout, a water content profile such as that shown in Figure 7.1 would be used.

For the partially wetted soils in the transition zone, it is necessary to modify the fully wetted oedometer data to determine the values of percent swell and swelling pressure to be used to predict heave. Figure 8.1 shows normalized percent swell plotted against degree of saturation for various values of initial degree of saturation. The normalized percent swell was determined by dividing the values of percent swell that occurred at a particular degree of saturation by the maximum values of percent swell at full wetting in an oedometer test. The data shown in Figure 8.1 can be used to calculate the reduced percent swell to use in calculations of predicted heave under conditions of partial wetting. A general form of the relationship between normalized percent swell and degree of saturation for soil was derived by performing regression analyses on the observed experimental data shown in Figure 8.1 (Chao 2007). The relationship is expressed by equation (8-7).

$$\varepsilon_{s\%N} = (850.17S_i^2 - 505.61S_i + 47.14)S_f^2$$
$$+ (-283.20S_i^2 + 250.35S_i - 21.94)S_f \qquad (8\text{-}7)$$
$$+ (-45.70S_i^2 - 9.57S_i + 1.19)$$

where:

   $\varepsilon_{s\%N}$ = normalized percent swell,
   S_i = initial degree of saturation, and
   S_f = final degree of saturation.

Although this is a complex form of equation, it has the advantage of being useful for analysis in computer modeling.

In addition to determining a value for the percent swell at partial wetting, the swelling pressure for a partially wetted soil must be determined. Reichler (1997) showed that curves for reloading the sample after swelling are nearly parallel even for the samples with different values of percent swell. This suggests a procedure for determining a value

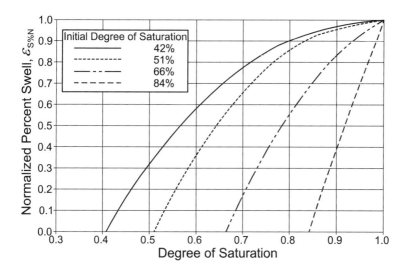

FIGURE 8.1. Normalized percent swell vs. degree of saturation (modified from Chao 2007).

of reduced swelling pressure, σ''_{cvN}, to be used for computing heave in a partially wetted soil. This procedure is shown in Figure 8.2. To determine the reduced swelling pressure a line is drawn parallel to the C_H line and passing through the value of normalized percent swell, $\varepsilon_{s\%N}$, that was determined by Figure 8.1 or equation (8-7). Where that line intersects the axis for zero swell is the reduced swelling pressure, σ''_{cvN}.

Example 8.4 shows computations for a site in which full wetting was predicted to extend through only a portion of the depth of potential

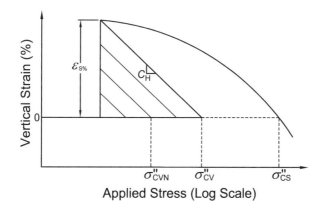

FIGURE 8.2. Procedure for determining swelling pressure for partially wetted soil.

heave. In that example, the water migration analysis predicted that at the end of the design life the wetting profile would be as shown in Figure E8.4-1. Full wetting was predicted to occur to a depth of 16 ft (4.9 m), below which the soil was partially wetted to a depth of 22 ft (6.7 m). Thus, the design active zone extends to a depth of 22 ft, with only partial wetting in the transition zone between 16 and 22 ft. The computations of heave to the depth of 16 ft are the same as for Example 8.1. Below a depth of 16 ft the values of $\varepsilon_{s\%}$ and σ''_{cv} were adjusted in accordance with the procedure just described.

It is shown in Example 8.4 that the design heave was computed to be 9.0 in. (229 mm) as compared to the ultimate heave of 10.1 in. (257 mm) calculated in Example 8.1. It is interesting that a reduction of 10 ft (3 m) in the depth of the design active zone resulted in a reduction of only 1 in. (25 mm) in the calculated heave. The reason for such a small change in the amount of heave is that because of the higher overburden stress in the lower 10 ft (3 m), most of the heave takes place in the shallower soils.

EXAMPLE 8.4

Given:

The same soil profile as in Example 8.1 with a design wetting profile as shown in Figure E8.4-1.

Find:

Design heave.

Solution:

Because wetting is predicted to occur to a depth of 22 ft, the soil profile is divided into 11 layers, each of 2 ft thickness. For the three lower layers the percent swell and the swelling pressure will be corrected to account for the fact that the soil in the transition zone between 16 and 22 ft is only partially wetted. The method for correcting the percent swell and the swelling pressure is shown in Figures E8.4-2 and E8.4-3. Figure E8.4-2 contains data from Figure 8.1 that contains data for values of initial degree of saturation, S_i, of 56.6 percent, and 66.0 percent. In Example 8.1 the initial degree of saturation for the soil was given to be 64.7 percent. The curve for an initial degree of saturation of 64.7 percent was determined by interpolation between the curves for S_i of 56.6 percent and 66 percent. At the midpoint of the soil layer from 16 to 18 ft (i.e., $z = 17$ ft). The predicted degree of saturation can be determined by interpolation in Figure E8.4-1. Thus,

$$(S_i)_{17ft} = 100 - \frac{1}{6} \times (100 - 64.7) = 94.1\%$$

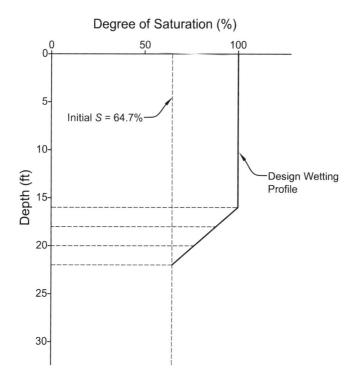

FIGURE E8.4-1. Design wetting profile for Example 8.4.

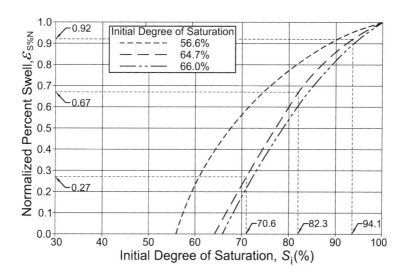

FIGURE E8.4-2. Determination of normalized percent swell for Example 8.4.

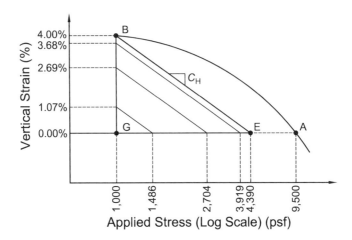

FIGURE E8.4-3. Determination of reduced swelling pressure for Example 8.4.

In Figure E8.4-2, the normalized percent swell corresponding to a value of $S_i = 94.1\%$ is 0.92. Thus, the value of percent swell to be used for that layer is

$$(\varepsilon_{s\%})_{17\text{ft}} = 0.92 \times 4.0 = 3.68\%$$

In Figure E8.4-3, a line is drawn parallel to the C_H line (BE) and passing through the point for $\varepsilon_{s\%} = 3.68\%$. This line intersects line GE at a value of $\sigma''_{cv} = 3,919$ psf.

In a similar fashion, the initial degree of saturation and normalized percent swell for the depths of 19 ft and 21 ft are computed to be $(S_i)_{19} = 82.3\%$, $(S_i)_{21} = 70.6\%$, $(\varepsilon_{s\%})_{N19} = 0.67$, and $(\varepsilon_{s\%})_{N21} = 0.27$. The values of $\varepsilon_{s\%}$ and σ''_{cv} for each layer are shown in Figure E8.4-3.

The three lower layers can be considered to be different soils. The soil profile to be analyzed is given in Table E8.4-1 with the three lower layers labeled as Claystones 1, 2, and 3. The corrected values of $\varepsilon_{s\%}$ and σ''_{cv} are shown in Table E8.4-1 and all other soil properties are the same as in Example 8.1.

TABLE E8.4-1   Revised Soil Profile for Partial Wetting in Example 8.4

Depth (ft)	Soil Type	CS % Swell, $\varepsilon_{s\%}$ @ 1,000 psf (%)	CV Swelling Pressure, σ''_{cs} (psf)
0–16	Claystone	4.00	4,390
16–18	Claystone 1	3.68	3,919
18–20	Claystone 2	2.69	2,704
20–22	Claystone 3	1.07	1,486

The computations are the same as for a soil profile with four layers, and are shown in Table E8.4-2.

TABLE E8.4-2   Free-Field Heave Computations for Partial Wetting in Example 8.4

Depth to Bottom of Layer (ft)	Depth Below Excavation (ft)	Interval (ft)	Soil Type	Total Unit Weight (pcf)	Incremental Stress @ midlayer (psf)	Overburden Stress (psf)	Incremental Heave (in.)	Cumulative Heave (in.)	Cumulative Heave from z_p (in.)
0.00	0.00								8.98
2.00	2.00	2.0	Claystone	136	136.22	136.22	2.25	2.25	6.73
4.00	4.00	2.0	Claystone	136	136.22	408.65	1.54	3.79	5.19
6.00	6.00	2.0	Claystone	136	136.22	681.09	1.21	5.00	3.98
8.00	8.00	2.0	Claystone	136	136.22	953.52	0.99	5.99	2.99
10.00	10.00	2.0	Claystone	136	136.22	1225.96	0.83	6.82	2.16
12.00	12.00	2.0	Claystone	136	136.22	1498.39	0.70	7.52	1.46
14.00	14.00	2.0	Claystone	136	136.22	1770.83	0.59	8.11	0.87
16.00	16.00	2.0	Claystone	136	136.22	2043.26	0.50	8.61	0.37
18.00	18.00	2.0	Claystone1	135	134.93	2314.41	0.34	8.95	0.03
20.00	20.00	2.0	Claystone2	132	132.42	2581.76	0.03	8.98	0.00
22.00	22.00	2.0	Claystone3	130	129.95	2844.13	0.00	8.98	

8.1.5 Discussion of Earlier Oedometer Methods Proposed to Compute Heave

8.1.5.1 Department of the Army (1983)

Technical Manual TM 5-818-7 by the US Department of the Army (1983) presents guidance and information regarding the geotechnical investigation necessary for the selection and design of foundations for heavy and light military-type buildings constructed in expansive clay soil areas. In the manual, two approaches are provided to predict heave for a soil layer using CS test results. In the first approach, the following equation is used:

$$\Delta H = \frac{\varepsilon_{s\%vo}H}{100} \tag{8-8}$$

in which $\varepsilon_{s\%vo}$ is the percent swell measured from a sample inundated at the overburden stress in the CS test, and other symbols are as previously defined.

Equation (8-8) is essentially the same as equation (8-2), and is based on the fundamental concept to predict heave of a soil layer. However, as noted previously, it does not consider the nonlinear nature of the variation of heave or applied stress throughout the thickness of the layer.

The second approach presented in TM 5-818-7 uses the percent swell and swelling pressure determined from the CS test to determine a parameter similar to C_H. Using the notation of this book, the equation for heave of a soil layer for the second approach is shown in equation (8-9).

$$\Delta H = \frac{C_{DA}}{1 + e_0} H \log \left[\frac{\sigma''_{cs}}{\sigma''_f} \right] \tag{8-9}$$

where:

    C_{DA} = Department of Army heave parameter, or DA heave index,
    e_0 = initial void ratio of the soil layer,
    σ''_{cs} = swelling pressure from CS test, and
    σ''_f = final vertical effective pressure.

Equation (8-9) has the same form as that of equation (8-5) except that the heave parameter, C_{DA}, in equation (8-9) is defined in terms of void ratio and is based on σ''_{cs} instead of σ''_{cv}. The DA heave index, C_{DA}, is the slope of the loading portion of the consolidation-swell curve between points B and A shown in Figure 8.3, and is expressed in equation (8-10). Figure 8.3 compares the heave parameters as determined by the different oedometer methods to be discussed. In that figure, the heave index,

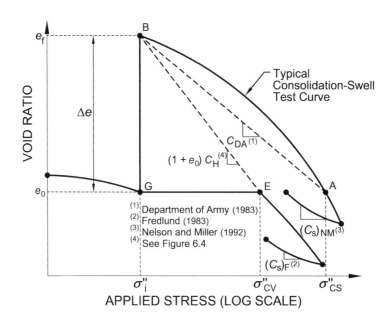

FIGURE 8.3. Comparison of different heave parameters.

C_H, is multiplied by $(1 + e_0)$ to reflect the fact that the vertical axis in Figure 8.3 is plotted in terms of void ratio and C_H is expressed in terms of percent strain.

$$C_{DA} = \frac{\Delta e}{\log \sigma_{cs}'' - \log \sigma_i''} = \frac{\Delta e}{\log \left[\dfrac{\sigma_{cs}''}{\sigma_i''} \right]} \tag{8-10}$$

Vertical strain is related to void ratio by equation (8-10).

$$\varepsilon_{s\%} = \frac{\Delta e}{1 + e_0} \times \frac{1}{100} \tag{8-11}$$

Thus, the similarity of C_{DA} to the heave index, C_H, is evident.

8.1.5.2   Fredlund (1983)
In 1983, Fredlund proposed a method for calculating heave based on data from the CV test (Fredlund 1983). Fredlund and Rahardjo (1993) and Fredlund, Rahardjo, and Fredlund (2012) presented a heave calculation method that is similar to the method presented in Fredlund (1983). The basic data required for this method are the corrected swelling pressure from the CV test and the swelling index, C_s, measured on the unloading portion of the curve shown in Figure 8.3.

The basic equation that they proposed is shown in equation (8-12). Its similarity to equation (8-9) is noted, except that equation (8-12) uses the slope of the rebound curve, C_s.

$$\Delta H = \frac{C_s}{1 + e_0} H \log \left[\frac{\sigma_{cv}''}{\sigma_f''} \right] \tag{8-12}$$

The symbols in equation (8-12) are as previously defined.

The total heave of the entire soil profile is equal to the sum of the heave for each layer.

8.1.5.3   Nelson and Miller (1992)
Nelson and Miller (1992) presented a method for computing heave that was based on the method presented by Fredlund (1983). They suggested that the swelling index, C_s, shown in equation (8-12) can be determined from the slope of the rebound curve of the CS test, as shown in Figure 8.3. As discussed in Section 8.1.6 it is more correct to use the parameter C_H.

8.1.6 Comments on the Heave Index

The evolution of heave prediction methods using oedometer tests has been largely related to the appropriate definition of the heave index, C_H. Burland (1962), Fredlund (1983), Nelson and Miller (1992), Fredlund and Rahardjo (1993), and Fredlund, Rahardjo, and Fredlund (2012) proposed using the slope of the unloading curve from either the CS test or the CV test. The method for determining C_H that is presented in chapter 6 is more rigorous and is based on consideration of both applied stress and suction changes. The method proposed by the US Department of the Army as shown in Figure 8.3 is similar to that developed in chapter 6, except that it uses σ''_{cs} instead of σ''_{cv}. These two methods inherently consider the influence of both matric suction and net normal stress on heave. This is in contrast to the use of the slope of an unloading curve from an oedometer test, C_s. That value is the slope of a stress–strain curve for a fully wetted soil that has exhausted its expansion potential. As such, the parameter C_s does not reflect the heave that will occur due to wetting. The parameter C_H, defined in chapter 6, is a rigorously determined parameter that considers the change in volume that occurs due to a reduction in soil suction under a particular applied stress. It is the value that should be used in calculating heave.

8.2 SOIL SUCTION METHODS

Soil response to suction changes can be predicted in much the same manner as soil response to saturated effective stress changes. It was shown in chapter 6 that the amount of swell that occurs under a constant inundation stress is the result of a decrease in matric suction.

Equation (5-11) is a general equation for the constitutive surface in terms of void ratio, applied stress, and matric suction. The term $C_m \Delta \log(u_a - u_w)$ in that equation represents the contribution to heave due to changes in matric suction. The parameter C_m is the *matric suction index*. If the total stress does not change (i.e., $C_t \Delta \log(\sigma - u_a) = 0$), then the first term is the governing stress state variable for calculating heave. For application of that equation to compute heave, the initial suction conditions may be determined by direct measurement, and the final suction conditions can be assumed or calculated. For full wetting conditions, it would be appropriate to assume that the matric suction would go to zero. For conditions other than full wetting, the assumed final suction distribution does not necessarily have to correspond to zero suction. For unsaturated conditions, the soil suction generally will be greater than zero.

8.2.1 McKeen (1992)

McKeen (1992) developed a method to compute potential heave using soil suction as the primary stress state variable. This method relates heave of soil directly to suction changes. The heave prediction equation proposed by McKeen (1992), in terms of symbols consistent with this book, is,

$$\Delta H = C_h \times \Delta \psi_m \times H \times f \times s \qquad (8\text{-}13)$$

where:

$$\Delta H = \text{vertical heave of the layer being considered,}$$
$$C_h = \text{suction compression index for the layer,}$$
$$\Delta \psi_m = \text{matric suction change in the layer in pF,}$$
$$H = \text{thickness of the layer,}$$
$$f = \text{lateral restraint factor, and}$$
$$s = \text{coefficient for load effect.}$$

McKeen (1992) used test data from several field monitoring sites to determine an equation to express the suction compression index, C_h, in terms of the slope of the SWCC (i.e., $\Delta \psi_m / \Delta w$), where Δw is the change in water content. The test data represented clays and shales from five different locations in the United States. From the test data, a linear relationship was presented to estimate C_h. Perko, Thompson, and Nelson (2000) extended the data beyond that used by McKeen by conducting tests on undisturbed samples of clay and claystone bedrock obtained from the Denver, Colorado, area. An empirical relationship shown in Figure 8.4 was developed for determining C_h over a broader range of materials. The equation developed by Perko, Thompson, and Nelson (2000) is expressed by equation (8-14):

$$C_h = (-10) \left(\frac{\Delta \psi_m}{\Delta w} \right)^{-2} \qquad (8\text{-}14)$$

McKeen (1992) proposed that the slope of the SWCC could be expressed as

$$\frac{\Delta \psi_m}{\Delta w} = \left(\frac{\psi_{mf} - \psi_{mi}}{0 - w_i} \right) \qquad (8\text{-}15)$$

in which ψ_{mf} is the soil suction at oven-dry water content and ψ_{mi} is the in situ soil suction in pF at a water content of w_i. McKeen (1992) considered the soil suction at zero water content to be near 6.25 pF (174 MPa).

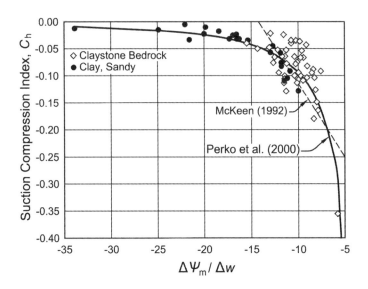

FIGURE 8.4. Suction compression index as a function of the slope of SWCC (Perko, Thompson, and Nelson 2000).

The value of ψ_{mf} at oven-dry water content was discussed in chapter 4. Thus, the McKeen (1992) method for calculating heave requires determination of the expected change in soil suction for each layer within the zone of wetting. This change in soil suction can be determined from the difference between the measured soil suction profile and an estimated final soil suction profile.

The application of McKeen's method is fairly straightforward. For a particular layer of soil, the suction in that soil and its water content are measured to determine ψ_{mi} and w_i. The value of ψ_{mf} corresponds to $w_f = 0$ and may be assumed to be 6.25 pF, as indicated by McKeen or another value as discussed in chapter 4. The value of $\Delta\psi_{mi}/\Delta w_i$ determined from equation (8-15) is used in equation (8-14) to determine C_h, which then can be used to determine heave using equation (8-13).

The complication of this method, however, lies in the determination of the lateral restraint factor, f, and the coefficient, s, for load effect on heave. The lateral restraint factor is used to convert volume change to vertical dimensional change. McKeen (1992) suggests that the factor f is equal to

$$f = \frac{1 + 2K_0}{3} \qquad (8\text{-}16)$$

where K_0 is the coefficient of earth pressure at rest.

TABLE 8.1 K_0 Values from Lytton (1994)

K_0	Condition of Soil
0.00	Badly cracked
0.33	Drying
0.67	Wetting
1.00	Cracks closed, wetting

The factor has been found to be 0.5 for highly fractured clays and 0.83 for massive clay with little or no fractures (Thompson 1997). Lytton (1994) back-calculated values of K_0 from field observations of heave to be as shown in Table 8.1.

McKeen suggests that the coefficient for load effect, s, can be determined using the equation

$$s = 1.0 - 0.01 \times (\%SP) \tag{8-17}$$

in which %SP is the percentage of swelling pressure represented by the total applied stress. Thus,

$$\%SP = \frac{\sigma_{vo}''}{\sigma_{cv}''} \times 100 \tag{8-18}$$

Thus, in order to determine a value for s it is necessary to also perform an oedometer test on a sample of the same soil on which the suction was measured in order to determine σ_{cv}''. The application of McKeen's method to predict heave is shown in Example 8.5.

EXAMPLE 8.5

Given:

An expansive soil site consists of 10 ft of silty clay, underlain by claystone to the maximum boring depth of 50 ft. The measured soil suction profile for the site is shown in Figure E8.5. The soil properties and assumed values are given in Table E8.5. Values of σ_{cv}'' were measured in oedometer tests.

Find:

Predicted heave using the McKeen (1992) soil suction method.

Solution:

Suction values shown in Figure E8.5 indicate that constant soil suction is reached at about 30 ft. Thus, the final soil suction at equilibrium may be considered to be 3 pF, and the depth of the active zone will be taken to be 30 ft. For this example, the soil was divided into three 10 ft thick layers.

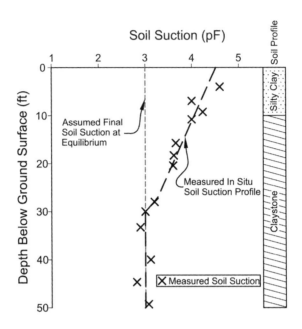

FIGURE E8.5. Soil suction profile for Example 8.5.

TABLE E8.5   Soil Profile for Example 8.5

Depth (ft)	Soil	Water Content, w_i (%)	Total Unit Weight (pcf)	Soil Suction at Zero Water Content (%)	CV Swelling Pressure, σ''_{cv} (psf)
0–10	Silty clay	18.0	120	6.44[*]	5,000
10–20	Claystone	22.0	125	6.40[**]	6,500
20–50	Claystone	25.0	125	6.40[**]	10,000

[*]per Cumbers et al. (2008)
[**]per Chao (2007)

Top layer: 0 to 10 ft

The three soil suction values shown in Figure E8.5 for the upper 10 ft layer are 4.6, 4.0, and 4.2 pF. Thus, the average measured soil suction, ψ_{s1}, for the first layer is,

$$(\psi_{s1})_{AVG} = \frac{4.6 + 4.0 + 4.2}{3} = 4.27 \text{ pF}$$

The suction change, $\Delta\psi_{s1}$, that will take place due to full wetting is,

$$\Delta\psi_{s1} = 4.27 - 3.0 = 1.27 \text{ pF}$$

The water content corresponding to the average soil suction of 4.27 pF is 18 percent. Thus, from equation (8-15), the slope of the SWCC, $(\Delta\psi/\Delta w)_1$, is

$$\left(\frac{\Delta\psi}{\Delta w}\right)_1 = \left(\frac{6.44 - 4.27}{0 - 0.18}\right) = -12.06$$

The value of C_h for the first layer can be read from Figure 8.4 or calculated using equation (8-14).

$$C_{h1} = -10(-12.06)^{-2} = -0.069$$

The effective stress at the midpoint of the top layer, which is 5 ft below the ground surface, is

$$\sigma''_{vo1} = 120 \text{ pcf} \times 5 \text{ ft} = 600 \text{ psf}$$

The lateral restraint factor, f, can be computed from equation (8-16). For both of the soils in this example, the cracks are closed and the soil is wetting, causing heave. For both the silty clay and the claystone, K_0 can be estimated from Table 8.1 as 1.0.

Thus, for both the silty clay and the claystone

$$f = \left(\frac{1 + 2 \times 1}{3}\right) = 1.0$$

From equation (8-17), the coefficient for load effect on heave, s, for the top layer is

$$s_1 = 1.0 - 0.01 \times \left(\frac{\sigma''_{vo1}}{\sigma''_{cv1}} \times 100\right) = 1.0 - 0.01 \times \left(\frac{600}{5,000} \times 100\right) = 0.88$$

From equation (8-13), the heave of the top layer is

$$\Delta H_1 = C_{h1} \times \Delta\psi_{s1} \times H_1 \times f \times s_1 = (-0.069) \times (1.27) \times 120 \text{ in.} \times 1 \times 0.88$$
$$= -9.25 \text{ in.}$$

The negative sign indicates upward movement.

Middle layer: 10 to 20 ft

The average measured soil suction for the middle layer is

$$(\psi_{s2})_{AVG} = \frac{4.0 + 3.7 + 3.6}{3} = 3.77 \text{ pF}$$

The suction change, $\Delta\psi_{s2}$, that will take place due to wetting is

$$\Delta\psi_{s2} = 3.77 - 3.0 = 0.77 \text{ pF}$$

The slope of the SWCC is

$$\left(\frac{\Delta\psi}{\Delta w}\right)_2 = \left(\frac{6.40 - 3.77}{0 - 0.22}\right) = -11.95$$

and

$$C_{h2} = (-10)(-11.95)^{-2} = -0.070$$

The effective stress at the midpoint of the middle layer, which is 15 ft below the ground surface, is

$$\sigma''_{vo2} = 120\,\text{pcf} \times 10\,\text{ft} + 125\,\text{pcf} \times 5\,\text{ft} = 1{,}825\,\text{psf}$$

and

$$s_2 = 1.0 - 0.01 \times \left(\frac{\sigma''_{vo2}}{\sigma''_{cv2}} \times 100\right) = 1.0 - 0.01 \times \left(\frac{1{,}825}{6{,}500} \times 100\right) = 0.72$$

Thus, the heave of the middle layer is

$$\Delta H_2 = C_{h2} \times \Delta\psi_{s2} \times H_2 \times f \times s_2 = (-0.070) \times (0.77) \times 120\,\text{in.} \times 1 \times 0.72$$
$$= -4.66\,\text{in.}$$

Bottom layer: 20 to 30 ft

The average measured soil suction, ψ_{s3}, for the bottom layer is

$$(\psi_{s3})_{\text{AVG}} = \frac{3.6 + 3.2 + 3.0}{3} = 3.27\,\text{pF}$$

The suction change, $\Delta\psi_{s3}$, is

$$\Delta\psi_{s3} = 3.27 - 3.0 = 0.27\,\text{pF}$$

Thus,

$$\left(\frac{\Delta\psi}{\Delta w}\right)_3 = \left(\frac{6.40 - 3.27}{0 - 0.25}\right) = -12.52$$

and

$$C_{h3} = (-10)(-12.52)^{-2} = -0.064$$

The effective stress at the midpoint of the bottom layer, which is 25 ft below the ground surface, is

$$\sigma''_{vo3} = 120\,\text{pcf} \times 10\,\text{ft} + 125\,\text{pcf} \times 15\,\text{ft} = 3{,}075\,\text{psf}$$

and

$$s_3 = 1.0 - 0.01 \times \left(\frac{\sigma''_{vo3}}{\sigma''_{cv3}} \times 100 \right) = 1.0 - 0.01 \times \left(\frac{3{,}075}{10{,}000} \times 100 \right) = 0.69$$

Thus, the heave of the bottom layer is

$$\Delta H_3 = C_{h3} \times \Delta \psi_3 \times H_3 \times f \times s_3 = (-0.064) \times (0.27) \times 120 \,\text{in.} \times 1 \times 0.69$$
$$= -1.43 \,\text{in.}$$

The total heave is the sum of the heave computed for each of the three layers.

$$\Delta H_{total} = -(9.25 + 4.66 + 1.43) = -15.3 \,\text{in.}$$

The negative sign indicates upward heave.

8.2.2 Department of the Army (1983)

A matric suction–water content relationship was proposed by the US Department of the Army (DA) in 1983 to evaluate the swell behavior of a particular soil. The DA (1983) Technical Manual (TM 5-818-7) presented the following equation:

$$\Delta H = \frac{C_m}{1 + e_0} H \log \left[\frac{(u_a - u_w)_0}{(u_a - u_w)_f} \right] \qquad (8\text{-}19)$$

in which all symbols are as previously defined.

The DA technical manual presented an equation to determine the suction index, C_m, as follows:

$$C_m = \frac{\alpha G_s}{100 B} \qquad (8\text{-}20)$$

where:

α = compressibility factor (slope of specific volume versus water content curve),

B = slope of matric suction versus water content curve, and

G_s = specific gravity of solids.

The compressibility factor, α, is the ratio of the change in volume to a corresponding change in water content. The specific volume is defined as γ_w/γ_d where γ_w is the unit weight of water and γ_d is the dry density of a soil. This parameter may be determined by a test similar to that for shrinkage limit such as the linear shrinkage test, or the CLOD test described in the next section (Nelson and Miller 1992). These equations, however, have not found widespread application.

8.2.3 Hamberg and Nelson (1984)

A simplified heave prediction procedure that uses the CLOD test was proposed by Hamberg and Nelson (1984) and was presented in Nelson and Miller (1992). The CLOD test provides a method to determine the relationship between a change in water content and a change in volume. From that relationship the heave of the soil can be determined. This method has the advantage that it uses undisturbed samples of any shape that can be extracted from test pits or exploratory holes. It has the disadvantage that it measures volume change during the drying process instead of the wetting process. It has particular advantages for assessing soil shrinkage during drying.

The original test method using the CLOD test procedure was the coefficient of linear extensibility (COLE) test that is used routinely by the USDA Natural Resource Conservation Service (NRCS) and National Soil Survey Laboratory (Brasher et al. 1966). The CLOD test is a modification of the COLE procedure. The basic difference between the CLOD test and the COLE test procedures is that in the CLOD test, volume changes are monitored along a gradually varying moisture change path. This results in a smooth shrinkage (or swelling) curve for each sample.

The CLOD test procedure develops a relationship between water content and void ratio (i.e., shrinkage), as the sample dries. The procedure is described in Nelson and Miller (1992). The slope of the water content–void ratio curve is designated the CLOD index, C_w, and is analogous to the matric suction index in equation (5-11). Thus, the CLOD index, C_w, is an index of volume change with respect to water content:

$$C_w = \frac{\Delta e}{\Delta w} \tag{8-21}$$

Noting that $\varepsilon_s = \Delta e/(1+e_0)$, total heave, ρ, can be determined by substituting equation (8-21) into equation (8-1).

$$\rho = \sum_{i=1}^{n} \frac{H_i \Delta e_i}{(1+e_0)_i} = \sum_{i=1}^{n} \frac{C_w \Delta w}{(1+e_0)_i} H_i \tag{8-22}$$

The void ratio can be computed from the volume readings if the specific gravity of the solids is known. Because water content is directly related to soil suction, the relationship between void ratio and water content is tantamount to expressing the effect of suction on void ratio. This is true only at water contents greater than the shrinkage limit, because changes in water content below the shrinkage limit are not accompanied by changes in volume. The relationship between water content and void ratio is generally linear over a range of water contents greater than the shrinkage limit.

This equation considers heave only due to suction changes with no net normal stress in any direction. To predict heave, initial water content profiles can be measured during the site investigation. The final water content profile after construction must be predicted on the basis of soil profiles, groundwater conditions, and environmental factors.

8.2.4   Lytton (1994)

Lytton (1994) proposed a method to predict volumetric strain in a soil mass that considered changes in three stress state variables. The equation he proposed, in terms of the notation used in this book, is

$$\frac{\Delta V}{V} = -\gamma_{\psi m}\log\left[\frac{\psi_{mf}}{\psi_{mi}}\right] - \gamma_{\sigma}\log\left[\frac{\sigma_f''}{\sigma_i''}\right] - \gamma_{\psi_o}\log\left[\frac{\psi_{of}}{\psi_{oi}}\right] \qquad (8\text{-}23)$$

where:

$\Delta V/V$ = volumetric strain,

ψ_{mi}, ψ_{mf} = initial and final matric suction,

σ_i'', σ_f'' = initial and final mean principal stress,

ψ_{oi}, ψ_{of} = initial and final osmotic suction,

$\gamma_{\psi m}$ = matric suction volumetric compression index,

γ_{σ} = mean principal stress volumetric compression index, and

γ_{ψ_o} = osmotic suction volumetric compression index.

The change in matric suction is the primary factor that generates the heave and shrinkage. Osmotic suction rarely changes appreciably. The mean principal stress increases by only small amounts except under footings. Thus, the last two terms in equation (8-23) can be neglected, and the major portion of the heave is generated by changes in matric suction.

Lytton (1994) proposed that the heave, as expressed by the vertical strain, $\Delta H/H$, can be estimated from the volumetric strain by using a crack fabric factor, f. He suggested that back-calculated values of the crack fabric factor, f, are 0.5 for soil drying conditions and 0.8 for soil wetting conditions. Thus,

$$\frac{\Delta H}{H} = f \left[\frac{\Delta V}{V} \right] \qquad (8\text{-}24)$$

8.3 EMPIRICAL METHODS

A number of empirical relationships for predicting heave have been proposed. These are usually based on limited test data collected from the particular geographic region in which they were developed. Table 8.2 summarizes several empirical relationships from the literature. As shown in Table 8.2, most of the empirical relationships are based on soil classification parameters, primarily the plasticity index. A major shortcoming of these methods is that they do not consider the in situ condition of the soil, and neglect initial density and water content.

Furthermore, they are based on a limited amount of data. Caution must be exercised if they are applied out of the area in which they were developed. They should be considered only as an indicator of the potential for the soil to be expansive, and not for quantitative prediction.

The potential vertical rise (PVR) method was developed by the Texas Department of Transportation. The most recent version of the procedure is presented in Tex-124-4 (Texas DOT 2014). The method purports to provide a quantitative value of heave. However, it has the same limitations as mentioned above since it is based primarily on the plasticity index and natural water content.

8.4 PROGRESSION OF HEAVE WITH TIME

In order to predict the amount of heave at a given time, such as the design life, the progression of heave must be considered. One method that has been used to predict the progression of heave is an observational method based on periodic elevation surveys and heave measurements, and extension of that database into the future by means of curve fitting. Another method is to analyze the migration of water in the zone of potential heave, and relate free-field heave to water content changes. Both of these methods are discussed in the following sections.

8.4.1 Hyperbolic Equation

The observational method for predicting heave is introduced in this section by fitting a hyperbolic equation to observed data to predict the

TABLE 8.2   Summary of Empirical Heave Prediction Methods

Source	Equation	Legend/Description
Seed, Woodward, and Lundgren (1962)	$\varepsilon_{s\%} = (2.16 \times 10^{-3}) \times PI^{2.44}$	$\varepsilon_{s\%}$ = percent swell (%) PI = plasticity index
van der Merwe (1964)	$\rho_i = f_i \times PE_i$	ρ_i = heave for layer i f_i = reduction factor for layer i PE_i = potential expansiveness for layer i
Ranganathan and Satyanarayana (1965)	$\varepsilon_{s\%} = (4.13 \times 10^{-4}) \times (SI)^{2.67}$	SI = shrinkage index = $LL - SL$ LL = liquid limit SL = shrinkage limit
Nayak and Christensen (1971)	$\varepsilon_{s\%} = 0.0229 \times PI^{1.45} \times (c/w) + 6.38$ P_s (psi) = $[(3.5817 \times 10^{-2}) \times PI^{1.12} \times c^2/w^2] + 3.7912$	w = initial water content P_s = swelling pressure c = clay content γ_d = dry density
Vijayvergiva and Ghazzaly (1973)	Log $\varepsilon_{s\%} = 1/12 \times (0.4LL - w + 5.5)$ Log P_s (tons/ft^2) = $1/19.5 \times (\gamma_d + 0.65LL - 139.5)$	
Schneider and Poor (1974)	Log $\varepsilon_{s\%} = 0.90(PI/w) - 1.19$ for $S = 0$ ft Log $\varepsilon_{s\%} = 0.65(PI/w) - 0.93$ for $S = 3$ ft Log $\varepsilon_{s\%} = 0.51(PI/w) - 0.76$ for $S = 5$ ft Log $\varepsilon_{s\%} = 0.41(PI/w) - 0.69$ for $S = 10$ ft Log $\varepsilon_{s\%} = 0.33(PI/w) - 0.62$ for $S = 20$ ft	S = surcharge
Weston (1980)	$\varepsilon_{s\%} = (4.11 \times 10^{-4}) \times LLw^{4.17} \times \sigma_{vo}^{-0.386} \times w^{-2.33}$	LLw = weighted liquid limit σ_{vo} = overburden stress
Chen (1988)	$\varepsilon_{s\%} = 0.2558e^{0.0838 \times PI}$	e = base of natural logarithm = 2.71828

progression of heave with time. The method is illustrated in the case study presented below.

The progression of heave of the slabs and piers can be analyzed by fitting measured survey data to a hyperbolic function of the form shown in equation (8-25).

$$\rho = \frac{t}{a + bt} \qquad (8\text{-}25)$$

where:

ρ = slab or pier heave at the time, t,

t = the time since movement began, and

a and b = curve fitting parameters.

The ultimate heave, ρ_{ult}, can be determined by taking the limit of equation (8-25) as t approaches infinity. Thus,

$$\rho_{ult} = \lim_{t \to \infty} \rho = \frac{1}{b} \qquad (8\text{-}26)$$

A method to determine the parameters a and b is to rewrite equation (8-25) in the form shown in equation (8-27).

$$\frac{t}{\rho} = a + bt \qquad (8\text{-}27)$$

In equation (8-27), t/ρ is a linear function of t. The parameters a and b are the intercept and slope of a plot of t/ρ versus t.

The rate of heave can be obtained by taking the derivative of equation (8-25). The first derivative of equation (8-25) is:

$$\frac{d\rho}{dt} = \frac{a}{(a + bt)^2} \qquad (8\text{-}28)$$

The application of this method is demonstrated in Example 8.6.

Care must be taken when applying the observational method if the soil or bedrock at some depth is significantly different from the shallower soils. In this case, the rate of heave in the deeper soils can differ from that in the shallower soils that were wetted during the observation period. If the expansion potential of the soil does not vary greatly throughout the depth of soil that will be wetted during the time period of interest, this method can be used effectively.

EXAMPLE 8.6

Given:

A building is experiencing distress of a slab-on-grade due to heaving of expansive soils. At the present time, the slab has heaved 8-4 in. since the end of construction. The slab has a design life of 20 years. The soil profile under the slab consists of an expansive clay to a depth of 40 ft. The depth of potential heave is 36 ft. The slab has been surveyed periodically since the end of construction. Results of the elevation surveys for a point on the slab are shown in Figure E8.6-1.

Find:

1. Expected additional heave that will occur by the end of the design life.
2. Ultimate heave of the slab.
3. Rate of slab heave at the end of the design life.
4. Plot heave as a function of time.

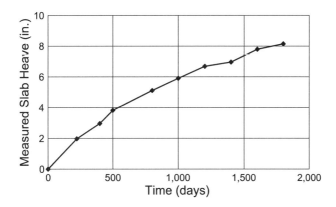

FIGURE E8.6-1. Measured survey data for the slab in Example 8.6.

Solution:

Part 1:
Because the soil is relatively uniform over the depth of potential heave, it is reasonable to use the hyperbolic curve fitting technique expressed by equation (8-25) to compute the expected total heave. The survey data given in Figure E8.6-1 were plotted in the form of equation (8-27) in Figure E8.6-2. The values of a and b were determined to be 100.6 days/in. and 0.07 in.$^{-1}$, respectively, as shown in Figure E8.6-2.

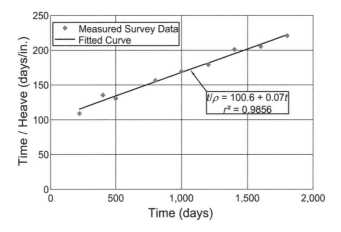

FIGURE E8.6-2. Plot of time/heave versus time for Example 8.6.

The heave, as expressed by equation (8-25) is,

$$\rho = \frac{t}{100.60 + 0.07t}$$

where t is in days and ρ is in inches.

At the end of the design life, $t = 20$ yrs $\times$ 365 days/yr $= 7,300$ days. Thus, the predicted heave at that time is

$$\rho_{(20 \text{ yrs})} = \frac{7,300}{100.60 + 0.07(7,300)} = 11.9 \text{ in.}$$

Therefore, the additional heave, $\Delta\rho$, that is expected to occur by the end of the design life, is

$$\Delta\rho_{(20 \text{ yrs})} = 11.9 - 8.4 = 3.5 \text{ in.}$$

Part 2:
The ultimate heave of the slab is

$$\rho_{\text{ult}} = \frac{1}{b} = \frac{1}{0.07} = 14.3 \text{ in.}$$

Part 3:
At $t = 20$ yrs, the rate of heave is given by equation (8-28). Thus,

$$\frac{d\rho}{dt} = \frac{100.60}{[100.60 + 0.07(7,300)]^2}$$

$$= 2.7 \times 10^{-4} \frac{\text{in.}}{\text{day}} = 0.10 \frac{\text{in.}}{\text{yr}}$$

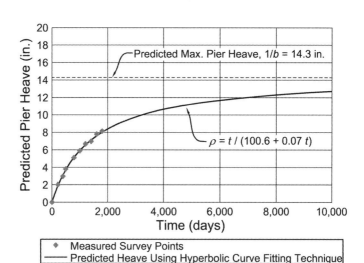

FIGURE E8.6-3. Predicted progression of slab heave with time using the hyperbolic curve fitting technique for Example 8.6.

Part 4:
Heave as a function of time is shown in Figure E8.6-3.

If the slab is replaced at the present time, it can be expected that additional heave in the amount of 3.5 in. will occur. By the end of the design life it may be necessary to replace the slab more than once. At the end of the design life the slab will be experiencing only 0.1 in. of heave per year, and that rate should be decreasing. Based on this information, the owner of the slab is able to make informed decisions regarding remediation and future maintenance costs.

CASE STUDY—DENVER INTERNATIONAL AIRPORT (SLAB HEAVE)

The hyperbolic curve fitting method was used to analyze slab movement at the DIA building that was discussed in Section 3.2. Construction of the building began in 1991 and continued into 1992. Heave due to the expansive soils was observed to be occurring even before construction was finished. Some elevation monitoring was undertaken at that time, but due to the lack of a stable benchmark for the elevation survey, those data were not reliable. Stable deep benchmarks were installed and detailed monitoring of the slab and pier movement was initiated in September 2000. The slab and pier heave that had occurred between the initiation of construction (1991) and September 2000 was estimated based on as-built floor elevations and design drawings. The highest values of slab and pier heave that were measured by the end of 2011 were 6.2 and 5.0 in., respectively.

The measured survey data were first analyzed by fitting it to equation (8-25) using data only from the first 11 months of monitoring. This is commonly the general period of time for which data are available in practice. A second analysis was also performed using a longer 22-month survey period.

Figure 8.5 compares the slab heave measured in June 2006 with values that were predicted for that time using the data from the first 11 months of monitoring (September 2000 to July 2001) (Chao 2007). The dashed line in Figure 8.5 represents the best-fit linear relationship to the data and shows that the hyperbolic function represented the data with an expected error rate of 19 percent. The envelope of the points shows a maximum error of about 43 percent. The percentage error appears to be the least at higher values of heave. This represents good accuracy for heave prediction. The accuracy is likely influenced by the drought cycle in Colorado, particularly the record dry year in 2002.

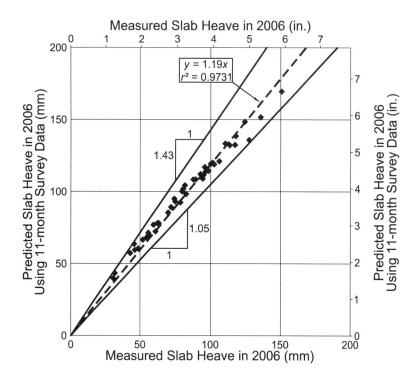

FIGURE 8.5. Measured slab heave versus predicted slab heave in 2006 based on initial 11-month survey data for the building at DIA (Chao 2007).

After the initial analysis using the 11-month set of data, the hyperbolic relationship was revised based on 22 months of survey data. Figure 8.6 shows the comparison between the slab heave measured in 2006 and that predicted using survey data taken over a period of 22 months (Chao 2007). The dashed line in Figure 8.6 shows an expected accuracy of about 4 percent. The envelope of the points shows a range of error of up to about 17 percent. The use of the 22-month set of data compared to the 11-month data increased the accuracy of prediction significantly. This is very good predictive accuracy and demonstrates that the longer the monitoring time period available to determine the curve-fitting parameters, the more accurate will be the results. The good correlation between the measured and predicted heave shows that, if a reasonable set of data can be obtained over a sufficiently long period of time, the use of the hyperbolic function to represent heave versus time is a valuable tool.

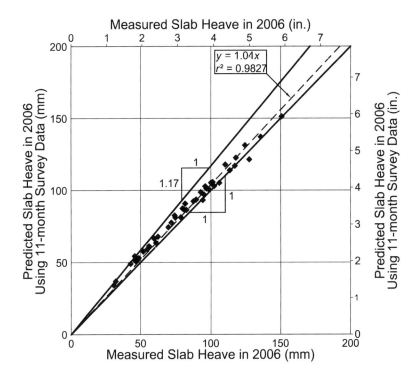

FIGURE 8.6. Measured slab heave versus predicted slab heave in 2006 based on 22-month survey data for the building at DIA (Chao 2007).

8.4.2 Use of Water Migration Modeling to Analyze Rate of Heave

Water migration modeling, as was discussed in chapter 7, is also useful in predicting the progression of heave. Water content profiles can be analyzed at various points in time and used to analyze predicted heave at those points. Example 8.4 illustrated the computation of heave for a design wetting profile, taking into account the partial wetting in the transition zone. Figure 7.11 shows water content profiles determined at different times by means of computer modeling. If it is desired to predict heave at various times within the design life, the heave can be computed for each of the groundwater profiles corresponding to those points in time.

8.5   FREE-FIELD SURFACE MOVEMENT FOR SHRINK–SWELL SOILS

In areas having dense, highly expansive claystones, the general pattern of heave is for heave to continue to increase with time, although it may occur more rapidly during wet periods and slow down during dry periods. Surface movement due to shrinkage is small or nonexistent. However, for soils where the expansive soils are less dense and heave is due mainly to the soils in the upper soil profile, both shrink and swell may be significant. Slab-on-grade foundations tend to be commonly used in locations with shrink-swell conditions. Heave will occur during wet periods and settlement will occur during dry periods. Design of those slabs, therefore, must consider the potential for both upward and downward movement.

Fityus, Cameron, and Walsh (2005) present a method for predicting potential slab movement based on the *shrink swell test*. This method has found fairly widespread use in Australia and is referenced in the Australian standard for the design of residential slabs and footings (AS 2870-2011) (Standards Australia 2011). The shrink swell test uses results from both an unrestrained shrinkage test and a confined oedometer swell test performed in parallel. In the shrinkage test a sample is allowed to first air dry and is subsequently oven-dried. At the end of each drying period the axial shrinkage strain, ε_{sh}, is measured. The swelling strain, ε_{sw}, is measured after wetting of a sample of the same soil in an oedometer test. The shrink swell test considers the two strains added together. Because the oedometer sample is confined, whereas the shrinkage sample is not, the value of ε_{sw} is adjusted by a factor of 2. Thus, the total range of strain, ε_T, is equal to

$$\varepsilon_T = \varepsilon_{sh} + \frac{\varepsilon_{sw}}{2} \tag{8-29}$$

The total suction is assumed to be equal to a value on the order of pF = 2.2 to 2.5 at the time when the swelling strain is completed. The total volume change is assumed to take place between the "wilting point" for trees and a water content close to saturation. The magnitude of suction change over which the volume change takes place is assumed to be equal to 1.8 pF (Fityus, Cameron, and Walsh 2005).

The shrink swell index, I_{ss} is thus defined as

$$I_{ss} = \frac{\varepsilon_{sh} + \dfrac{\varepsilon_{sw}}{2}}{1.8} \tag{8-30}$$

An instability index, I_{pt}, is then defined as

$$I_{pt} = \alpha \cdot I_{ss} \tag{8-31}$$

where α is an adjustment factor to account for effects of overburden stress and lateral confinement on a sample in the in situ condition. The adjustment factor is defined as

$$\alpha = \begin{cases} 1.0 \text{ for cracked zone} \\ 2 - \frac{z}{5} \text{ for uncracked zone} \end{cases} \tag{8-32}$$

where z = depth in meters.

This formulation of a value for α assumes that, in an upper zone where the soil is cracked, there is no lateral confinement. Below the cracked zone, the confinement is assumed to increase the vertical strain, but at a depth of 30 ft (10 m) it is assumed that no swell occurs. This is tantamount to assuming that the depth of potential heave, as discussed in chapter 7, is equal to 30 ft (10 m) for all soils.

The net ground surface movement, y_s, considering effects of both shrinkage and swelling, is given by equations (8-33a) and (8-33b).

$$y_s = \sum_{i=1}^{n} I_{pt,i} \cdot \Delta(\psi_m)_i \cdot \Delta z_i \tag{8-33a}$$

$$y_s = \sum_{i=1}^{n} \alpha \cdot I_{ss,i} \cdot \Delta(\psi_m)_i \cdot \Delta z_i \tag{8-33b}$$

To apply equations (8-33a) and (8-33b), the soil layer is divided into n layers, I_{ss} and α are determined for the soil in each layer, and the expected suction change, $\Delta\psi_m$, in pF is determined or assumed.

There are a number of assumptions involved in the application of this method, but it has found acceptance in Australia to the point where it is referenced in the Australian standard AS 2870. A more complete discussion of the assumptions made in development of the method and a more detailed description of the testing procedures is given by Fityus, Cameron, and Walsh (2005).

8.6 DISCUSSION OF HEAVE PREDICTION

Heave prediction is generally accomplished using the oedometer method. The oedometer method is used because the CS test is commonly conducted in general geotechnical engineering practice. Different forms of equations to predict heave based on the oedometer method have been presented in this chapter. These approaches to heave prediction are all based on the same fundamental principle. The differences in these methods are related to the manner in which the heave index parameter is determined.

Various methods of predicting heave using the soil suction method were presented. Although the testing methods required for these methods are not complex, they are not normally conducted in most geotechnical engineering laboratories. Filter paper is commonly used to measure soil suction of samples for use in predicting heave. An advantage of the filter paper method is the wide range of soil suction over which it can be used and its simplicity. A disadvantage for the use of this method is the degree of accuracy required for weighing the filter paper. Filter paper measurements tend to be relatively unreliable and unrepeatable compared to the results of oedometer tests. Oedometer tests are routinely conducted by most geotechnical engineering laboratories practicing in areas with expansive soils.

Heave can be calculated using the oedometer method with good accuracy. However, to achieve a good degree of accuracy requires careful drilling, sampling, and soil testing along with rigorous analyses. Along with that, the accuracy with which we can predict changes in the environmental factors must be kept in mind. Where uncertainty exists, we must assume reasonably conservative scenarios and attempt to apply the most rigorous theoretical concepts available. Our goal must always be accurate prediction of heave. Overprediction of heave can lead to costly and unreasonable foundations, while underprediction of heave can lead to excessive and costly mitigation and litigation measures. Thus, it is important to carefully select the design parameters used in the analyses to make the most accurate prediction possible. Reasonably conservative values should be used if exact values are not known.

References

Brasher, B. R., D. P. Franzmeier, V. Valassis, and S. E. Davidson. 1966. "Use of Saran Resin to Coat Natural Soil Clods for Bulk-Density and Water-Retention Measurements." *Soil Science* 101(2): 108.

Burland, J. B. 1962. "The Estimation of Field Effective Stresses and the Prediction of Total Heave Using a Revised Method of Analyzing the Double Oedometer Test." *The Civil Engineer in South Africa*. Transactions of the South African Institute of Civil Engineering.

Chao, K. C. 2007. *Design Principles for Foundations on Expansive Soils*. PhD dissertation, Colorado State University, Fort Collins, CO.

Chen, F. H. 1988. *Foundations on Expansive Soils*. New York: Elsevier Science.

Cumbers, J. M., J. D. Nelson, K. C. Chao, and D. D. Overton. 2008. "An Evaluation of Soil Suction Measurements Using the Filter Paper Method and Their Use in Volume Change Prediction." *Proceedings of the 1st European Conference on Unsaturated Soils*, Durham, UK, 389–395.

Fityus, S. G., D. A. Cameron, and P. F. Walsh. 2005. "The Shrink Swell Test." *Geotechnical Testing Journal, ASTM* 28(1): 1–10

Fredlund, D. G. 1983. "Prediction of Ground Movements in Swelling Clays." *Proceedings of the 31st Annual Soil Mechanics and Foundation Engineering Conference*, Minneapolis, MN.

Fredlund, D. G., and H. Rahardjo. 1993. *Soil Mechanics for Unsaturated Soils*. New York: John Wiley and Sons.

Fredlund, D. G., J. U. Hasan, and H. Filson. 1980. "The Prediction of Total Heave." *Proceedings of the 4th International Conference on Expansive Soils*, Denver, CO, 1–11.

Fredlund, D. G., H. Rahardjo, and M. D. Fredlund. 2012. *Unsaturated Soil Mechanics in Engineering Practice*. Hoboken, NJ: John Wiley and Sons.

Hamberg, D. J., and J. D. Nelson. 1984. "Prediction of Floor Slab Heave." *The 5th International Conference Expansive Soils*, Adelaide, South Australia, 137–217.

Jennings, J. E. B., and K. Knight. 1957. "The Prediction of Total Heave from the Double Oedometer Test." *Proceedings of Symposium on Expansive Clays, South African Institute of Civil Engineers, Johannesburg*, 7(9): 13–19.

Lytton, R. L. 1994. "Prediction of Movement in Expansive Clay." In *Vertical and Horizontal Deformations of Foundations and Embankments*, edited by A. Yeung and G. Feaalio. Geotechnical Special Publication No. 40, ASCE New York, 1827–1845.

McKeen, R. G. 1992. "A Model for Predicting Expansive Soil Behavior." *Proceedings of the 7th International Conference on Expansive Soils*, Dallas, TX, 1, 1–6.

Nayak, N. V., and R. W. Christensen. 1971. "Swell Characteristics of Compacted Expansive Soils." *Clay and Clay Minerals* 19(4): 251–261.

Nelson, J. D., and D. J. Miller. 1992. *Expansive Soils: Problems and Practice in Foundation and Pavement Engineering*. New York: John Wiley and Sons.

Nelson, J. D., D. B. Durkee, and J. P. Bonner. 1998. "Prediction of Free Field Heave Using Oedometer Test Data." *Proceedings of the 46th Annual Geotechnical Engineering Conference*, St. Paul, MN.

Nelson, J. D., D. K. Reichler, and J. M. Cumbers. 2006. "Parameters for Heave Prediction by Oedometer Tests." *Proceedings of the 4th International Conference on Unsaturated Soils*, Carefree, AZ, 951–961.

Palit, R. M. 1953. "Determination of Swelling Pressure of Black Cotton Soil." *Proceedings of the 3rd International Conference on Soil Mechanics and Foundation Engineering*, Switzerland, 170.

Perko, H. A., R. W. Thompson, and J. D. Nelson. 2000. "Suction Compression Index Based on CLOD Test Results." In *Advances in Unsaturated Geotechnics,* edited by C. D. Shackelford, S. L. Houston, and N. Chang. Reston, VA: ASCE Press, 393–408.

Ranganathan, B. V., and B. Satyanarayana. 1965. "A Rational Method of Predicting Swelling Potential for Compacted Expansive Clays." *Proceedings of the 6th International Conference on Soil Mechanics and Foundation Engineering*, International Society for Soil Mechanics and Geotechnical Engineering, London, 1, 92–96.

Reichler, D. K. 1997. "Investigation of Variation in Swelling Pressure Values for an Expansive Soil." Master's thesis, Colorado State University, Fort Collins, CO.

Salas, J. A. J., and J. M. Serratosa. 1957. "Foundations on Swelling Clays." *Proceedings of the 4th International Conference on Soil Mechanics and Foundation Engineering*, London, 1, 424–428.

Schneider, G. L., and A. R. Poor. 1974. "The Prediction of Soil Heave and Swell Pressures Developed by an Expansive Clay." Research Report Number: TR-9-74, Construction Research Center, University of Texas.

Seed, H. B., R. J. Woodward Jr., and R. Lundgren. 1962. "Prediction of Swelling Potential for Compacted Clays." *Journal of the Soil Mechanics and Foundation Division*, ASCE 88(SM3): 53–87.

Standards Australia. 2011. Australian Standard, Residential Slabs and Footings-Construction, as 2870-2011. Standards Australia International Ltd., Sydney, NSW.

Texas Department of Transportation. 2014. "Test Procedure for Determining Potential Vertical Rise." TxDOT Designation Tex-124-E.

Thompson, R. W. 1997. "Evaluation Protocol for Repair of Residences Damaged by Expansive Soils." *Proceedings for the Unsaturated Soils Conference*, ASCE Geotechnical Special Publication No. 68.

US Department of the Army (DA). 1983. "*Technical Manual TM 5-818-7, Foundations in Expansive Soils.*" Washington, DC.

van der Merwe, D. H. 1964. "The Prediction of Heave from the Plasticity Index and the Percentage Clay Fraction of Soils." *The Civil Engineer in South Africa* 6, 103–107.

Vanapalli, S. K., and L. Lu. 2012. "A State-Of-the Art Review of 1-D Heave Prediction Methods for Expansive Soils." *International Journal of Geotechnical Engineering* 6, 15–41.

Vijayavergiya, V. N., and O. I. Ghazzaly. 1973. "Prediction of Swelling Potential for Natural Clays." *Proceedings of the 3rd International Conference on Expansive Soils*, Haifa, Israel, 1, 227–236.

Weston, D. J. 1980. "Expansive Roadbed, Treatment for Southern Africa." *Proceedings of the 4th International Conference on Expansive Soils*, Denver, CO, 1, 339–360.

9

General Considerations for Foundation and Floor Design

Design of foundations and floors for structures is much more challeng-ing when dealing with an expansive soil site than a nonexpansive soil site. The cost of inadequate design will have serious and costly effects. The site investigation and design of foundations and floors will be more complex than for nonexpansive soil sites, and the construction will be more complex as well. Furthermore, although soil sampling methods are generally similar to those for other sites, the laboratory testing, analysis, and design are more extensive. Failure to recognize the need for appropriate robust foundation design and attempts to design foundations for the same cost as ordinary sites has resulted in serious economic losses on the part of both builders and owners. This and following chapters will discuss alternative types of foundations and floors for expansive soils and methods for design of such.

Soil treatment and moisture control are discussed in chapter 10. Foundation and building elements including shallow foundations, deep foundations, and floor systems are discussed in chapters 11 through 13. Figure 9.1 shows a number of different foundation types that have been used on expansive soil sites, some with more success than others. This chapter will discuss the design of the various systems and various factors that need to be considered for expansive soil applications. The quantification of risk associated with the design and construction on an expansive soil site will be addressed.

The design of foundation and slab systems can apply one, or more, of the following three design concepts.

1. The structure can be designed to be stiff enough to provide the rigid-ity necessary for the structure to perform its intended function as the soil heaves.
2. The foundation can be designed to isolate the structure from the expansive soil.
3. The soil can be stabilized sufficiently to reduce its expansion poten-tial to the degree where it will not produce intolerable movement.

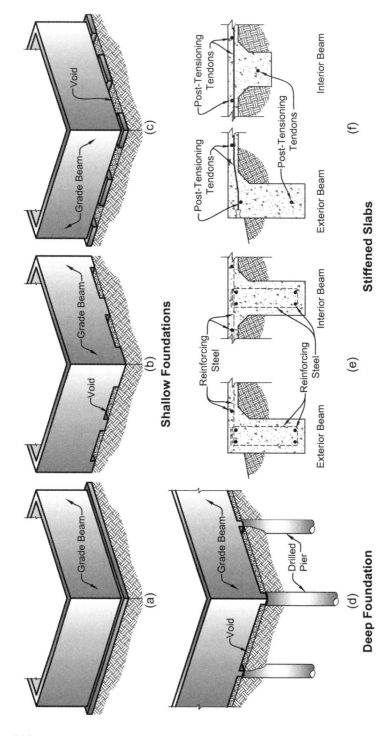

FIGURE 9.1. Foundation types typically used on expansive soil sites: (a) continuous footing; (b) voided stem wall; (c) isolated pad; (d) pier and grade beam; (e) reinforced mat; (f) post-tensioned slab.

228

These three design concepts can also be used in combination with each other to provide a satisfactory foundation design.

Shallow foundations should not be used on expansive soil sites except those with low expansion potential. Stiffened mat foundations or reinforced slab-on-grade foundations are primarily used to provide rigidity to withstand expansive soil heave. These foundation types are most effective on sites where soil movement is not excessive. They are used in the southwestern United States, South Africa, Australia, and other parts of the world, as well. Post-tensioned slabs are a common type of stiffened mat foundation. However, unless these types of foundations are very stiff they do not function well in moderately to highly expansive soils.

A deep foundation system is used primarily to isolate the structure from the expansive soil. Such a system generally consists of a deep foundation that supports a grade beam on which the structure is constructed. The deep foundation is founded at a depth such that future heave will not exceed the tolerable limits of the structure. The grade beam is then isolated from the expansive soil to prevent surface heave from affecting the foundation. For structures where basements are constructed, the grade beam also serves as the basement wall. Deep foundation systems are also used for the remediation of distressed foundations. Typically, those systems use a type of patented foundation element such as helical piles, micropiles, or push piers.

Soil treatment usually involves either moisture-conditioning and recompacting the expansive soil or the use of chemical admixtures to reduce the expansion potential of the soil. The admixture can be mixed with the soil during moisture-conditioning or it can be injected into the in situ soil. The successful application of soil treatment procedures requires evaluation of the pertinent properties of the soils and their reaction with the additive to be used. The limitations of the methods, and correct implementation of the treatment procedures, must be understood. This method is often used in conjunction with other foundation and slab systems to reduce the design requirements of those systems.

Moisture control alternatives are a form of soil treatment. Moisture control alternatives mainly involve installation of moisture barriers and/or subsurface drains to control the rate of moisture increase and to minimize seasonal fluctuations of the water content of the soil. It must be realized that it is virtually impossible to prevent an increase in water content of the foundation soils by using moisture control alternatives.

Floor slab alternatives for structures on expansive soils include slabs-on-grade or structural floors. If a mat type foundation such as a post-tensioned slab is used, it also functions as the floor slab. Structural floor slabs are supported on the grade beam or foundation wall and are isolated from the soil. Construction of slab-on-grade floors on highly expansive soils should be avoided because of the historical high failure rate of such floor systems.

9.1 RISK AND LIFE CYCLE COSTS

Many factors, not all of which are technical, influence the selection of foundation type, design methods, and soil treatment practices. Some of those factors include differences in climate, soil conditions, costs, design life, mortgage lending practices, and legal standards. In the northern and central areas of North America, for example, most residences are constructed with basements. Pier and grade beam foundation construction is common in these areas. In the southern Atlantic, Gulf Coastal Plain, and West Coast states, basements are uncommon and stiffened slab-on-grade construction is popular.

The initial cost of foundation systems on expansive soils is typically the driving factor in the selection of foundation type. However, life cycle costs can be even more important than initial costs. Expansion potential may not manifest itself until months or years after construction. This fact, in itself, is the root cause for much of the litigation between owners and builders. For this reason, risk assessment must be a part of the selection of a design alternative. Funding agencies and owners need to become involved in the design process at an early stage because they are the ones that will bear the cost of failures. All parties involved in a project must make every effort to keep risks low and develop a mutual understanding of what the risks are. This is true for remediation projects as well as original construction.

Financing practices and regulatory issues may have a significant influence on the selection of design alternatives. Often, the foundation design alternative that is optimal for minimizing risk of damage involves costs that exceed the anticipated level of funding for the project. As a result, the owners or financiers may prescribe the use of a lower cost foundation system. A lower-cost system may be a viable alternative, but it may carry with it a higher degree of risk to the structure. It is not appropriate for the design engineer to assume those risks on behalf of the client without making the client aware of the nature and magnitude of risk involved.

9.1.1 Classification of Expansion Potential

One particularly perplexing aspect of risk assessment is the lack of a standard method of quantifying *expansion potential,* also termed *swell potential.* Various methods of classification of expansion potential have been proposed on the basis of soil tests. However, not only do soil sample conditions vary in the different swell tests that have been used to evaluate swell potential (i.e., remolded or undisturbed), but surcharge loading and other testing factors vary over a wide range of values.

In all cases the term *expansion potential* refers to the relative capacity for expansion of the different soils. The amount of swell that may be realized in the field, however, is a function of the environmental conditions as well. This will include whether the soil is undisturbed or has been recompacted. Two soils may have the same expansion potential according to their laboratory classification, but may exhibit very different amounts of swell (Nelson, Chao, and Overton 2007; Nelson et al. 2011).

A number of methods have been developed to classify the swell potential of a soil. Most of these provide a type of rating to provide an assessment of the degree of probable expansion. These ratings are usually qualitative terms such as "low," "medium," "high," and "very high."

The use of Atterberg limits to classify the swell potential is a common approach. That approach, however, does not directly address the swelling characteristics of the soil and ignores the in situ condition of the soil. It has found some success in limited areas, but the use of these relationships outside of the local area where the relationship has been developed is usually not successful.

Holtz and Gibbs (1956) developed a classification system in terms of "low" to "very high" based on Atterberg limits and colloid content. Their system is shown in Table 9.1. The criteria listed in Table 9.1 were based on tests on undisturbed soil samples. This classification system was developed for the US Bureau of Reclamation (USBR) for application in the design of canal linings over expansive soil. It was adopted by the USBR in its *Earth Manual* (USBR 1998). In Table 9.1 it is seen that

TABLE 9.1   Classification of Expansion Potential Based on Colloid Content, Plasticity Index, and Shrinkage Limit (USBR 1998)

Data from Index Tests				
Colloid Content (% less than 0.001 mm)	Plasticity Index (%)	Shrinkage Limit (%)	Probable Expansion* (% Total Volume Change)	Degree of Expansion
< 15	< 18	> 15	< 10	Low
13–23	15–28	10–16	10–20	Medium
20–31	25–41	7–12	20–30	High
> 28	> 35	< 11	> 30	Very high

*Applied vertical stress equal to 1.0 psi (144 psf or 7 kPa)

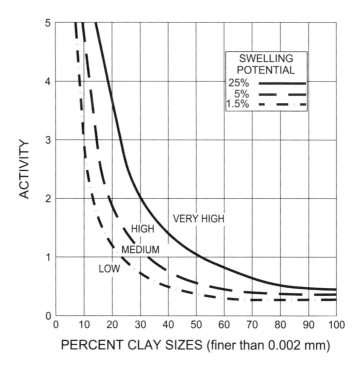

PERCENT CLAY SIZES (finer than 0.002 mm)

FIGURE 9.2. Swelling potential for compacted clays based on activity and percent clay (Seed, Woodward, and Lundgren 1962).

the values of the different parameters overlap between groups. This is necessary to be able to apply both indices and colloid content together. The *Earth Manual* emphasizes that the in-place moisture and density of the soil should be taken into account when identifying expansion potential.

Seed, Woodward, and Lundgren (1962) studied the swelling characteristics of compacted clay and developed the chart shown in Figure 9.2. The chart relates percentage of clay and the activity of the clay to swelling potential. Activity was defined by Skempton (1953) and is shown in equation (2-1). Swelling potential was defined on the basis of amount of swell in an oedometer test. The samples tested in the oedometer test were compacted to standard Proctor dry density at optimum water content and inundated under a stress of 1 psi (144 psf or 7 kPa).

Most classification systems overlap to some degree. Snethen, Johnson, and Patrick (1977) evaluated a number of published criteria for classifying expansion potential. They concluded that liquid limit and plasticity index are the best indicators of potential swell if natural

TABLE 9.2   Expansion Potential Classified on Basis of Liquid Limit, Plasticity Index, and In Situ Suction (Modified from Snethen, Johnson, and Patrick 1977)

Liquid Limit (%)	Plasticity Index (%)	Natural Soil Suction		Potential Swell* (%)	Classification of Potential Swell
		(pF)	(kPa)		
< 50	< 25	< 3.1	143	< 0.5	Low
50–60	25–35	3.1–3.6	143–383	0.5–1.5	Marginal
> 60	> 35	> 3.6	>383	> 1.5	High

*Inundated at in situ overburden stress

conditions are taken into account as well. A statistical analysis of laboratory data correlating expansion potential to a large number of independent variables resulted in the classification system shown in Table 9.2. This approach includes consideration of the in situ soil suction, which is an indicator of the natural conditions and environment. The US Department of the Army (1983) and the American Association of State Highway and Transportation Officials (AASHTO 2008) both adopted the criteria proposed by Snethen, Johnson, and Patrick (1977) for classification of expansion potential.

Chen (1988) developed a correlation between the percent fines, liquid limit, and standard penetration test blow counts to expansion potential, as shown in Table 9.3. In this method the blow count is used as a measure of in situ conditions as opposed to soil suction as shown in Table 9.2. He suggested that this classification can be used as a guide for the choice of foundation type.

TABLE 9.3   Expansive Soil Classification Based on Percent Passing No. 200 Sieve, Liquid Limit, and Standard Penetration Resistance for Rocky Mountain Soils (Modified from Chen 1988)

Laboratory and Field Data					
Percentage Passing No. 200 Sieve (%)	Liquid Limit (%)	Standard Penetration Resistance (Blows/ft)	Probable Expansion* (% Total Volume Change) (%)	Swelling Pressure (ksf**)	Degree of Expansion
< 30	< 30	< 10	< 1	1	Low
30–60	30–40	10–20	1–5	3–5	Medium
60–95	40–60	20–30	3–10	5–20	High
> 95	> 60	> 30	> 10	> 20	Very high

*$\sigma_i'' = 1,000$ psf (48 kPa)
**ksf = kips per square foot

TABLE 9.4   Recommended Representative Swell Potential Descriptions and Corresponding Slab Performance Risk Categories (CAGE 1996)

Representative Percent Swell (500 psf Surcharge) (%)	Representative Percent Swell (1,000 psf Surcharge) (%)	Slab Performance Risk Category
0 to < 3	0 to < 2	Low
3 to < 5	2 to < 4	Medium
5 to < 8	4 to < 6	High
≥ 8	≥ 6	Very high

The Colorado Association of Geotechnical Engineers presented a guideline for risk evaluation of slab performance (CAGE 1996). This classification system is shown in Table 9.4 and defines risk on the basis of the percent swell in an oedometer test. CAGE noted in the guideline that the variability of soil and bedrock conditions and other factors, such as depth of wetting, existing and anticipated groundwater conditions, surface grading and drainage, must also be considered when evaluating the slab performance risk.

9.1.2 Risk Factor

All foundation and slab systems constructed on expansive soil have some associated risk of movement. The risk, of course, depends on the type of foundation, the adequacy of the site investigation, and the competency of the design. However, all of these factors being equal, the risk associated with a site is largely a function of the amount of free-field heave that can occur. It is convenient to have a method for classifying and quantifying the risk associated with a site based on the laboratory data prior to performing detailed analyses.

The expansion potential classification systems that were presented here attempted to classify risk based largely on index properties. The CAGE (1996) method provides some degree of quantification by using the percent swell in an oedometer test. However, it does not consider the depth and extent of the expansive soil in the soil profile. Furthermore, none of the systems described so far takes into account the swelling pressure. Also, the inundation stress will influence the percent swell that will occur. The inundation stress, the percent swell, and the swelling pressure must all be considered in evaluating the risk associated with an expansive soil site. The risk factor, *RF*, is defined below and takes into account all three of these parameters (Nelson, Chao, and Overton 2007; Nelson et al. 2011).

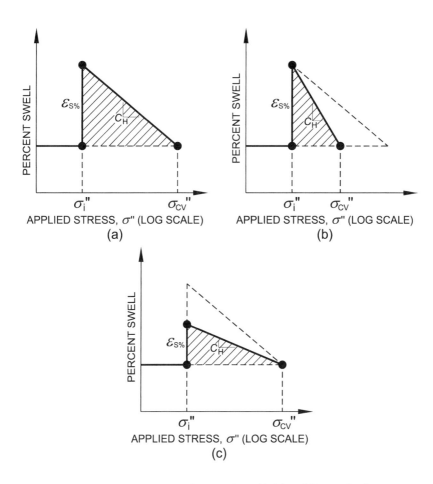

FIGURE 9.3. Oedometer test results for samples exhibiting different risk factor: (a) base case for comparison; (b) same $\varepsilon_{s\%}$ as in (a) but with smaller σ''_{cv}; (c) same σ''_{cv} as in (a) but with lower $\varepsilon_{s\%}$.

Figure 9.3 depicts the rationale behind the definition of *RF*. Figures 9.3a and 9.3b show oedometer test results for two samples that exhibited the same value of percent swell but different values of swelling pressure. From equation (8-6) it is evident that the one with the larger value of swelling pressure will heave more. Similarly, for two samples that exhibit the same value of swelling pressure, but different amounts of percent swell as depicted in Figures 9.3a and 9.3c, the one exhibiting the higher value of percent swell will heave more. A reasonable system on which to quantify risk that considers both percent swell and swelling pressure would be based on the area under the C_H line. The risk factor, RF, is therefore defined as

$$RF = \varepsilon_{s\%} \, (\log \, \sigma_{cv}'' - \log \sigma_i'') = \varepsilon_{s\%} \log \frac{\sigma_{cv}''}{\sigma_i''} \qquad (9\text{-}1)$$

where $\varepsilon_{s\%}$ is expressed in percent.

The test results shown in Figure 9.3 all relate to samples inundated under the same inundation stress, σ_i''. For the same soil having the same value of swelling pressure but tested at a higher inundation stress, a lower value of RF would be computed if only the shaded area is considered. Therefore, in comparing the values of RF for different soils, it is important that they refer to the same inundation stress. The percent swell was expressed as a function of the inundation stress in terms of the heave index, C_H, that was defined in chapter 6. The percent swell measured at different values of inundation stress can be adjusted to a common value of inundation stress as shown in Figure 9.4. In that figure, the percent swell, $\varepsilon_{s\%2}$, is adjusted back to correspond to the percent swell that would have occurred at an inundation stress equal to σ_{i1}'' by extending the C_H line backwards or forwards to the value of σ_{i1}''. The relationship shown in equation (9-2) can be developed by reference to Figure 9.4. The RF value for a sample inundated at an inundation stress of σ_{i2}'' can be calculated using the adjusted value of $\varepsilon_{s\%1}$ in equation (9-1).

$$\varepsilon_{s\%1} = \frac{\log(\sigma_{cv}'') - \log(\sigma_{i1}'')}{\log(\sigma_{cv}'') - \log(\sigma_{i2}'')}\varepsilon_{s\%2} = \frac{\log\left(\frac{\sigma_{cv}''}{\sigma_{i1}''}\right)}{\log\left(\frac{\sigma_{cv}''}{\sigma_{i2}''}\right)}\varepsilon_{s\%2} \qquad (9\text{-}2)$$

For purposes of consistency the risk factor will be defined for a common value of inundation stress of 1,000 psf (48 kPa). This value is one that is commonly used for oedometer tests.

The application of RF to quantify risk was demonstrated by computing heave using actual data taken from a database collected from a large number of soil reports that have been reviewed by the authors. The database contains results of more than 3000 CS tests conducted on a variety of material types with a wide range of expansion potential, ranging from low to very high. Many of the samples were for highly overconsolidated and highly expansive claystones.

For purposes of illustration, free-field heave was computed for a single stratum of soil or bedrock extending to a large depth. For these

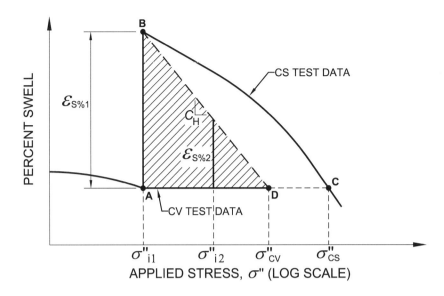

FIGURE 9.4. Effect of inundation stress on percent swell.

computations, data from 40 samples chosen at random were used. It was assumed that the soil had the same properties throughout its depth.

Figure 9.5 shows the computed values of free-field heave plotted as a function of risk factor, RF. Figure 9.5 shows a good correlation between RF and predicted heave. When the same data were plotted as a function of only percent swell or swelling pressure, the plots showed a wide range of scatter (Nelson et al. 2011).

For a soil profile having multiple soil layers a weighted value of RF, RF_w, can be computed by applying a weighting factor based on the thickness of the individual strata relative to the depth of the design active zone. Thus,

$$RF_w = \sum_{i=1}^{n} \frac{z_i}{z_{AD}} RF_i \qquad (9\text{-}3)$$

where:

$RF_i = RF$ for layer i,
  z_i = soil layer thickness for layer i, and
  z_{AD} = depth of design active zone.

Sample calculations for RF_w are shown in Examples 11.1 and 11.2.

The weighted risk factor, RF_w, and the predicted value of free-field heave for the fully wetted soil profile were computed for a number of actual soil profiles that have been observed in the Front Range area of Colorado. Figure 9.6 shows the computed heave as a function of the

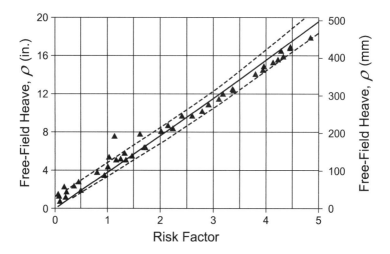

FIGURE 9.5. Free-field heave as a function of risk factor for a single soil layer (modified from Nelson et al. 2011). (Reproduced with permission of ASCE.)

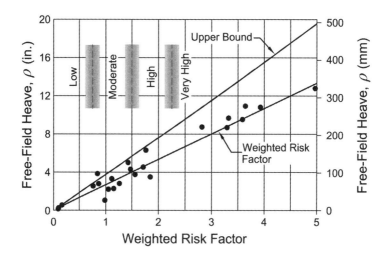

FIGURE 9.6. Free-field heave as a function of weighted risk factor for sites with multiple soil layers.

weighted risk factor for the soil profiles used in the computations. The data shown in Figure 9.6 correlate fairly well with computed free-field heave, but are lower than the correlation line from Figure 9.5. Thus, the correlation line from Figure 9.5 is considered to be an upper boundary for the weighted risk factor.

In Figure 9.6 the weighted risk factors are identified with descriptors of low, moderate, high, and very high. The rationale for the dividing lines between the zones is as follows. For a site that could experience a free-field heave of about 2 in. (50 mm), a relatively simple foundation system could be designed and slabs-on-grade could be designed to accommodate that amount of movement. Thus, there would not be the need for intensive site investigation or out-of-the-ordinary design methods, and the risk of damage to a structure would be considered low. That amount of free-field heave corresponds to a risk factor of 0.75 on the weighted average line shown in Figure 9.6.

As the predicted free-field heave increases, the complexity of the required design and the sophistication of the foundation system increase. Also, there is less room for error during construction. At what point the severity of the associated risk becomes what one would describe as "very high" is not a clearly defined point. However, for a site that could exhibit about 6 in. (150 mm) of free-field heave, the risk associated with damage to the flatwork and slabs would be "very high." Deep foundations would be required and detailed site investigation along with careful analysis of the design would be needed. A value of about 6 in. (150 mm) of free-field heave corresponds to a risk factor of 2.25, and therefore, that is marked as the lower limit for "very high."

Two zones identified as "moderate" and "high" are divided at a value of $RF = 1.5$. The assignment of RF values to the zones is to some extent subjective and for that reason, shaded areas are shown adjacent to the dividing lines in Figure 9.6. That allows for some engineering judgment to be exercised in classifying the risk of a site. It also allows for the inclusion of other site-specific factors such as environmental conditions or intended use of the site.

It is important to note that the weighted average line presented in Figure 9.6 was developed using soil data and stratigraphy particular to the Front Range of Colorado. Differences in the weighted average may occur for different regions of the world; therefore, the readers are encouraged to develop a weighted average line for their particular region. In the absence of empirical data, it is recommended that the reader use the upper limit line as a conservative estimate to compare free-field heave to risk factor.

It is important to understand what the risk classification, based on RF, means. Having a rating of very high does not necessarily mean that any structure constructed on that site has a very high probability of failure. Instead, it means that because it has a high value of RF, more extensive measures are warranted for site investigation, foundation design, and construction quality control. For a site with nonexpansive soil, having an *RF* of 0.0, ordinary methods of exploration and conventional design methods will suffice, provided no other site-specific geotechnical issues exist. Required construction quality assurance and control will be the same as normally provided. However, for a site with a high *RF* rating, additional exploratory borings at closer spacing and deeper depths, continuous core sampling, and good-quality sampling procedures are called for. Rigorous design methods using up-to-date theoretical concepts with conservative assumptions should be applied. During construction frequent inspections should be carefully performed. Experienced and knowledgeable technicians should be employed for the inspections. They should be aware of the special conditions that exist, and close communication must be maintained with the project engineer. Only well-qualified and highly reputable contractors and subcontractors should be used. In basic terms, the higher the *RF* rating, the lower the tolerance for errors, whether they are inadvertent or the result of ignoring the soils engineer's recommendations.

Two case studies are presented next to demonstrate the application of the risk factor to the potential for slab movement. Both cases involved houses located in the Front Range area of Colorado.

CASE STUDY—LOW *RF* SITE

This case study involved a two-story single family house with a slab-on-grade basement floor and a spread footing foundation. The soils report indicated that the risk of slab movement at this site was low. The design level geotechnical investigation at the site consisted of one boring drilled to a depth of 25 ft within the anticipated footprint of the structure. The soil profile consisted of 17 ft of sandy clay overlying claystone bedrock. Laboratory tests were performed on samples from this site and nearby lots. Soil properties are listed in Table 9.5.

TABLE 9.5   Soil Properties for Low *RF* Site

Depth (ft)	Soil Type	LL (%)	PI (%)	<#200 (%)	$\varepsilon_{s\%}$ * (%)	σ_{cs}'' * (psf)
0–10	Sandy clay	39	19	65	0	0
10–17	Sandy clay	—**	—	—	0.8	2,000
17–25	Claystone	—	—	—	2.9	7,400

* $\sigma_i'' = 1,000$ psf
** — = Not measured

Approximately eight years after construction of the structure, an elevation survey of the basement concrete slab measured the maximum out-of-levelness of

TABLE 9.6   Comparison of Level of Risk for Slab Movement Using Various Classification Systems for Low *RF* Site

Classification System	Soil Properties Used for Classification	Range of Value	Level of Risk
Holtz and Gibbs (1956)	Plasticity index	19%	Medium
Snethen, Johnson, and Patrick (1977)	Liquid limit	39%	Low
Chen (1988)	Plasticity index	19%	
	% passing #200 sieve	65%	High
	Liquid limit	39%	
	Plasticity index	19%	
	Percent swell	0.8–2.9%	
	Swelling pressure	2,000–7,400 psf	
CAGE (1996)	Percent swell	0.8–2.9%	Medium
RF Method	Weighted *RF*	0.6	Low

the slab to be 1.2 in. Some minor hairline cracks were observed in the basement slab. No significant distress was observed throughout the entire structure.

Table 9.6 shows the classification of risk for this site performed in accordance with the classification systems that were discussed in the previous section. As indicated in Table 9.6, the risk of the slab movement was determined to be either "low," "medium," or "high," depending on which classification system is used. The risk of the slab movement was determined to be "medium" in accordance with the majority of the classification systems. The weighted risk factor, RF_w, was calculated to be 0.6 for this site. Using the weighted average line in Figure 9.6, this would correspond to a low risk of movement. This risk potential agrees with the actual performance of the foundation and the floor slab.

CASE STUDY—VERY HIGH *RF* SITE

This case study involved a two-story single-family house founded on a pier and grade beam foundation with a slab-on-grade basement floor. The drilled piers were 23 ft long. The forensic investigation included one boring drilled to a depth of 40 ft in the front yard of the structure. The soil profile consisted of 9 ft of silty clay, underlain by 5 ft of weathered claystone, and claystone bedrock. Soil properties are listed in Table 9.7.

Distress to the house included drywall cracks and differential movement of the floor slabs beginning within the first two years after construction. Four years after construction, the maximum out-of-levelness of the basement slab was reported to be approximately 3.5 in. Significant drywall crack widths up to $1/4$ in. were observed at the second level of the residence. Structural inspections performed four and 5 years after construction indicated that the structure was continuing to

TABLE 9.7   Soil Properties for Very High *RF* Site

Depth (ft)	Soil Type	LL (%)	PI (%)	$\varepsilon_{s\%}$ (%)	σ''_{cs} (psf)
0–9	Silty clay	48	28	0.7*	1,378*
9–14	Weathered claystone	54	31	4.1**	9,720**
14–19	Claystone	51	33	3.4**	10,048**
19–29	Claystone	48	27	2.6**	5,534**
29–40	Claystone	48	28	1.4**	2,463**

*$\sigma''_i = 1,600$ psf
**$\sigma''_i = 1,000$ psf

experience vertical and lateral movements. An example of the slab distress at the residence is shown in Figure 9.7.

FIGURE 9.7. Basement concrete slab for Very High *RF* Site.

The risk of the slab movement was classified using the engineering index properties obtained from the forensic investigation for the site in accordance with the classification systems that were previously discussed. Table 9.8 shows that the risk of the slab movement was determined to be either "marginal," "high," or "very high," depending on which classification system is used. The weighted risk factor, RF_w, was calculated to be 2.3 for this case. In accordance with Figure 9.6,

TABLE 9.8   Comparison of Level of Risk for Slab Movement Using Various Classification Systems for Very High *RF* Site

Classification System	Soil Properties Used for Classification	Range of Value	Level of Risk
Holtz and Gibbs (1956)	Plasticity index	27–33%	High
Snethen, Johnson, and Patrick (1977)	Liquid limit	48–54%	Marginal
	Plasticity index	27–33%	
Chen (1988)	Liquid limit	48–54%	High
	Plasticity index	27–33%	
	Percent swell	0.7–4.1%	
	Swelling pressure	1,378–10,048 psf	
CAGE (1996)	Percent swell	0.7–4.1%	High
RF Method	Weighted *RF*	2.3	Very high

this corresponds to a very high risk of slab movement. This value of risk factor matches with the actual performance of the basement slab and the structure.

These two cases demonstrate what the *RF* means. For the low *RF* site, ordinary soil investigation and foundation design measures were adequate. For the very high *RF* site, the geotechnical analysis performed during a forensic investigation showed that the drilled piers were too short. It also was concluded that a slab-on-grade basement floor was the wrong floor system to use. If longer piers had been designed and if a structural basement floor had been constructed, the house would not have experienced distress. Thus, the high value of *RF* would have alerted the designer to the need for a more extensive site investigation and more conservative design.

9.2   FOUNDATION ALTERNATIVES

Table 9.9 lists the various aspects of the design alternatives that were discussed above, and shows their similarities and differences. Table 9.10 lists the design principles, procedures, and details, along with construction quality control and remedial construction measures.

9.3   FACTORS INFLUENCING DESIGN OF STRUCTURES ON EXPANSIVE SOILS

9.3.1   Tolerable Foundation Movement

The amount of heave that the structure can tolerate is the most important design parameter with regard to foundation design. More

TABLE 9.9 Foundation Design Alternatives (Modified from Nelson and Miller 1992)

Foundation Type	Design Objective	Advantages	Disadvantages	Risk Assessment
Modified continuous perimeter spread footings	Provide rigid foundation to resist differential soil movement Increase bearing pressure to reduce heave	Simple construction No specialized equipment	Only effective in soils with very low expansion potential	Low to moderate risk for soils with very low expansion potential High risk for soils with moderate to high expansion potential
Drilled pier and grade beam foundation	Isolate structure from expansive soil	Has a long and successful track record Can be used in a variety of soil types Provides reliable system for soils with high expansion potential	Need for rigorous design and construction Potentially higher construction cost	Low risk if designed and installed properly
Patented piers	Isolate structure from expansive soil	Commonly used for remediation of structures on expansive soils Can be installed in low-headroom applications in limited access areas Provide reliable system for soils with high expansion potential Can be installed with batter inclination to resist lateral loads on foundation	Need for rigorous design and construction Requires specialty contractors May require on-site test pier to verify installability and load bearing capacity Vertically installed patented piers provide little resistance to lateral forces	Low risk if designed and constructed properly
Stiffened slabs-on-grade (includes post-tensioned slabs)	Provide rigid foundation to resist differential soil movement	Faster to construct than deep foundations Provides a reliable system for soils with low to moderate expansion potential	History of poor performance if proper design and construction quality control are not provided Configuration of building must be relatively simple If a tendon or anchorage in post-tensioned slab fails, additional operations are required for repair	Low to moderate risk for soils of low to moderate expansion potential Moderate to high risk for soils with high expansion potential

TABLE 9.10 Design Summary for Major Foundation Systems Used in Expansive Soils (Modified from Nelson and Miller 1992)

Foundation Type	Design Principle	Design Procedures	Design Details	Quality Control	Remedial Measures
Modified continuous perimeter spread footings	Provide sufficiently high bearing pressure to counteract swell pressures Provide foundation wall or beam capable of spanning differential heave between footings	Empirical Minimal design Generally follows local rule of thumb	Can provide void space under portions of foundation wall (to concentrate loads over small areas)	Standard construction inspection	Post-tension foundation wall Add adjacent grade beam (sister wall) Underpin with patented piers
Drilled pier and grade beam foundation	Provide sufficient anchorage in deep zones where expansion is limited	Establish pier length based on design active zone Provide reinforcing steel to resist tension in pier Isolate grade beam from soil	Extend reinforcing steel over entire pier length and into grade beam Provide detail to preserve void space beneath structural slabs and beneath grade beam between piers Provide separation between grade beam and floating floor slab	Avoid "mushroom" at top of pier Assure clean pier bottom Maintain void space and soil retainer under grade beam and structural slabs Assure cleanliness and proper length of reinforcing steel Place concrete by tremie method to avoid segregation of concrete when placing pier Assure isolation between grade beam and floor slab	Excavate void space under grade beam if inadequate void space was initially provided Remove improper piers and underpin foundation Post-tension or install new grade beam

continued

TABLE 9.10 (*Continued*)

246

Foundation Type	Design Principle	Design Procedures	Design Details	Quality Control	Remedial Measures
Patented piers	Provide sufficient anchorage	Select an appropriate type of patented pier Design pier length based on design active zone	For micropiles: Extend reinforcing steel over entire pier length Provide adequate connection to grade beam to resist design tensile loads Can provide low-friction casing in upper zones Provide void space beneath structural slabs and beneath grade beam between piers Provide for resistance to lateral loads	Check connection to grade beam Other aspects are the same as for the drilled pier and grade beam foundation	Same as for the drilled pier and grade beam foundation
Stiffened slab-on-grade (includes post-tensioned slabs)	Provide sufficiently stiff foundation to minimize structural distortions	Compute free-field heave Define free-field soil heave pattern (edge or center heave mound shape) Define slab loading conditions Evaluate seasonal volume changes of soil Provide adequate structural design of slab details	Provide adequate shear connection to ensure interaction between stiffening beam and slab Provide adequate reinforcing or post-tensioning in stiffening beams Adjust beam spacing to coincide with walls or concentrated loads	Standard reinforced concrete inspection Avoid over-compaction or undercompaction of soil beneath slab Assure adequate placement depth for anchors in post-tension construction	Underpin around edges and interior stiffening beams Epoxy cracks

important than total heave is differential foundation movement and the distance over which differential movement will take place. This is termed the *angular distortion,* $\Delta\rho/L$, where $\Delta\rho$ is the differential movement and L is the distance over which $\Delta\rho$ has occurred. The tolerable amount of total heave and angular distortion will depend on the nature of the building and the tolerance for distress in the building. For example, a steel frame building with steel siding that is used for storing nonperishable goods would be much more tolerant of heave and angular distortion than would be a safety-critical building such as a hospital or a public occupancy building.

The structure should be able to perform its functions and require only normal maintenance. The ability for a structure to tolerate deformation depends on the brittleness of the building materials, length to height ratio, relative stiffness of the structure in shear and bending, and mode of deformation where heave or settlement occurs. The US Army Corps of Engineers recommended the tolerable angular distortion shown in Table 9.11 for structures constructed of different types of building materials. Some superstructures, such as masonry walls or glass blocks, are intolerant of even relatively small differential movement. For structures with those construction elements, a rigid foundation, such as a heavily stiffened mat, or one that will isolate the structure, such as a pier and grade beam foundation, should be used. Other superstructures, such as those of timber or steel construction, can tolerate relatively greater differential movement.

The tolerance for differential foundation movement and severity of distress depends on a number of factors. As noted above, the owners or tenants of a building used for storage of industrial materials may be more tolerant of out-of-levelness or wall cracks than would be the owners of an expensive, upscale residence. Geographic location can also be a factor. People in areas where differential movement has been historically common are more tolerant of cracking and out-of-levelness than are people from areas where such things are not common. On the other hand, movement that impairs the function of the structure cannot be tolerated. An example would be movements that cause doors, windows, or other points of egress to be difficult to open. This would be a safety issue in case of fire, particularly if a child needed to use that as an emergency exit. Day (1998) categorized distress to buildings caused by soil movements into the three general groups presented in Table 9.12.

Day (1998) categorized the severity of cracking damage based on approximate crack widths, typical values of maximum differential movement, and maximum angular distortion of the foundation. The various categories and the nature of distress pertinent to each are

TABLE 9.11   Tolerable Angular Distortion of Superstructure Systems (US Department of the Army 1983)

Superstructure System	Tolerable Angular Distortion, $\Delta\rho/L^*$	Description
Rigid	1/600 to 1/1,000	Precast concrete block, unreinforced brick, masonry or plaster walls, slab-on-grade
Semirigid	1/360 to 1/600	Reinforced masonry or brick reinforced with horizontal and vertical tie bars or bands made of steel bars or reinforced concrete beams; vertical reinforcement located on sides of doors and windows; slab-on-grade isolated from walls
Flexible**	1/150 to 1/360	Steel, wood framing; brick veneer with articulated joints; metal, vinyl, or wood panels; gypsum board on metal or wood studs; vertically oriented construction joints; strip window or metal panels separating rigid wall sections with 25-ft spacing or less to allow differential movement; all water pipes into structure with flexible joints; suspended floor or slab-on-grade isolated from walls (heaving and cracking of slab-on-grade probable and accounted for in design)
Split construction**	1/150 to 1/360	Walls or rectangular sections heave as a unit (modular construction); joints at 25-ft spacing or less between units and in walls; suspended floor or slab-on-grade isolated from walls (probable cracking of slab-on-grade); all water pipes and drains equipped with flexible joints; construction joints in reinforced and stiffened mat slabs at 150-ft spacing or less and cold joints at 65-ft spacing or less

$^*\Delta\rho$ = differential movement, and L = horizontal distance.
**A $\Delta\rho/L$ value exceeding 1/250 is not recommended for normal practice; $\Delta\rho/L$ value exceeding 1/150 often leads to structural damage.

listed in Table 9.13. Architectural damage is generally associated with a damage category of at least "very slight," shown in Table 9.13. Although Table 9.13 categorizes fine cracks as "very slight," it should be noted that a high frequency of hairline cracks can indicate foundation movement. Thus, a high frequency of even "very slight" damage can be an indication of more severe distress. Functional damage is generally associated with a damage category of "slight" or "moderate." Structural damage commonly relates to the "severe" to "very severe" damage categories. Depending on the nature of the structure, structural damage could also be present in the "moderate" damage category.

TABLE 9.12   Categorization of Types of Distress (Modified from Day 1998)

Type of Distress	Description
Architectural	Also referred to as *cosmetic damage*. Affects the appearance of the building and is usually related to minor cracks in the walls, floors, and finishes. Cracks in plaster walls greater than 0.02 in. (0.5 mm) wide and cracks in masonry walls greater than 0.04 in. (1 mm) wide are considered to be typical threshold values that would be noticed by the building occupants (Burland, Broms, and Demello 1997).
Functional (or Serviceability)	Affects the use of the building. Examples include jammed doors and windows, extensively cracked and falling plaster, and tilting of wall and floors. Ground movements may cause cracking that leads to premature deterioration of construction materials or leaking roofs and facades.
Structural	Compromises the structural integrity of the building, and affects the stability. Examples include cracking or distortions of supporting members such as beams, columns, or load-bearing walls. Damages in this category could lead to structural collapse.

Guidelines for the Evaluation of Foundation Movement for Residential and Other Low-Rise Buildings, published by the Foundation Performance Association (FPA 2007) contains allowable criteria for foundation movement. The FPA 2007 guidelines state that the onset of excessive distress in the superstructure appears to occur when the angular distortion exceeds 1/240 to 1/480. There may be exceptions to this rule where the distress is minimal at larger deflection ratios. However, the opposite has also been found to occur. FPA (2007) proposed the allowable criteria for foundation movement shown in equation (9-4) based on review of projects in the Houston, Texas, area. They proposed that the maximum allowable distortion, measured from edge-to-edge of the foundation in any direction, is given as

$$\begin{cases} \text{Deflection Limit (in inches)} = \frac{kL}{360} \\ \text{Maximum Allowable Tilt} = 1.0\% \end{cases} \tag{9-4}$$

where:

k = modification factor to adjust the deflection limit when the profile being considered is in a direction that is not parallel to one of the foundation's principal axes. The k value varies from 1.000 (if along a principal axis) to 1.414 (if diagonally across a square), and

L = effective horizontal length in inches.

TABLE 9.13 Categorization of Severity of Cracking Damage (Modified from Day 1998)

Damage Category	Description of Typical Damage	Approximate Crack Width	Maximum Differential Movement, Δ	Angular Distortion, Δρ/L
Negligible	Hairline cracks.	<1/64 in. (<0.1 mm)	< 1.2 in. (<30 mm)	< 1/300
Very slight	Fine cracks that can be easily treated during normal decoration; perhaps an isolated slight fracture in building; cracks in external brickwork visible on close inspection.	1/32 in. (1 mm)	1.2–1.5 in. (30–40 mm)	1/300–1/240
Slight	Cracks that can be easily filled; redecoration would probably be required. Several slight fractures may appear showing the inside of the buildings; cracks might be visible externally and some repointing may be required; doors and windows may stick.	1/8 in. (3 mm)	1.5–2.0 in. (40–50 mm)	1/240–1/175
Moderate	Cracks that require some opening up and that can be patched by a mason; recurrent cracks that can be masked by suitable lining; repointing of external brickwork. Where possible, a small amount of brickwork replacement may be required. Doors and windows stick; service pipes may fracture; weather-tightness is often impaired.	1/4–1/2 in. (5–15 mm) or a number of cracks > 1/8 in. (3 mm)	2.0–3.0 in. (50–75 mm)	1/175–1/120
Severe	Large cracks requiring extensive repair work involving breaking-out and replacing sections of walls (especially over doors and windows). Window frames and door frames are distorted. Also noticeably sloping floors; leaning or bulging walls; some loss of bearing in beams; disrupted service pipes.	1/2–1 in. (15–25 mm), but also depends on number of cracks	3.0–5.0 in. (75–130 mm)	1/120–1/70
Very severe	Requires major repair involving partial or complete rebuilding; beams lose bearing; walls lean and require shoring; windows are broken with distortion. There is danger of structural instability.	Usually > 1 in. (25 mm), but also depends on number of cracks	> 5.0 in. (>130 mm)	> 1/70

**TABLE 9.14   Maximum Tolerable Angular Distortion for Slab-on-Grade Foundation
Systems to Limit Damage to Superstructure**

Type of Superstructure	Maximum Tolerable Angular Distortion, $\Delta\rho/L$	Reference
Wood frame	1/200	BRAB (1968)
Nonmasonry, timber or prefabricated	1/200	Woodburn (1979)
Unplastered masonry or gypsum wallboard	1/300	BRAB (1968)
Stucco or plaster	1/360	BRAB (1968)
Brick veneer (articulated)	1/300	Woodburn (1979)
Brick veneer (standard)	1/500	Woodburn (1979)
Masonry (articulated)	1/800	Woodburn (1979)
Masonry (solid)	1/2,000	Woodburn (1979)

Equation (9-4) does not consider the nature of the superstructure placed on the foundation. Table 9.14 lists the maximum tolerable angular distortion for slab-on-grade foundation systems to limit damage to certain commonly used superstructure types (Wray 1978). In Table 9.14, the maximum tolerable angular distortion for a structure should be determined by the weakest exposed finish material in the superstructure (BRAB 1968).

Measurements of post-construction deflection or angular distortion are often used for quantification of distress. However, without an as-built or previous elevation survey such measurements are only indications of distress.

Original construction tolerances have to be considered. As-built conditions of slabs are shown in Table 9.15. The average angular distortion for newly constructed slabs is approximately 1/340. However, actual foundation movement cannot be estimated simply by subtracting the average values shown in Table 9.15 from the measured values, because the foundation may have moved such that it has tilted in a different direction. A series of elevation surveys at different times is valuable in demonstrating ongoing movement. Other indications of movement must also be considered, such as cracking of exterior and/or interior walls, racked doors and windows, or other signs of distress.

9.3.2 Design Life

The design life of a foundation will depend on the structure and its use. In some cases, particularly for commercial or government buildings, the design life is specified to provide a common basis for design of various building elements. It does not mean that the building will be discarded at the end of the design life, but it might be put to some other use. Also, it

TABLE 9.15   Out-of-Levelness and Angular Distortion for Newly Constructed Slabs

References	Location	Out-of-Levelness (in.)		Angular Distortion	
		Range	Average	Range	Average
Koenig (1991)	San Antonio, TX	0.125–1.0	0.54	—	—
Marsh and Thoeny (1999)	Southern CA	0.6–1.0	0.75	—	—
Walsh, Bashford, and Mason (2001)	Phoenix, AZ	0.25–1.18	0.53	1/857–1/101	1/334
Noorany et al. (2005)	CA	0.2–2.2	0.53	1/1,000–1/71	1/346

is important to realize that the design life of the foundation will be different than that of other elements (e.g., the roofing material).

The Housing Facts, Figures, and Trends published by National Association of Home Builders (1997) indicated that the design life for residential foundations should be 200 years. The US Department of Housing and Urban Development (2002) considers the design life for residential foundations to be 100 years. Therefore, a minimum design life of 100 years for residential foundations should be considered. According to Schmalz and Stiemer (1995), the design life may be 30 to 60 years for commercial and industrial buildings, 60 to 120 years for public sector buildings, and more than 120 years for civic and high-quality buildings.

9.3.3   Design Active Zone and Degree of Wetting

The water content profile that will exist throughout the design life of the foundation forms the basis for computation of free-field heave and design of foundation elements to limit movement to within tolerable amounts. The profile that exhibits the greatest amount of wetting for design usually occurs at the end of the design life, but that is not always the case. The design active zone has been defined in chapter 1 and discussed in chapter 7.

9.3.4   Site Grading

Although site grading is not a structural foundation design detail, it is of paramount importance in controlling surface water to minimize infiltration into the soil and must be considered in the overall plan. The required grade should be specified in the soils and foundation report. Site grading will be discussed in detail in chapter 10. Various codes specify the grades to be achieved for the first several feet away from the foundation. Specified values of grade range from about 5 to 10 percent. A minimum grade of 10 percent for the first 10 ft away from the foundation has been the industry standard for expansive soil sites in Colorado and other areas of the western United States.

Problems frequently arise in achieving sufficient grade if the engineer who develops the grading plan is not made aware of that requirement or decides to disregard it. It is important that the first-floor elevations for the structure be set sufficiently high relative to the curbs and streets so that the required grade can be achieved. A statement to that effect should be included in the foundation and soils report when the required grade is specified.

9.4 REMEDIAL MEASURES

The previous sections of this chapter have been directed primarily to initial design of foundations and floors. When problems arise, remedial measures need to be designed. The remediation plan designs should generally consider all of the factors discussed above, but there are some additional factors to be considered. Nelson and Miller (1992) have discussed the application of remediation plan designs. The following discussion is taken largely from that reference.

The design of a foundation on expansive soils always involves a certain degree of risk. Usually, a lower cost of the design alternative will be associated with a higher degree of risk. The same principle applies to the design of remedial measures. The maximum cost of a remedial measure is one that would require the removal of the damaged structure and reconstructing a new one. Even this procedure requires considerable engineering design. Even though the foundation soils will have caused movement of the foundation, it will be necessary to assess the current state of expansion and to predict future expansion potential that may still exist. The new foundation will need to be capable of withstanding this future movement.

Nelson and Miller (1992) interviewed engineers at locations from Canada to Texas, from the midwestern United States to the West Coast, and in several countries such as Israel and South Africa. They noted that in almost all locations, there existed some case in which the cost of remedial repairs would exceed the cost of removing the structure and rebuilding it. That has also been the authors' experience. Not all cases are that severe, however, and remediation techniques have been developed for cost-effective remedial measures.

Prior to undertaking remedial measures a number of important issues must be discussed among the owner, the engineer, and other parties that may be financially involved. Nelson and Miller (1992) noted the following questions that should be considered:

- What is the cause and extent of the damage?
- What remedial measures are applicable?
- Should remedial measures be undertaken at this time? If damage is not severe and continued future movement is anticipated, it may be better to wait until the rate of movement has slowed.
- Should remedial measures be undertaken at all? If damage is not severe and remedial measures offer no significant improvement, it may be appropriate to make cosmetic repairs and continue maintenance.
- Who is financially responsible for the repair?
- What criteria should be used to select the remedial measure and scope of repairs to be employed? These should include considerations of cosmetic and structural benefits, as well as actual costs.
- What residual risk will exist after the repairs have been completed? It should be expected that future movement will continue and the cost/risk relationship of the selected remedial measures must be considered.

The economic responsibility to implement a remedial measure may fall on the owner, developer, builder, contractor, or insurer. The party responsible for the repair generally makes the decisions regarding the choice of a remedial measure. However, that party may have only a limited period of liability, whereas the owner's interest is usually for an indefinite long-term period. It is important that the party or parties who will benefit most from the repairs be in a position to at least influence the selection of the remedial measures to be undertaken. The role of the engineer is to define the cause of the damage, make recommendations for corrective action, and possibly approve or inspect the completed remedial work.

The selection and application of technically effective and cost-effective remedial measures are influenced by the differing interest groups. An agreement must be reached to determine how much future risk, both actual and economic, is acceptable as compared to immediate costs and present values. An important point of contention is often the definition of what constitutes a failure, or whether remedial measures are warranted and necessary. Since each interest group has differing objectives, it is often necessary to reach a compromise to avoid future controversy. The ultimate liability of each party should be made clear.

The process of applying remedial measures will be greatly improved if all parties can agree on performance standards and assignment of responsibilities for the assumptions of residual risks. Realistically, this rarely occurs. Long-term guarantees on remedial repairs are virtually impossible to provide realistically. It may be difficult to adequately evaluate the economic soundness of a particular remedial measure, but applying only cosmetic repairs (leveling, patching, redecoration) without appropriate analysis of causes and damage is often an unrewarding investment.

In the selection of a remedial measure, each case should be evaluated based on the pertinent questions listed at the beginning of this section. It is important that the advice of a professional engineer be sought to outline possible alternatives, initial costs, residual risks, and benefits. The success of the remedial measures employed will depend on the knowledge and skill of the engineer and contractor and the degree of quality control that is exercised.

Alternative remedial measures discussed in chapters 11 through 13 will give the reader an indication of some methods that are available. Frequently, local contractors have developed innovative techniques that can be very successful in their area. Each case should be evaluated by a competent engineer to provide the owner with the necessary information on which to base a decision.

There are no standard procedures in the application of remedial measures. In general, innovation is necessary in the design of different systems, and the application of newly developed techniques should be encouraged even if they have not been implemented elsewhere. A complete listing of procedures that have been used would require the discussion of case studies too numerous to be included here.

References

American Association of State Highway and Transportation Officials (AASHTO). 2008. "AASHTO T 258 Standard Method of Test for Determining Expansive Soils." Washington, DC.

Building Research Advisory Board (BRAB). 1968. "Criteria for Selection and Design of Residential Slabs-On-Ground." Publication 1571, National Academy of Sciences Report Number 33 to Federal Housing Administration, Washington, DC.

Burland, J. B., B. B. Broms, and V. F. B. Demello. 1997. "Behavior of Foundations and Structures: State of the Art Report." *Proceedings of the 9th International Conference on Soil Mechanics and Foundation Engineering*, Japanese Geotechnical Society, Tokyo, Japan, 495–546.

Chen, F. H. 1988. *Foundations on Expansive Soils*. New York: Elsevier Science.

Colorado Association of Geotechnical Engineers (CAGE). 1996. "Guideline for Slab Performance Risk Evaluation and Residential Basement Floor System Recommendations (Denver Metropolitan Area) Plus Guideline Commentary." CAGE Professional Practice Committee, Denver, CO.

Day, R. W. 1998. "Settlement Behavior of Posttensioned Slabs-On-Grade." *Journal of Performance of Constructed Facilities*, ASCE 12(2): 56–61.

Foundation Performance Association (FPA). 2007. "Guidelines for the Evaluation of Foundation Movement for Residential and Other Low-Rise Buildings." Document No. FPA-SC-13-0, FPA, Houston, TX.

Holtz, W. G., and H. J. Gibbs. 1956. "Engineering Properties of Expansive Clays." *Transactions ASCE* 121, 641–677.

Koenig, A. S. 1991. "Relative Elevation Data Analysis on Newly Constructed Residential Foundations." *Proceedings of the ASCE Texas Section Spring Meeting, Expansive Clay Research Institute*, San Antonio, TX.

Marsh, E. T., and S. A. Thoeny. 1999. "Damage and Distortion Criteria for Residential Slab-on-Grade Structures." *Journal of Performance of Constructed Facilities*, ASCE 13(3): 121–127.

National Association of Home Builders. 1997. *Housing Facts, Figures, and Trends*. Washington, DC: National Association of Home Builders.

Nelson, J. D., and D. J. Miller. 1992. *Expansive Soils: Problems and Practice in Foundation and Pavement Engineering*. New York: John Wiley and Sons.

Nelson, J. D., K. C. Chao, and D. D. Overton. 2007. "Definition of Expansion Potential for Expansive Soil." *Proceedings of the 3rd Asian Conference on Unsaturated Soils*, Nanjing, China, 587–592.

Nelson, J. D., K. C. Chao, D. E. Overton, and J. S. Dunham-Friel. 2011. "Evaluation of Level of Risk for Structural Movement Using Expansion Potential." *Proceedings of the Geo-Frontiers Conference*, Dallas, TX, 2404–2413.

Noorany, I. M., E. D. Colbaugh,, R. J. Lejman, and M. J. Miller. 2005. "Levelness of Newly Constructed Posttensioned Slabs for Residential Structures." *Journal of Performance of Constructed Facilities, ASCE* 19(1): 49–55.

Schmalz, T. C., and S. F. Stiemer. 1995. "Consideration of Design Life of Structures." *Journal of Performance of Constructed Facilities*, ASCE 9(3): 206–219.

Seed, H. B., R. J. Woodward, Jr., and R. Lundgren. 1962. "Prediction of Swelling Potential for Compacted Clays." *Journal of the Soil Mechanics and Foundation Division*, ASCE 88(SM3): 53–87.

Skempton, A. W. 1953. "The Colloidal 'Activity' of Clays." *Proceedings of the 3rd International Conference on Soil Mechanics and Foundation Engineering*, Switzerland, 1, 57–61.

Snethen, D. R., L. D. Johnson, and D. M. Patrick. 1977. "An Evaluation of Expedient Methodology for Identification of Potentially Expansive

Soils." Soils and Pavements Laboratory, US Army Engineers Waterways Experiment Station, Vicksburg, MS, Report No. FHWA-RE-77-94.

US Bureau of Reclamation (USBR). 1998. *Earth Manual* (3rd ed.). Denver, CO: US Department of the Interior, Water Resources Technical Publication.

US Department of the Army (DA). 1983. *Technical Manual TM 5-818-7, Foundations in Expansive Soils*. Washington, DC.

US Department of Housing and Urban Development. 2002. "Durability by Design, A Guide for Residential Builders and Designers." Washington, DC.

Walsh, K. D., H. H. Bashford, and B. C. A. Mason. 2001. "State of Practice of Residential Floor Slab Flatness." *Journal of Performance of Constructed Facilities, ASCE* 15(4): 127–134.

Woodburn, J. A. 1979. "Interaction of Soils, Footings, and Structures." In *Footings and Foundations for Small Buildings in Arid Climates, with Special Reference to Australia*, edited by P. J. Fargher, J. A. Woodburn, and J. Selby Institution of Engineers, Australia. South Australia Division, 79–89.

Wray, W. K. 1978. "Development of a Design Procedure for Residential and Light Commercial Slabs-on-Ground Constructed over Expansive Soils." PhD dissertation, Texas A&M University, College Station, TX.

10

Soil Treatment and Moisture Control

As an alternative or supplement to the design of a structure or pavement it may be desirable, and economical, to either alter the properties of the soil to reduce its expansion potential or to stabilize the moisture regime around a structure or pavement. Various techniques have been used to alter the characteristics of the soil or stabilize the moisture regime around the foundation or pavement. These techniques can be grouped into one or more of the following categories:

- Removal (overexcavation) and replacement
- Prewetting
- Chemical admixtures
- Water content control

Some methods may be used alone or in conjunction with specific foundation or pavement alternatives. Depending on the situation a particular method may be applied before or after construction. Choice of the method to be used will rely on experience and sound engineering judgment. A program of site investigation, laboratory testing, and consideration of construction techniques is necessary to evaluate the possible alternatives.

The choice of a particular method will depend on whether the application is for a foundation system or a pavement. The methods most commonly used to treat expansive pavement subgrade include:

- Overexcavation and replacement with nonexpansive soil
- Overexcavation and replacement with moisture-conditioned on-site soils
- Chemical admixtures

Pavement systems such as gravel and asphalt cement are more flexible than rigid portland cement roadways. In general, flexible pavement

258

is more tolerant of differential movement than are rigid pavements or building foundations. Flexible pavements tend to perform better than rigid pavements in moderate to highly expansive soils.

Due to different performance requirements and bearing pressures between pavements and foundations, the soil classification methods and the techniques used for the determination of soil expansion potential in pavement subgrades are typically modified from those used for foundations. Also, the depth of investigation and treatment are generally much shallower for pavements than for a foundation system in the same soil type.

Test methods and analyses vary considerably between different states and municipalities. For example, the Colorado Department of Transportation (DOT) design manual (2014) requires the use of a CS test performed under an inundation stress of 200 psf (10 kPa), while the Texas DOT (2014) uses the PVR method to evaluate the swell potential of a pavement subgrade. Appropriate regulatory agencies should be consulted to identify appropriate methods for their area.

Some of the currently used soil treatment procedures will be discussed in this chapter in terms of effectiveness, economy, and ease of implementation. Limitations of each procedure are also discussed.

10.1 OVEREXCAVATION AND REPLACEMENT

Overexcavation of expansive soils and replacement with nonexpansive or treated soils has been used to mitigate soil heave under a foundation or subgrade. In this method, the expansive soil is excavated to an appropriate depth to minimize heave to an appropriate amount, and then appropriately treated and compacted fill is placed to bring the soil up to grade. Factors that need to be considered are the required depth of removal, and the amount, location, and cost of the fill. The necessary depth of soil that is required to be removed will depend on the overall soil profile, the nature of the soil that will be used for fill, and the allowable heave. An additional advantage provided by a stiff layer of compacted low to nonexpansive fill is that it will tend to even out the differences in heave of the underlying native soil, thereby reducing the differential heave.

If nonexpansive fill is not available, the swell characteristics of the on-site expansive soils can be altered by moisture-conditioning and compaction control. Compacting the material wet of optimum and to a lower density will reduce the expansion potential, but care must be taken to ensure that the recompacted soil has adequate strength and will not cause intolerable settlement. In some instances, chemical additives can be used together with moisture-conditioning.

Generally, the expansive soil layer extends to a depth that is too great to economically allow complete removal and replacement. Appropriate soil testing and analyses should be conducted to design the overexcavation and evaluate the expected potential heave after the overexcavation and recompaction process. The design depth of overexcavation must take into account the predicted heave.

Chen (1988) recommended a minimum of 3 ft to 4 ft (1 m to 1.3 m), but these depths were found to be ineffective for sites with highly expansive soils. Thompson (1992a and 1992b) conducted an analysis of insurance claims and concluded that if 10 ft (3 m) or more of nonexpansive soil existed below the footings, the frequency of claims was lower than for shallower depths. Based on Thompson's papers, depths of overexcavation of 10 ft (3 m) or more have frequently been specified in the Front Range area of Colorado. In some cases, the upper 20 ft (6 m) of soil across an entire subdivision has been excavated, moisture-conditioned, and recompacted in place.

Overexcavation and replacement also helps to control water content fluctuations in the underlying expansive soil layer. Most of the seasonal water content fluctuation will take place in the upper several feet of soil. However, if there is a high potential for expansion in the underlying soil, the overexcavated zone may not adequately prevent surface heave or shrinkage. If the underlying expansive soil becomes wetted intolerable movement may result. Some of the potential water sources cannot be controlled or predicted. Therefore, the design engineer must assume that events may occur during the life of the structure and make proper design decisions accordingly. Potential water sources and water migration was discussed in chapter 7.

Care must be taken to avoid the collection of water in the overexcavation zone. Permeable granular fill is not recommended for the replacement fill. Highly permeable fill will provide an easy pathway for water, and create a reservoir for this water to accumulate. This is commonly referred to as a *bathtub effect*. Such a condition will ultimately result in seepage into expansive subgrades or foundation soils. Impermeable, nonexpansive fill is more satisfactory. If granular material must be used, permanent, positive drainage and moisture barriers, such as geomembranes, should be provided to minimize moisture infiltration into this zone.

Even without granular soil, the removed and recompacted material is likely to have a higher hydraulic conductivity than the underlying in situ soils and bedrock. An underdrain system placed at the bottom of the overexcavation zone can be effective in intercepting the flow of ground-water. Care must be taken to provide positive drainage to the drain and to make sure that the drain does not just concentrate water in an area where it will result in greater soil wetting and heave.

Some advantages of treatment by overexcavation and replacement include the following:

- The cost of soil replacement can be more economical than other treatment procedures, since it does not require special construction equipment.
- Mixing with soil treatment additives can be done more uniformly resulting in some degree of soil improvement.
- Overexcavation and replacement may require less delay to construction than other procedures that require a curing period.

Disadvantages of overexcavation and replacement methods include the following:

- If the fill must be imported, the cost of nonexpansive fill with low permeability can be significant.
- Removal and recompaction of the on-site expansive soils may not sufficiently reduce the risk of foundation movement if the recompacted on-site soils exhibit intolerable expansion potential.
- The required thickness of the recompacted backfill material may be too great to be practical or economical.
- If the backfill material is too permeable the overexcavation zone may serve as a reservoir and provide a long-term source of water to the foundation soils and bedrock.

If overexcavation and replacement is not effective by itself, it may be used in conjunction with other foundation alternatives. If the potential heave can be reduced adequately, it might be possible to use a stiff mat foundation instead of a more costly deep foundation. Also, when used together with a deep foundation, the required length of the piers may be reduced.

Example 10.1 illustrates the effect that overexcavating and replacing a layer of the expansive surface soil with a nonexpansive compacted soil has on free-field heave. In this example, it is assumed that if the on-site soils are moisture-conditioned and recompacted, no expansion potential remains after recompaction.

EXAMPLE 10.1

Given:

The same soil profile and data as for Example 8.1, with the exception that the upper 10 ft of expansive soil will be removed and replaced with the reworked and compacted soil that was excavated. It is assumed that by adjusting the water content and recompacting the soil it will be nonexpansive. The fill has a saturated unit weight of 129.9 pcf.

Find:

Ultimate free-field heave for conditions of full wetting after removal of the expansive soil and replacement with nonexpansive soil.

Solution:

The depth of potential heave can be computed in the same manner as in Example 8.1, but using the appropriate unit weight for the fill. Note that, as in Example 8.1, the saturated unit weight of claystone is used in the computations. Thus,

$$129.9 \times 10 + 136.2 \times (z_p - 10) = 4,390 \text{ psf}$$

$$z_p = 32.68 \text{ ft}$$

The computations of heave are similar to those for Example 8.1. The computations are shown in Table E10.1. In these computations the upper three layers were replaced with nonexpansive soil. The computed heave is 3.7 in. Hence, the overexcavation and replacement with nonexpansive fill reduced the heave from 10.1 in. to 3.7 in.

Example 10.1 shows that replacement with nonexpansive fill can have a significant influence on the heave. However, even a small amount of remaining expansion potential in the replacement soil can have a serious impact on the ultimate free-field heave. That is illustrated by Example 10.2. This example uses the same soil properties for the claystone as Example 10.1, but the recompacted fill has some expansion potential.

TABLE E10.1 Free-Field Heave Computations for Example 10.1

Depth to Bottom of Layer (ft)	Depth Below Excavation (ft)	Interval (ft)	Soil Type	Total Unit Weight (pcf)	Incremental Stress @ midlayer (psf)	Overburden Stress (psf)	Incremental Heave (in.)	Cumulative Heave (in.)	Cumulative Heave from z_p (in.)
0.00	0.00								3.65
3.27	3.27	3.27	Nonexpansive fill	129.9	212.4	212.4	0.00	0.00	3.65
6.54	6.54	3.27	Nonexpansive fill	129.9	212.4	637.3	0.00	0.00	3.65
9.81	9.81	3.27	Nonexpansive fill	129.9	212.4	1,062.1	0.00	0.00	3.65
13.08	13.08	3.27	Claystone	136.2	222.7	1,497.2	1.14	1.14	2.51
16.35	16.35	3.27	Claystone	136.2	222.7	1,942.5	0.87	2.01	1.64
19.62	19.62	3.27	Claystone	136.2	222.7	2,387.9	0.65	2.66	0.99
22.89	22.89	3.27	Claystone	136.2	222.7	2,833.2	0.46	3.12	0.53
26.16	26.16	3.27	Claystone	136.2	222.7	3,278.6	0.31	3.43	0.22
29.43	29.43	3.27	Claystone	136.2	222.7	3,723.9	0.17	3.60	0.05
32.70	32.70	3.27	Claystone	136.2	222.7	4,169.3	0.05	3.65	

EXAMPLE 10.2

Given:

Same soil profile and data as in Example 10.1 except that after testing the fill in the upper 10 ft it was found that it exhibited 0.7 percent swell with a CS swelling pressure of $\sigma''_{cs} = 750$ psf. The inundation stress was $\sigma''_i = 500$ psf.

Find:

Ultimate free-field heave for conditions of full wetting after placement of 10 ft of the low expansive fill.

Solution:

The depth of potential heave will be the same as for Example 10.1. The CV swelling pressure of the fill soil was not measured. Therefore, equation (6-3) was used to estimate a value of σ''_{cv} for the fill soil, as in Example 8.2. The value of m for remolded fill was assumed to be 0.6.

Thus, from equation (6-3),

$$\log \sigma''_{cv} = \frac{\log \sigma''_{cs} + m \times \log \sigma''_i}{1 + m} = \frac{\log 750 + 0.6 \times \log 500}{1 + 0.6} = 0.771$$

$$\sigma''_{cv} = 644 \text{ psf}$$

The value for C_H for this value of σ''_{cv} is then

$$C_H = \frac{0.007}{\log \left[\dfrac{644}{500}\right]} = 0.064$$

The computations are similar to those in the previous examples and are presented in Table E10.2. The resulting free-field heave was calculated to be 4.9 in. It is seen that even the small amount of expansion potential in the removed and replaced compacted fill increased the predicted heave over that for nonexpansive fill. This demonstrates the importance of testing the fill when the *overexcavation and replacement* method is used. It is not appropriate to just assume that reconditioning the soil will totally eliminate the potential for heave in that layer.

10.2 PREWETTING METHOD

Prewetting of expansive soils has been used with varying levels of success to reduce the expansion potential. In previous decades, it has been applied fairly commonly, but it is being used less frequently in

TABLE E10.2 Free-Field Heave Computations for Example 10.2

Depth to Bottom of Layer (ft)	Depth Below Excavation (ft)	Interval (ft)	Soil Type	Total Unit Weight (pcf)	Incremental Stress @ midlayer (psf)	Overburden Stress (psf)	Incremental Heave (in.)	Cumulative Heave (in.)	Cumulative Heave from z_p (in.)
0.00	0.00								4.86
3.27	3.27	3.27	Moisture-conditioned fill	129.9	212.3	212.3	1.20	1.20	3.66
6.54	6.54	3.27	Moisture-conditioned fill	129.9	212.3	637.0	0.01	1.21	3.65
9.81	9.81	3.27	Moisture-conditioned fill	129.9	212.3	1,061.7	0.00	1.21	3.65
13.08	13.08	3.27	Claystone	136.2	222.7	1,496.7	1.14	2.35	2.51
16.35	16.35	3.27	Claystone	136.2	222.7	1,942.0	0.87	3.22	1.64
19.62	19.62	3.27	Claystone	136.2	222.7	2,387.4	0.65	3.87	0.99
22.89	22.89	3.27	Claystone	136.2	222.7	2,832.7	0.46	4.33	0.53
26.16	26.16	3.27	Claystone	136.2	222.7	3,278.1	0.31	4.64	0.22
29.43	29.43	3.27	Claystone	136.2	222.7	3,723.4	0.17	4.81	0.05
32.70	32.70	3.27	Claystone	136.2	222.7	4,168.8	0.05	4.86	

recent years. The concept behind this treatment method assumes that increasing the water content in the expansive foundation soils will cause heave to occur prior to construction, thereby reducing post-construction expansion potential. Results of the prewetting method are unreliable and its use is not encouraged. Nevertheless, it warrants some discussion.

There have been some reports of success in using this method if the problem soils have sufficiently high hydraulic conductivity such that water penetration can occur within a reasonable time frame. However, this procedure can have serious drawbacks. The hydraulic conductivity of expansive soils is typically so low that the time required for adequate wetting can be up to several years or decades. In some cases, after the water has been applied for long periods of time, serious loss of soil strength can result causing a reduction in bearing capacity. Furthermore, even after a prolonged period of surface ponding, the wetting front of the infiltrating water may not have advanced to the full depth of the design active zone. Even after surface ponding has been discontinued, redistribution of water can continue throughout the active zone for many years. Continued migration of water into lower more expansive layers can result in continued heave.

Commonly applied procedures for prewetting are to construct dikes or small earth berms to impound water in the flooded area, or to construct trenches below the foundation that can be flooded (Dawson 1959). Alternatively, holes can be drilled around the site to provide access for water. Installation of vertical sand drains installed on a grid pattern can reduce the time needed for water penetration and heave by reducing the length of the flow path (Blight and deWet 1965). The main factors to be considered are the coefficient of permeability, the depth of the expansive clay layer, the area affecting the construction, the radius of the sand drains, and the number of drains installed. Other soil properties, such as clay mineralogy and fabric, may influence the rate of swell. The in situ coefficient of permeability may vary from that calculated from laboratory tests, especially with desiccated, highly fissured soil. Some approaches to expedite the prewetting process are to inject water under pressure or to add a surfactant to the water (Das 2011).

Although the prewetting method can take very long periods of time to accomplish, it reportedly has been successful if sufficient time exists to allow the soil to swell. Teng, Mattox, and Clisby (1972) described a procedure used to reduce the swell potential of a highly fissured Yazoo clay formation in Mississippi. Sand drains 20 ft (6 m) deep were used and ponding lasted 140 days. However, it also must be noted that an adjacent remolded fill section that was also flooded did not exhibit the same positive results as the in situ soil. The lack of success in the fill area can be attributed primarily to the fact that the compacted soil likely had a much lower hydraulic conductivity than did the in situ soil.

Soil wetting was also employed as a remedial measure in clays of the Hawthorne Formation in Gainesville, Florida (Schmertmann 1983). Shallow sand-filled wells approximately 8 ft to 10 ft (2.4 m to 3.1 m) deep were installed around a house foundation. A constant low head of water was maintained in the wells for a period of about four years. After that time, the magnitude of the differential elevation of the foundation had decreased to about half of what had existed when the remediation plan was implemented. At that point, the water contents had stabilized to the extent that cosmetic repairs could be completed and the structure was put back into use.

In some cases, a deflocculant such as household laundry detergent has been added to the flooding water to increase the permeability of the soils. There have been some reports of successful application of this method in the past (Felt 1953; Holtz and Gibbs 1956; Blight and deWet 1965; McDowell 1965; van der Merwe, Hugo, and Steyn 1980; Steinberg 1981).

The frequency of successful applications being documented in the literature can be very misleading. There have been many failures resulting from the loss of soil strength and the creation of unsuitable working conditions due to excessive moisture in the upper soil layers. Unfortunately, only the successes are reported in the literature and not the failures.

10.3   CHEMICAL ADMIXTURES

Chemical admixtures that are available for the treatment of expansive soil may be divided into two groups. These include traditional materials such as hydrated lime, portland cement, and fly ash, and nontraditional chemical agents such as potassium compounds, sulfonated oils, ammonium chloride, and others. The traditional materials rely mainly on calcium exchange and pozzolanic reactions to effect treatment. The nontraditional chemical agents rely on various proprietary and unpublished chemical reactions. Potassium-based compounds reportedly rely on potassium ions to modify the clay lattice of montmorillonite thereby altering the mineral structure to that of illite. The types and applications of the two groups of chemical admixtures are discussed in this section.

10.3.1   Lime Treatment

Lime treatment has been used successfully on many projects to minimize swelling and improve soil plasticity and workability. Many state highway departments have researched lime treatment and frequently use this method. Generally, from 3 to 8 percent by weight of hydrated lime is added to the soil. It is also used as a follow-up treatment over ponded areas to add strength to the surface and to provide a working surface for equipment (McDowell 1965; Teng, Mattox, and Clisby 1973).

The primary reactions in the lime reaction include cation exchange, flocculation-agglomeration, lime carbonation, and pozzolanic reaction. (Thompson 1966 and 1968; Little 1995; NLA 2004). The strength characteristics of a lime-treated soil depend primarily on soil type, lime type, lime percentage, and curing conditions such as time and temperature.

Lime is not an effective treatment for all types of soils. In general, clay soils with a minimum of 25 percent passing the No. 200 sieve and a plasticity index greater than 10 percent are considered to be good candidates for soil treatment (NLA 2004). Caution must be exercised, however, with the use of lime. Some soil components such as sulfates, organics, and phosphates can cause reactions that can have serious adverse effects.

10.3.1.1   Type of Lime
Table 10.1 lists several types of lime used as additives. Care must be taken to assure that industrial lime is used. Agricultural limes will not provide suitable soil treatment results. Quicklime is manufactured by chemically transforming calcium carbonate ($CaCO_3$) into calcium oxide (CaO) by heating. Quicklime will react with water to form hydrated lime. Either quicklime or hydrated lime can be used as an agent for soil treatment. If quicklime is used, the first water that is introduced will be used in the chemical reaction to form hydrated lime, which then reacts with the soil. Caution must be exercised when using quicklime. It can cause serious burns to skin and eyes if personnel come into contact with it. Modern spreading equipment can reduce the potential safety hazards associated with using quicklime.

Most lime used for soil treatment is "high calcium" lime, which contains 5 percent or less magnesium oxide or hydroxide (NLA 2004).

TABLE 10.1   Lime Materials Used in Soil Treatment

Type	Formula
Quicklime	CaO
Hydrated lime	$Ca(OH)_2$
Dolomitic lime	$CaO \bullet MgO$
Normal hydrated or monohydrated dolomitic lime	$Ca(OH)_2 \bullet MgO$
Pressure hydrated or dehydrated dolomitic lime	$Ca(OH)_2 \bullet Mg(OH)_2$

However, sometimes dolomitic lime, which contains 35 to 46 percent magnesium oxide or hydroxide can be used. Dolomitic lime can also perform well when used for soil treatment, but the magnesium fraction of the lime requires more time to react than does calcium. The type of lime that is used can influence the strength of the treated soil. Dolomitic lime generally will be more effective in increasing strength.

10.3.1.2   Soil Factors
Factors influencing the lime reactivity of a soil include the following:

- A soil pH greater than about 7 indicates good reactivity.
- Organic carbon greatly retards lime-soil reactions.
- Poorly drained soils tend to have higher lime reactivity than well-drained soils.
- Calcareous soils have good reactivity.
- The presence of soluble sulfate salts in the soil can react with lime to cause ettringite-induced heave.

The production of strong cementing agents can occur from reactions between lime, water, and aluminous or siliceous substances. The high pH environment created by the addition of lime increases the solubility of silica in the soils. The lime supplies a divalent calcium cation that can form calcium silicates and calcium aluminum hydrates, which can form physical bonds between particles to increase soil strength.

Soils that are highly weathered and are better drained are less reactive than poorly drained soils. One indicator of the degree of weathering is the Ca/Mg ratio. Weathering causes calcium ions to be leached from the system, thereby reducing the Ca/Mg ratio. A low pH also indicates weathering. Well-drained soils might have a high iron content, which might disrupt pozzolanic reactions (Currin, Allen, and Little 1976). In tropical or subtropical regions, the silica-alumina ratio can be used as an index of lime reactivity (Harty 1971).

Lime treatment is not recommended for sandy soils with no fine fraction. Also, it is not very effective with silt-loam soils. Fly ash or other pozzolanic material can be added to most granular soils to improve the gradation and reactivity of the soil. Clay-gravel material has been successfully stabilized for use as pavement base (Winterkorn 1975).

10.3.1.3   Ettringite Formation
The formation of the mineral ettringite can be particularly problematic if lime is introduced into soils containing soluble sulfates. Ettringite results from the reaction of the lime and the sulfates with

alumino-hydroxides that are formed by the dissolution of clay minerals in the high pH environment created by the lime. The aluminum and calcium react with the sulfates to form the mineral ettringite. The ettringite molecule contains 26 molecules of water in a hydrated state, thereby increasing the volume of the treated soil accordingly. This larger volume results in serious heaving of the soils. The authors have observed instances where the introduction of lime into sulfate-bearing subgrade soils has resulted in heave of pavements in amounts of about 2 ft (300 mm). This phenomenon has also been observed by others and is referred to as sulfate-induced heave (Dermatas 1995; Mitchell and Soga 2005), lime-induced heave (Hunter 1988; Perrin 1992), or ettringite-induced heave (Puppala, Intharasombat, and Vempati 2005; Puppala, Talluri, and Chittoori 2014).

The formation of ettringite is temperature dependent. As long as the temperature is higher than 60°F, the ettringite molecule can continue to grow. Below that temperature, ettringite is transformed through a series of intermediate steps to another mineral, thaumasite.

The temperature dependence can delay the formation of ettringite, only to result in post-construction problems. An example of this occurred under a foundation of a building in western Colorado. Lime had been added to treat the fill beneath and around the foundation. The construction took place at lower temperatures, but after the temperature warmed, the formation of ettringite caused heave. Careful testing of the soils prior to the addition of lime is important. The same is true for testing of additives used to treat the soil. In one instance, the soil did not contain a sufficient amount of sulfates to be a problem, but the additive did. Fly ash was added to the soils beneath floor slabs in a shopping center being constructed in Louisiana. The fly ash contained sulfate, which provided the necessary components for significant ettringite-induced heave.

The National Lime Association (NLA 2004) indicates that sulfate concentrations of less than 3,000 ppm are unlikely to cause ettringite-induced heaving problems. Soils with sulfate concentrations of 3,000 to 5,000 ppm can be treated if a sufficient amount of water is used and sufficient time is allowed for the lime and soil to cure after mixing during construction.

If soils with higher concentrations of sulfates are treated with lime, the lime-treated soil should be prewetted to a water content of 3 to 5 percent above the optimum water content and allowed to remain at this water content for a sufficient period of time to allow full formation of the ettringite prior to compaction. This time period could be as much as 7 to 14 days. After that curing period, the soil is generally remixed and then compacted in accordance with the specifications. Once the ettringite is formed, it is relatively stable, but it must be recognized that additional ettringite mineral might continue to form as long as sufficient amounts of reactants are present in the soil.

10.3.1.4 *Testing for Reactivity and Required Lime Content*
Eades and Grim (1966) developed an easily applied test to determine if a soil is lime reactive and how much lime, in percent by weight, is necessary to achieve a desired volume change reduction. The test considers that when a calcium-based compound such as lime is added to clay soil, a reaction occurs that is based on soil-silica and soil-alumina solubility at a high pH. The pH of the lime/soil mixture can be used to identify the optimum lime content. The procedure is simple and can be completed within 1 hour. The results of the test are plotted on a graph showing percent lime versus pH. As the lime content is increased, the PI of the lime-treated soil will decrease. The point where the addition of more lime produces little, or no, decrease in PI is the point of maximum effectiveness. Also, as the lime content is increased, the pH of the soil increases. The higher pH is also associated with a lower PI. In the Eades and Grim test the lowest percent lime content to produce a pH of 12.4 is termed the *lime modification optimum* (or LMO) and is the approximate lime content to be used for treating the soil. A drawback of the Eades and Grim method is that a pH of 12.4 does not always ensure lime-soil reactivity (Currin, Allen, and Little 1976). Other factors must be considered to assess the effectiveness of lime in reducing the expansion potential. ASTM D6276 and ASTM C977 provide more detail on the determination of the LMO.

10.3.1.5 *Curing Conditions*
Higher temperature and longer curing time improve the gain in strength for lime-treated soils. They have some influence on expansion potential as well. It is advisable to schedule construction in order to obtain maximum benefit of summer temperatures before the onset of cold weather. If the soil temperature is less than 60° to 70°F (16° to 21°C) and is not expected to increase for 1 month, then the chemical reactions will be deterred and the benefits will be minimal (Currin, Allen, and Little 1976). Alternate methods of treatment should be used if cold environmental conditions are expected. The curing period should extend for at least 10 to 14 days before heavy vehicles are allowed on the lime-treated soil.

10.3.1.6 *Application Methods*

Perhaps the most commonly used method of lime application is to mix it mechanically. With this method it is difficult to mix deeper than about 12 in. (300 mm), and therefore it is used mainly for treatment of subgrades below pavements or slabs. For deeper depths of treatment the soil needs to be treated in several lifts. Lime can be applied dry, which is fastest, or in a lime slurry. If quicklime is used it must have water added for a proper reaction to occur. It is important that specifications be established and procedures be followed to ensure that the reaction requirements are met. Quality control of lime content, pulverization, mixing, and compaction must be maintained. After the lime is mixed into the desired depth of the subgrade, it should not be compacted until the size of the clumps, or clods, of the treated soil is small enough to allow sufficient mixing and adequate compaction. The top layer can then be sealed by rolling and compacting to prevent fluffing due to carbonation reactions or precipitation. After 3 or 4 days, the material can be mixed for the last time and then compacted to the specified moisture and density values. If the surface is sealed with an asphalt emulsion, the water content will stabilize in the lime-treated material over a period of time and pozzolanic reactions can take place.

Lime can also be injected in slurry form through holes drilled into the soil. This method has been used both for remedial measures and for preconstruction purposes. Results of the drill-hole technique are erratic, and its use is not encouraged. One factor that limits the effectiveness of the method is the inability to uniformly distribute the lime in the soil mass (Thompson and Robnett 1976). The diffusion of the slurry occurs very slowly unless the soil has an extensive network of fissures. Differences in various factors such as soil type and texture or quality control during construction undoubtedly account for the disparity in results using the drill-hole method. In general, one must expect a lower degree of confidence for success of the drill-hole technique than for other techniques. The same limitations that apply for prewetting also apply to this method. A higher level of success can be expected if the lime slurry can be injected under pressure.

Even with the high-injection pressures, field tests have indicated that penetration of injected lime slurry occurs only along planes of weaknesses or fissures and the slurry will not penetrate soil pores (Higgins 1965; Ingles and Neil 1970; Lundy and Greenfield 1968; Wright 1973). The procedure is most effective at times of maximum desiccation of the soil mass. If the soils are highly fissured, and are lime reactive, the injection of the lime can seal off zones of clay between the fissures and produce a stabilizing effect on the moisture content between fissures.

10.3.2 Cement Treatment

Portland cement has been found to be effective in treating a wide variety of soils, including granular soils, silts, and clays. The result of mixing cement with clay soil is similar to that of lime. It reduces the liquid limit, the plasticity index, and the potential for volume change. It increases the shrinkage limit and shear strength and improves resilient modulus (Chen 1988; Petry and Little 2002). However, it is not as effective as lime in treating highly plastic clays. Some clay soils have such a high affinity for water that the cement may not hydrate sufficiently to produce a complete pozzolanic reaction. When cement is used, it is generally because the soils are not lime reactive (Mitchell and Raad 1973).

For clays in which portland cement treatment is effective, mixing procedures similar to those used for lime treatment can be used. One difference in technique is that the time between cement addition and final mixing should be shorter than that used for lime treatment. Portland cement has a shorter hydration and setting time. Because of the strength increase that can be generated by the use of cement, the soil-cement mixture can increase pavement and slab strength significantly.

Generally, the amount of cement required to treat expansive soils ranges from 2 to 6 percent by weight (Chen 1988). A cement content of 2 to 6 percent can produce a soil that acts as a semirigid slab. This will aid in reducing differential heave throughout the slab. Little et al. (2000) indicated that cement treated materials may be prone to cracking as a result of hydration and moisture loss. Shrinkage cracks can compromise performance if they become wide and admit significant water. However, if proper construction procedures are followed, the effects of shrinkage cracks can be minimized. Techniques that have been used to minimize cracking problems include the following (Petry and Little 2002).

- Compaction of the cement-treated soil at a water content slightly drier than optimum water content.
- Precracking through inducement of weakened planes or early load applications.
- Delayed placement of surface hot mix, reduced cement content, and use of interlayers to absorb crack energy and prevent further propagation.

Mixing methods, and hence, processing costs are nearly the same for lime and cement. Overall treatment costs may be similar for both (PCA 1970).

10.3.3 Fly Ash Treatment

Fly ash has been used as an additive in the treatment of soils with lime. Its main purpose is to increase the pozzolanic reaction and improve the gradation of granular soils. Fly ash is a fine-grained residue that results from the combustion of pulverized coal in power plant boilers, and which is transported from the combustion chamber by exhaust gases (NLA 2004). Fly ash consists of spherical noncrystalline silica, aluminum, and iron oxides and unoxidized carbon. It is classified as nonplastic fine silt in accordance with the Unified Soil Classification System.

There are two major classes of fly ash: (1) Class F fly ash, which is produced from burning anthracite or bituminous coal, and (2) Class C fly ash, which is produced from burning lignite and subbituminous coal. Both classes of fly ash are pozzolans (Cokca 2001). However, Class C ash is cementitious and often contains higher levels of sulfate than Class F.

The use of fly ash reduces the plasticity index, permeability, and expansion potential of the soil. It increases the stiffness, strength, and freeze-thaw resistance. There are a wide variety of types of fly ash having different mechanical and chemical properties. Therefore, for a specific application, a comprehensive testing program should be performed to determine the design criteria necessary for optimum treatment. An understanding of the influence of both compaction and water control of the treated material is essential to achieving the optimum benefit from fly ash addition (Little et al. 2000).

Cokca (2001) showed that the swelling potential of expansive soils was reduced by approximately 68 percent when Class C fly ash was added. Puppala, Punthutaecha, and Vanapalli (2006) showed that treatment with Class F fly ash reduced the swelling pressures of expansive soils by as much as 65 percent.

10.3.4 Chemical Injection

Various chemical solutions have been used in injection processes to reduce the expansion potential of soils. The reaction of the chemical solutions with the soil depends on the nature of the cation in the dissolved chemicals. These reactions typically are more complex than just hydration alone (Pengelly et al. 1997)

Sodium chloride and calcium chloride have been used in soil treatment. The effect of sodium chloride on soil properties is variable. It generally has a greater effect in soils having a high liquid limit. Depending on the soil type, sodium chloride may increase the shrinkage limit and shear strength. For soils that react with sodium chloride, there may also be some beneficial control of frost heave.

Calcium chloride has the effect of stabilizing water content changes in soils, thereby reducing the potential for volume change. It has long been used to control frost heave in soils. Disadvantages of using calcium chloride are that it is easily leached from the soil, and the relative humidity must be at least 30 percent before it can be used.

Potassium chloride has also been used for treatment of expansive soils for many years. Potassium-based chemical injection works by saturating the soil mass with potassium ions, which change the mineralogical characteristics of the soil. As noted in chapter 2, the potassium ions have a size that is compatible with the spacing of the silicon atoms in the montmorillonite building blocks. The potassium forms a chemical bond between the crystal lattice, thereby modifying the montmorillonite to illite and reducing the expansion potential of the soil significantly. Laboratory and field studies by Addison and Petry (1998) and Pengelly and Addison (2001) have shown that the expansion potential of expansive soils treated with potassium ions is reduced significantly. However, the treatment of soils by injection methods is subject to the same limitations noted previously for prewetting. Injection of chemicals into most clay soils is difficult to achieve because of the relatively impervious nature of most expansive soils. Where it has been possible to directly mix the soil with the potassium chloride, the expansion potential can be reduced almost entirely. However, controlling the injection of chemical to achieve an even and complete distribution in an in situ soil deposit is questionable.

10.4 MOISTURE CONTROL ALTERNATIVES

Because the root cause of soil expansion is an increase in water content of the foundation soils, it would appear that in order to eliminate problems of heave one would only have to control the water content of the soil. However, it is virtually impossible to prevent an increase in water content of the foundation soils after site development. Nevertheless, it is possible to exercise some control over the rate of increase and the magnitude of seasonal fluctuations.

Research at Colorado State University by Porter (1977), Nelson and Edgar (1978), and Goode (1982) has shown that even if a distinct water table does not exist in the immediate depth below the zone of seasonal fluctuation, an increase in water content in the upper 10 ft to 20 ft (3 m to 6 m) of soil can occur when evapotranspiration is eliminated. In view of the futility of eliminating the increase in water content within foundation soils, moisture control can only be expected to stabilize water contents over time and minimize fluctuations.

Water content fluctuations beneath slabs can be reduced by means of horizontal barriers. Horizontal barriers consist of impermeable surfaces that extend outward for a considerable distance around the edges of the foundation or floor slab. Examples of large horizontal barriers are parking lot slabs or pavements constructed around commercial buildings such as convenience stores. Although these aprons do not totally eliminate an increase in water content, they do provide for development of a more uniform heave pattern under the building. The large areal extent of these barriers reduces the edge effects to a considerable distance away from the structure.

Vertical barriers have also been used to stabilize water contents under slabs or foundations. Goode (1982) placed vertical moisture barriers to a depth of 8 ft (2.5 m) around simulated floor slabs placed on natural ground. Measurements were made of the surface heave and water content profiles over a period of several years. The results indicated that, as for horizontal barriers, the water content of the soil under the slab still increased and the total heave was not decreased appreciably. However, the heave was significantly more uniform for the slabs with vertical barriers than for slabs without vertical barriers. Also, seasonal fluctuations were less. This is due primarily to the reduction in edge effects by the barriers. Moisture barriers are discussed in more detail in the next section.

Perimeter drains are also considered effective in moisture control around foundations. The primary function of perimeter drains is to avoid the presence of free water in the soil. Generally, for expansive clays, soil suction is high and permeability is very low. Consequently, perimeter drains are ineffective in eliminating heave below foundations. In fact, the presence of the drain may provide for a source of water if long-term positive drainage is not ensured. On the other hand, if the foundation soils have been overexcavated and replaced, a subsurface drain must be placed at the bottom of the overexcavation zone to reduce the potential for development of a perched water zone. Subsurface drains are discussed in Section 10.4.2.

Appropriate surface grading and drainage is another important factor to control moisture. Proper surface grading and drainage and appropriate landscaping practices are discussed in a following section.

10.4.1 Moisture Barriers

Site development in general will reduce evapotranspiration from the ground surface and increase the potential for infiltration. These effects increase the water content of the foundation soils and affect the distribution of water in the soils. Moisture barriers are an attempt to control changes in water content over time and make the water content distribution more uniform. For several reasons, however, moisture barriers cannot eliminate total heave. These reasons include the following:

- An increase in water content of the foundation soils will still occur because of water migration caused by changes in soil temperature. Chen (1988) noted that even in the case of a perfect impervious barrier, water migration due to thermal transfer will still introduce additional moisture into the foundation soils. Seasonal effects due to temperature change have been observed beneath simulated slabs at the Colorado State University Expansive Soils Field Test Site (Durkee 2000).
- The water content of foundation soils underneath moisture barriers will increase due to capillary rise from underlying soils. This will occur in arid climates even if a groundwater table, or phreatic surface, is not present. Seasonal effects in areas with arid climate conditions will generally result in a moisture deficiency in the upper soils. When the evapotranspiration from the surface is eliminated, water will migrate into the upper soils just due to suction gradients until hydrostatic conditions are reached. This effect has been discussed in chapter 7. However, if water contents are greater than equilibrium conditions at the time of construction, the migration of water may cause a decrease in water content, resulting in shrinkage.
- Water movement can occur through cracks, fissures, slickensides, or permeable layers below the barriers.

The primary effect of moisture barriers is to extend edge effects away from the foundation and minimize seasonal fluctuations of water content directly below the structure. The time period over which moisture changes occur is longer than that without barriers because the barrier increases the path length for water migration under the structure. This allows for water content to be more uniformly distributed due to suction gradients in the soil, causing heave to occur more slowly and in a more uniform fashion.

Various materials have been used to act as moisture barriers. Perhaps the most commonly used materials consist of geomembranes. Steinberg (1998) presented a comprehensive discussion of the use of geomembranes based on many years of experience with a number of actual applications. Moisture barriers can be placed as horizontal or vertical barriers. They are discussed in the following sections.

10.4.1.1   Horizontal Moisture Barriers

Horizontal barriers can be of a flexible nature or a rigid slab. Flexible horizontal barriers generally use an impermeable membrane consisting of a material such as polyethylene, polyvinyl chloride (PVC), polypropylene, high-density polypropylene, and other types of geomembranes. Information on longevity and long-term chemical stability should be obtained from the manufacturer prior to installation.

Membranes with thicknesses below 20 mil require special care to avoid puncture during placement. Care must be taken to prepare the surface prior to placement of the membrane. Vegetation, organic material, and all sharp, projecting materials must be removed. The surface should be even and uniformly compacted. A final rolling with a steel drum roller will aid in eliminating sharp projecting particles.

Most membranes (e.g., polyethylene sheeting) will degrade if exposed to sunlight. Therefore, barriers should be covered to protect them against ultraviolet radiation unless UV-resistant membranes are used. Membranes need to be protected from damage by chemical attack as well. Chemical compatibility of the membrane with the soil must be checked. Care should be taken when dumping and spreading soil cover to avoid damage from equipment or the dumped material. When membranes are used around the perimeter of a building, they should be protected from the environment by a layer of earth 6 in. to 12 in. (150 mm to 300 mm) thick (Johnson 1979; US Department of the Army 1983). They also should be placed deep enough to prevent damage from root growth.

The edges of the membrane should be secured so that leakage does not occur next to the foundation or at joints where the membrane overlaps. Batten boards, or waterproof glue or mastic, are recommended to attach the membrane to the foundation. There may be a tendency for moisture to accumulate at the outer edge of the membrane. Moisture migration into the foundation soils can be reduced by providing a subdrain at the membrane's edge, which can discharge water. The drains should be constructed with ample slope to avoid any chance of them backing up water. Surface water coming off the roof of a structure should be directed well away from the edge of the horizontal barrier.

Concrete aprons or sidewalks around building foundations form rigid moisture barriers. A flexible membrane or asphalt can be coupled with the concrete. Several reports have been published on the use of such barriers to establish constant uniform moisture contents (Mohan and Rao 1965; Lee and Kocherhans 1973; Najder and Werno 1973).

The construction of joints and seals is important when using a rigid horizontal barrier. Heaving can occur at the edge of the apron, and if it tilts toward the structure, surface water will be directed to the edge of the foundation and into foundation soils. A flexible sealed joint should be provided between the structure and the barrier. Sidewalks and aprons should have adequate slope away from the foundation so that even with some distortion flow will occur outward from the structure. Horizontal barriers will be more effective if surface drainage is provided to prevent ponding.

10.4.1.2 *Vertical Moisture Barriers*

Vertical moisture barriers function in much the same way as horizontal barriers in terms of slowing the rate of heave and causing the water content distribution to be more uniform below the structure. Vertical barriers are more effective than horizontal barriers in minimizing lateral moisture migration and maintaining long-term uniform water content distribution (Chen 1988 and 2000). They have been used effectively in the remediation of highway pavements as well as in new construction (Steinberg 1998).

A vertical moisture barrier usually consists of an excavated trench lined with any impermeable membrane such as polyethylene, concrete, asphalt, or impervious semihardening slurries. Membranes should be durable enough to resist puncturing or tearing during placement. Figure 10.1 shows one configuration of a vertical moisture barrier.

Equipment used to excavate trenches for construction of vertical moisture barriers is advancing rapidly in capability. Trenching machines have been observed to be more efficient for excavating than backhoes.

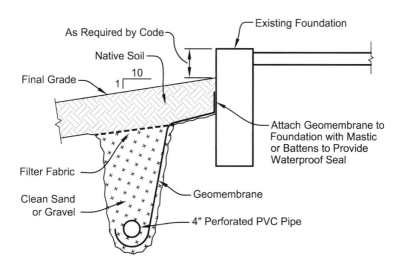

FIGURE 10.1. Configuration of a vertical moisture barrier.

Various materials have been used to backfill the vertical barrier trenches. Engineers in Australia have developed a flowable backfill material consisting of cement, fly ash, local sand, and water that can be used together with a geomembrane to create a deep vertical moisture barrier. The backfill material should be as impermeable as possible, such as grout or clay. The use of sand backfill in the trenches may provide a reservoir that eventually can leak through the membrane (Newland 1965). On the other hand, if saturated conditions are unlikely to exist, a granular material can act as a *capillary break*. The effectiveness of the barrier will be improved if adequate surface drainage away from the structure is provided.

The depth of the vertical barrier should be determined based on an assessment of the zone of water migration (Steinberg 1998). Chen (1988 and 2000) and Nelson and Miller (1992) indicated that vertical barriers should be installed at least as deep as the zone that is most affected by seasonal moisture change. It is generally not practical to place them to the full depth of the zone of seasonal moisture change, but a depth of one-half to two-thirds of that depth is recommended. Barriers less than 2 ft to 3 ft (0.6 m to 1 m) deep in general will not be sufficiently effective to warrant their placement (Snethen 1979). On highway projects, there are substantial reductions in maintenance costs where vertical membranes have been placed to a depth of 8 ft (2.4 m) below the ground surface (Steinberg 1998). If the barriers cannot be installed at the time of construction, they generally must be placed a distance of at least 3 ft (1 m) away from the structure to allow space for the excavating equipment to maneuver and to avoid disturbance of the foundation soil (US Department of the Army 1983). In that case, horizontal barriers must be installed to connect the vertical barrier to the outer edge of the structure.

Goode (1982) constructed field test plots to evaluate vertical moisture barriers. They were installed at the Colorado State University Expansive Soils Field Test Site and consisted of impermeable membranes to simulate a slab-on-grade. Four test plots were constructed, two of which had vertical barriers and two of which did not.

The soils consisted of highly expansive clay of the Pierre Shale formation. Previous investigations at the site indicated that the depth of the zone in which seasonal effects occurred extended to a depth of between 12 ft and 17 ft (4 m and 5 m) (Nelson and Edgar 1978).

The test results are described in detail in Goode (1982). In general, they indicated that for the first year or two, the heave under the slabs with vertical barriers was much less than that under the slabs with no barriers. After 4 to 5 years the total heave in both the slabs with and without vertical barriers was nearly the same. However, the differential heave observed under the slabs without the vertical barriers was greater than that observed in the slabs with vertical barriers (Hamberg 1985). The vertical barriers were effective in slowing the rate of heave and decreasing the amount of differential heave even though the eventual total heave was not affected.

10.4.2   Subsurface Drains

The placement of perimeter drains around the foundations is a common practice to avoid the presence of free water in the soil. Subsurface drains placed within the backfill zone, in the recompacted zone of overexcavated soil, or in relatively higher permeable material above claystone bedrock will aid in avoiding development of perched free water. However, perimeter drains are not very effective in totally avoiding heave below foundations.

A perimeter drain system generally consists of a trench with a perforated pipe, coarse clean gravel, and a geotextile drainage fabric. The perimeter drain can be placed either inside or outside of the foundation wall, as shown in Figure 10.2. When properly constructed, an exterior drain is more effective than an interior drain. An interior drain will not intercept seepage infiltrating through the backfill zone. This reduces its effectiveness in keeping free water away from the foundation. For remediation repair plans, the decision as to whether to use an exterior or an interior drain will depend on the ease of construction in conjunction with the overall plan.

It must be kept in mind that an improperly constructed drain may serve an opposite effect and actually bring water into the foundation soils. Thus, proper design and construction of the drain is of paramount importance. The highest point of the perimeter drain pipe should be placed several inches below the bottom of the foundation wall and/or the level of the floor slab. The perimeter drain should preferably have a slope of $1/4$ in. per ft (2 percent slope). A rigid perforated pipe should be used in the drain to provide a constant slope and avoid low spots from sagging pipe.

The perimeter drain must discharge into a sump, an area drain, or a suitable gravity outlet. If a sump is the discharge point for the drain, it must be equipped with a pump to pump out the drain water flowing into the sump. The authors' experience from inspection of many perimeter drain systems has shown that builders do not always install a pump. Water pumped from the sump must be discharged well away from the backfill zone and the foundation. Preferably, it should discharge into an

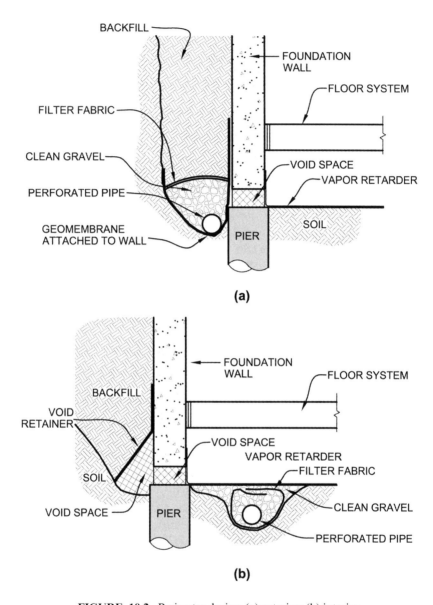

FIGURE 10.2. Perimeter drains: (a) exterior; (b) interior.

area-wide drainage system. If water is discharged near the foundation it will infiltrate into the more permeable backfill zone and be recycled in the perimeter drain.

Chen (2000) emphasized that when designing a subsurface drain, the possibility of the development of a perched water condition at a later date must be considered. This is especially true for a shallow foundation constructed on overexcavated and recompacted expansive soils.

10.4.3 Surface Grading and Drainage

Grading throughout the site and, in particular, around the perimeter of the building is one of the most important factors to minimize infiltration of surface water into the backfill and foundation soils (Houston et al. 2011; Nelson et al. 2011). Adequate slope of the ground surface is essential. In the western United States, a slope of 10 percent away from the structure for the first 10 ft (3 m) out from the foundation is generally recommended. A slope of 2 percent should be maintained beyond that distance. The *International Building Code* (2012) and the *International Residential Code* (2012) specify the surface grading be 5 percent or more whereas the *Uniform Building Code* (1997) requires it to be at least 8.3 percent. The *International Building Code* (2012) requires that impervious surfaces within 10 ft (3 m) of the building foundation be sloped a minimum of 2 percent away from the building. The table included in Figure 10.3 illustrates the application of those requirements. It is important to note that this slope is perpendicular to the foundation, not parallel to it. Slopes that are parallel to the foundation simply distribute water along the foundation instead of removing it from the backfill zone.

Site drainage is also influenced by pavements and roofs. The roof drainage system comprises gutters, downspouts, and splash blocks and serves to collect water from precipitation to carry it beyond the backfill zone. The downspouts should extend at least 5 ft (1.5 m) and preferably 10 ft (3 m) beyond the perimeter of the foundation. They should discharge to an area where the surface drainage is adequate to carry away the water. A swale or splash block should be provided at the end of the downspout extension to convey water away from the house.

In addition to influencing the site grading, landscaping practices may have a significant influence on the wetting of the foundation soils. Xeriscape landscaping should be used within the first 5 ft to 10 ft (1.5 m to 3 m) adjacent to buildings to eliminate the need for supplemental water from irrigation. Lawn sprinkler systems should not spray water any closer than 5 ft (1.5 m) from the foundation. Figure 10.3 shows the implementation of good drainage and landscaping practices.

Landscaping practices must be careful to maintain the slope of the ground surface. Settlement of the backfill zone can change the nature of the slope and can even result in a negative grade (i.e., a slope toward the foundation) or ponding of water.

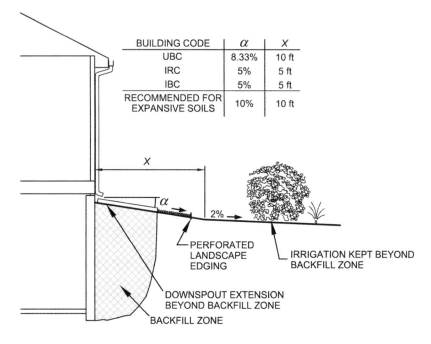

FIGURE 10.3. Landscaping and downspout locations.

The effect of ground surface slope and infiltration is difficult to model analytically or numerically. If the ground surface is modeled merely as a flat surface, the effect of slope is minimal. However, it has long been known in agrisciences and irrigation engineering that water ponding in small irregularities in the ground surface, a phenomenon known as *micro-ponding,* has a large influence on infiltration. These irregularities provide for storage of surface water and reduce runoff. Virtually all of the water stored in those irregularities will seep into the soil between precipitation and irrigation events. Earlier analyses that did not consider micro-ponding have erroneously come to the conclusion that slopes as flat as 5 percent were adequate to limit infiltration. Nelson et al. (2011) showed that when micro-ponding is considered, there is a major influence on infiltration.

The effect of micro-ponding can be demonstrated by means of a simple model as shown in Figure 10.4. In this model, surface irregularities are modeled as a rectangular indentation in the surface with depth, A, and length, L. For purposes of this analysis, it was assumed that after an irrigation or precipitation event, sufficient water will be introduced such that the irregularity will fill with water, as shown by the shaded areas in Figure 10.4a. For a unit width of slope, the volume of water

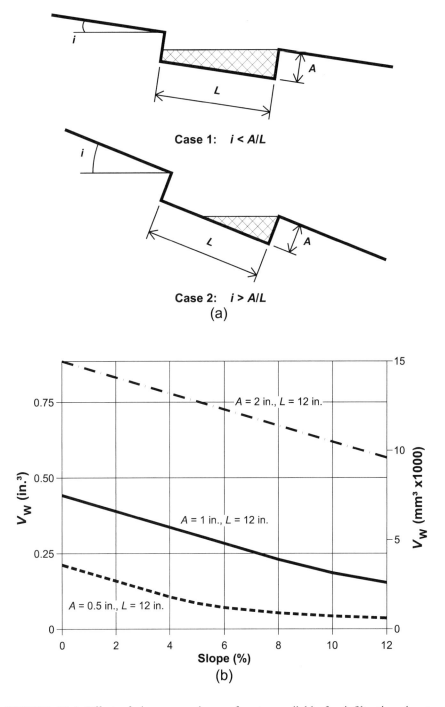

FIGURE 10.4. Effect of slope on volume of water available for infiltration due to micro-ponding: (a) simplified surface irregularity; (b) volume of water, V_w, ponded in surface irregularity.

collected in the irregularity for Cases (1) and (2) would be calculated by equation (10-1).

$$V_a = L\left(A - \frac{iL}{2}\right), \ i < \frac{A}{L} \qquad \text{Case (1)} \qquad \text{(10-1a)}$$

$$V_b = \frac{AL}{2i}, \ i > \frac{A}{L} \qquad \text{Case (2)} \qquad \text{(10-1b)}$$

where:

V_a = the volume of water collected in the irregularity for Case (1),

V_b = the volume of water collected in the irregularity for Case (2),

i = the slope of the surface as shown in Figure 10.4a,

A = the depth of the irregularity, and

L = the length of the irregularity.

At the end of an irrigation or precipitation event, the total volume of water collected in the irregularity will infiltrate into the ground. Thus, the amount of infiltration can be represented by the amount of water ponded in the irregularity. Whether the water in the irregularity will extend across the entire length, as shown for Case (1) in Figure 10.4a, or only extend part way, as shown for Case (2), would depend on the ratio A/L and the slope, i.

The actual total volume of water that the micro-ponding makes available to infiltrate into the ground depends on the actual size of the irregularity. As shown in Figure 10.4b, for a 2-in. (50-mm) -deep irregularity, the amount of micro-ponding is about 20 percent greater at a 5 percent slope than at a 10 percent slope. However, for the shallow irregularity, there is an increase of more than 100 percent at the 5 percent slope than at the 10 percent slope.

This simplified analysis demonstrates that as a result of micro-ponding, the slope of the general ground surface will have a profound effect on the infiltration of surface water into the ground.

To evaluate the effects of actual surface irregularities on the potential surface storage and infiltration, measurements were taken on four actual cross-sections at locations adjacent to two different houses. The cross-sections were surveyed in detail in order to detect small irregularities of the ground surface. Figure 10.5 shows cross-sections of several areas that were surveyed. The surface geometry is shown to be very irregular and areas where micro-ponding can occur are evident even for slopes greater than 10 percent. The amount of potential storage

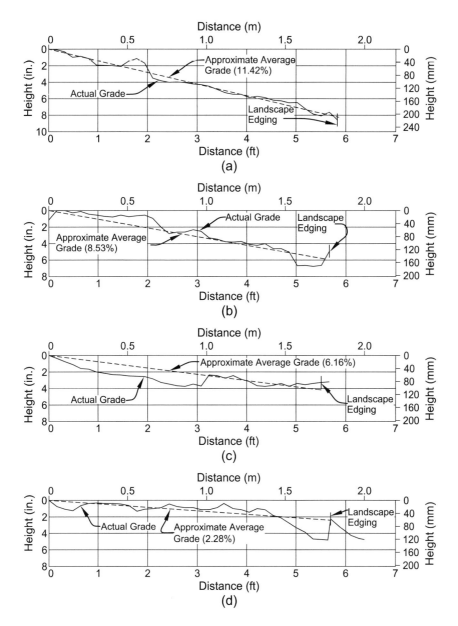

FIGURE 10.5. Measurements of surface geometry on four cross-sections: (a) south; (b) east; (c) north; (d) CB (Nelson et al. 2011).

TABLE 10.2   Storage Volume of Cross-Sections at Existing Slope (Nelson et al. 2011)

Cross-Section	Existing Slope (%)	Storage Volume (in.³)	
		Without Edging	With Edging
South	11.4	3.6	3.6
East	8.5	2.1	7.7
North	6.2	7.5	8.4
CB	2.3	11.1	22.3

Notes: South cross-section has no ponding adjacent to landscape edging.
Multiply cubic inches by 1.64×10^4 to convert to cubic mm.

TABLE 10.3   Storage Volume of Cross-Sections with Various Slopes (Nelson et al. 2011)

	Storage Volume (in.³)							
	Slope = 0%		2%		5%		10%	
Cross-Section	Without Edging	With Edging	Without Edging	With Edging	Without Edging	With Edging	Without Edging	With Edging
South	22.6	22.6	15.1	15.1	8.7	8.7	4.2	4.2
East	13.2	26.4	9.7	18.6	5.0	11.6	1.6	7.0
North	28.4	41.1	18.8	23.1	9.3	10.7	4.9	5.1
CB	30.1	43.1	13.0	24.3	6.0	15.6	1.8	9.4

Notes: South cross-section has no ponding adjacent to landscape edging.
Multiply cubic inches by 1.64×10^4 to convert to cubic mm.

for each cross-section is compounded by the installation of nonperforated landscape edging where the ground cover transitions from gravel to grass. The storage volume was calculated for a unit width of slope for each of the four cross-sections at their existing slopes with and without landscape edging. The volumes calculated are shown in Table 10.2.

The computed micro-ponding volume that would exist if the overall slopes shown in Figure 10.5 were different was also calculated. Table 10.3 presents the computed storage volume for the four cross-sections at varying slopes for conditions both with and without the landscape edging being present. For most cases, both with and without landscape edging, a decrease in slope from 10 percent to 5 percent results in an increase in potential infiltration of 100 percent or more. Decreasing the slope to 2 percent increases the volume of micro-ponding by as much as a factor of four.

It is evident from this case that simply maintaining proper drainage around a structure has a significant effect on the amount of water available for infiltration.

10.5   SUMMARY OF SOIL TREATMENT METHODS

The expansive soil treatment alternatives presented here may be used to control heave, either singly or in combination. The decision of whether to use some form of soil treatment to control heave, and the choice of the technique to be used, should be assessed with respect to the following:

- Economic factors
- Relative expected control of volume changes by implementing different treatment alternatives
- Site-specific conditions such as potential for volume change, moisture variations, degree of fissuring, and permeability
- Nature of the project
- Necessary strength of the foundation soils
- Tolerable movement of the foundation
- Time frame available for treatment

Geotechnical site investigations and testing programs are important in making a suitable selection of a treatment method. It is important that test conditions should duplicate field conditions. Some factors of particular interest include the following:

- Expansion potential for the site
- Design active zone
- Degree of soil fracturing
- Heterogeneity or uniformity of soils on-site
- Chemical reactivity of the soil
- Presence of undesirable chemical compounds
- Heterogeneity of water content and hydraulic conductivity of the soil
- Required strength of the soil

The decision to use soil treatment methods and the implementation of moisture barriers must be weighed against the advantages to be gained over other foundation design alternatives. The relative costs will be an important factor, but the reliability of the different methods and associated risk must also be considered. If soil treatment is used, the selection of which method to use remains a matter of applying judgment to comparisons of pros and cons for the selected alternatives, with special reference to the various factors involved.

References

Addison, M. B., and T. M. Petry. 1998. "Optimizing Multi-Agent, Multi-Injected Swell Modifier." Transportation Research Record No. 1611, Transportation Research Board, National Research Council, 38–45.

ASTM D6276-99. 1999. "Standard Test Method for Using pH to Estimate the Soil-Lime Proportion Requirement for Soil Stabilization." West Conshohocken, PA: ASTM International.

ASTM C977-10. 2010. "Standard Specification for Quicklime and Hydrated Lime for Soil Stabilization." West Conshohocken, PA: ASTM International.

Blight, G. E., and J. A. DeWet. 1965. "The Acceleration of Heave by Flooding. Moisture Equilibria and Moisture Changes Beneath Covered Areas." *Moisture Equilibria and Moisture Changes in Soils beneath Covered Areas, a Symposium in Print*, edited by G. D. Aitchison, Butterworths, Australia, 89–92.

Chen, F. H. 1988. *Foundations on Expansive Soils*. New York: Elsevier Science.

———. 2000. *Soil Engineering: Testing, Design, and Remediation*. Boca Raton, FL: CRC Press.

Cokca, E. 2001. "Use of Class C Fly Ashes for the Stabilization of an Expansive Soil." *Journal of Geotechnical and Geoenvironmental Engineering, ASCE* 127(7): 568–573.

Colorado Department of Transportation (DOT). 2014. *2014 Pavement Design Manual*.

Currin, D. D., J. J. Allen, and D. N. Little. 1976. "Validation of Soil Stabilization Index System with Manual Development." US Air Force Technical Report SRL-TR-76-0006.

Das, B. M. 2011. *Geotechnical Engineering Handbook*. Fort Lauderdale, FL: J. Ross Publishing.

Dawson, R. F. 1959. "Modern Practices Used in the Design of Foundations for Structures on Expansive Soils." *Colorado School of Mines Quarterly* 54(4): 66–87.

Dermatas, D. 1995. "Ettringite-induced Swelling in Soils: State-of-the-Art." *Applied Mechanics Reviews* 48(10): 659–673.

Durkee, D. B. 2000. *Active Zone Depth and Edge Moisture Variation Distance in Expansive Soils*. PhD dissertation, Colorado State University, Fort Collins, CO.

Eades, J. L., and P. E. Grim. 1966. "Quick Test to Determine Lime Requirements for Lime Stabilization." *Highway Research Record* 139, 61–72.

Felt, E. J. 1953. "Influence of Vegetation on Soil Moisture Contents and Resulting Soil Volume Changes." *Proceedings of the 3rd International Conference on Soil Mechanics and Foundation Engineering 1*, 24–27.

Goode, J. C. 1982. "Heave Prediction and Moisture Migration beneath Slabs on Expansive Soils." Master's thesis, Colorado State University, Fort Collins, CO.

Hamberg, D. J. 1985. "A Simplified Method for Predicting Heave in Expansive Soils." Master's thesis, Colorado State University, Fort Collins, CO.

Harty, J. R. 1971. "Factors Influencing the Lime Reactivity of Tropically and Subtropically Weathered Soils." PhD dissertation, University of Illinois, Urbana, IL.

Higgins, C. M. 1965. "High Pressure Lime Injection." Louisiana Department of Highways, Research Report 17, Interim Report 2.

Holtz, W. G., and H. J. Gibbs. 1956. "Engineering Properties of Expansive Clays." *Transactions ASCE* 121, 641–677.

Houston, S. L., H. B. Dye, C. E. Zapata, K. D. Walsh, and W. N. Houston. 2011. "Study of Expansive Soils and Residential Foundations on Expansive Soils in Arizona." *Journal of Performance of Constructed Facilities, ASCE* 25, 31–44.

Hunter, D. 1988. "Lime-Induced Heave in Sulfate-Bearing Clay Soils." *Journal of Geotechnical Engineering* 114(2): 150–167.

Ingles, O. G., and R. C. Neil. 1970. "Lime Grout Penetration and Associated Moisture Movements in Soil." CSIRO, Australia, Division of Applied Geomechanics, Research Paper 138.

International Building Code (IBC). 2012. *International Building Code*. Falls Church, VA: International Code Council.

International Residential Code (IRC). 2012. *International Residential Code for One-and Two-Family Dwellings*. Falls Church, VA: International Code Council.

Johnson, L. D. 1979. "Overview for Design of Foundations on Expansive Soils." US Army Engineers Waterways Experiment Station, Vicksburg, MS, Misc. Paper Gl-79-21.

Lee, L. J., and J. G. Kocherhans. 1973. "Soil Stabilization by Use of Moisture Barriers." *Proceedings of the 3rd International Conference on Expansive Soils*, Haifa, Israel, 1, 295–300.

Little, D. N. 1995. *Handbook for Stabilization of Pavement Subgrades and Base Courses with Lime*, Lime Association of Texas.

Little, D. N., E. H. Males, J. R. Prusinski, and B. Stewart. 2000. "Cementitious Stabilization." 79th Millennium Report Series, Transportation Research Board, Washington, DC.

Lundy, H. L., Jr., and B. J. Greenfield. 1968. "Evaluation of Deep in Situ Soil Stabilization by High Pressure Lime Slurry Injection." *Highway Research Record* 235, 27–35.

McDowell, C. 1965. "Remedial Procedures Used in the Reduction of Detrimental Effects of Swelling Soils." *Engineering Effects of Moisture Changes in Soils, Proceedings of the International Research and*

Engineering Conference on Expansive Clay Soils. College Station: Texas A&M University Press, 239–275.

Mitchell, J. K., and L. Raad. 1973. "Control of Volume Changes in Expansive Earth Materials." *Proceedings of Workshop on Expansive Clays and Shales in Highway Design and Construction*, Vol. 2, edited by D. R. Lamb and S. J. Hanna. Denver, CO: Federal Highway Administration, 200–219.

Mitchell, J. K., and K. Soga. 2005. *Fundamentals of Soil Behavior* (3rd ed.). Hoboken, NJ: John Wiley and Sons.

Mohan, D., and B. G. Rao. 1965. "Moisture Variation and Performance of Foundations in Black Cotton Soils in India." *Proceedings of Moisture Equilibria and Moisture Changes in Soils beneath Covered Areas.* Symposium, Soil Mechanics Section, Commonwealth Scientific and Industrial Research Organization, Butterworths, Australia, 175–183.

Najder, J., and M. Werno. 1973. "*Protection of Buildings on Expansive Clays.*" *Proceedings of the 3rd International Conference on Expansive Soils,* Haifa, Israel, *1*, 325–334.

National Lime Association (NLA). 2004. "Lime-Treated Soil Construction Manual: Lime Stabilization and Lime Modification." Bulletin 326, National Lime Association.

Nelson, J. D., and V. T. Edgar. 1978. "Moisture Migration Beneath Impermeable Membranes." *Proceedings of the 15th Annual Symposium Engineering Geology and Soil Engineering*, Idaho Department of Highways, Boise, ID.

Nelson, J. D., and D. J. Miller. 1992. *Expansive Soils: Problems and Practice in Foundation and Pavement Engineering*. New York: John Wiley and Sons.

Nelson, E. J., J. D. Nelson, K. C. Chao, and J. B. Kang. 2011. "Influence of Surface Grading on Infiltration." *Proceedings of the 5th Asia-Pacific Conference on Unsaturated Soils*, Dusit Thani Pattaya, Thailand, II, 785–789.

Newland, P. L. 1965. "The Behaviour of a House on a Reactive Soil Protected by a Plastic Film Moisture Barrier." *Engineering Effects of Moisture Changes in Soils, Proceedings of International Research and Engineering Conference on Expansive Soils*. College Station: Texas A&M University Press, 324–329.

Pengelly, A. D., and M. B. Addison. 2001. "In-Situ Modification of Active Clays for Shallow Foundation Remediation." *Expansive Clay Soils and Vegetative Influences on Shallow Foundations, Proceedings of the ASCE Geo-Institute Shallow Foundation and Soil Properties Committee Sessions at the 2001 Civil Engineering Conference*, 192–214.

Pengelly, A. D., D. W. Boehm, E. Rector, and J. P. Welsh. 1997. "Engineering Experience with In-Situ Modification of Collapsible and Expansive Soils." *Unsaturated Soil Engineering Practice*, ASCE Geotechnical Special Publication No. 68, 277–298.

Perrin, L. L. 1992. "Expansion of Lime-Treated Clays Containing Sulfates." *Proceedings of the 7th International Conference on Expansive Soils*, Dallas, TX, 1, 409–414.

Petry, T. M., and D. N. Little. 2002. "Review of Stabilization of Clays and Expansive Soils in Pavements and Lightly Loaded Structures—History, Practice, and Future." *Journal of Materials in Civil Engineering* 14(6): 447–460.

Porter, A. A. 1977. "The Mechanics of Swelling in Expansive Clays." Master's thesis, Colorado State University, Fort Collins, CO.

Portland Cement Association (PCA). 1970. *Recommended Practice for Construction of Residential Concrete Floors on Expansive Soil,* vol. 2. Los Angeles, CA: Portland Cement Association.

Puppala, A. J., N. Intharasombat, and R. K. Vempati. 2005. "Experimental Studies on Ettringite-Induced Heaving in Soils." *Journal of Geotechnical and Geoenvironmental Engineering*, ASCE 131(3): 325–337.

Puppala, A. J., K. Punthutaecha, and S. K. Vanapalli. 2006. "Soil-Water Characteristic Curves of Stabilized Expansive Soils." *Journal of Geotechnical and Geoenvironmental Engineering* ASCE 132(6): 736–751.

Puppala, A. J., N. Talluri, and B. C. S. Chittoori. 2014. "A Review of Calcium Based Stabilizer Treatment of Sulfate Bearing Soils: Challenges and Solutions." *Ground Improvement Journal*, London: Thomas Telford.

Schmertmann, J. H. 1983. Private communication with John D. Nelson.

Snethen, D. R. 1979. "Technical Guidelines for Expansive Soils in Highway Subgrades." US Army Engineers Waterways Experiment Station, Vicksburg, MS, Report No. FHWA-RD-79-51.

Steinberg, M. L. 1981. "Deep Vertical Fabric Moisture Barriers in Swelling Soils." *Transportation Research Record* 790, Transportation Research Board, 87–94.

Steinberg, M. 1998. *Geomembranes and the Control of Expansive Soils in Construction.* New York: McGraw-Hill.

Teng, T. C. P., R. M. Mattox, and M. B. Clisby. 1972. "A Study of Active Clays as Related to Highway Design." Research and Development Division, Mississippi State Highway Dept., Engineering and Industrial Research Station, Mississippi State University, MSHD-RD-72-045, 134.

———. 1973. "Mississippi's Experimental Work on Active Clays." *Proceedings of Workshop on Expansive Clays and Shales in Highway Design and Construction*, Laramie, WY, 1–17.

Texas Department of Transportation. 2014. "Test Procedure for Determining Potential Vertical Rise." TxDOT Designation Tex-124-E.

Thompson, M. R. 1966. "Lime Reactivity of Illinois Soils." *Journal of the Soil Mechanics and Foundation Division*, ASCE 92(SM5): 67–92.

———1968. "Lime Stabilization of Soils for Highway Purposes." Final Summary Report, Civil Engineering Studies, Highway Series No. 25, Illinois Cooperative Highway Research Program, Project IHR-76, 26.

Thompson, M. R., and Q. L. Robnett. 1976. "Pressure-Injected Lime for Treatment of Swelling Soils." *Transportation Research Record* 568, Transportation Research Board, 24–34.

Thompson, R. W. 1992a. "Swell Testing as a Predictor of Structural Performance." *Proceedings of the 7th International Conference on Expansive Soils,* Dallas, TX, *1*, 84–88.

———. 1992b. "Performance of Foundations on Steeply Dipping Claystone." *Proceedings of the 7th International Conference on Expansive Soils,* Dallas, TX, *1*, 438–442.

Uniform Building Code (UBC). 1997. *Uniform Building Code.* Vol. 1. Whittier, CA: International Conference of Building Officials.

US Department of the Army (DA). 1983. "Technical Manual TM 5-818-7, Foundations in Expansive Soils." Washington, DC.

van der Merwe, D. H., F. Hugo, and A. P. Steyn. 1980. "The Pretreatment of Clay Soils for Road Construction." *Proceedings of the 4th International Conference on Expansive Soils*, Denver, CO, 4, 361–382.

Winterkorn, H. F. 1975. "Soil Stabilization." Chapter 8. In *Foundation Engineering Handbook*. New York: Van Nostrand Reinhold.

Wright, P. J. 1973. "Slurry Pressure Injection Tames Expansive Clays." *Civil Engineering*, ASCE 43(10): 42–45.

11

Design Methods for Shallow Foundations

For sites where the heave is predicted to be of sufficiently small magnitude, shallow foundation systems such as conventional footings or stiffened slabs-on-grade may sometimes be used. Shallow foundations are often used in conjunction with overexcavation and replacement for expansive soil sites. The overexcavation and replacement method was discussed in chapter 10. The design and application of spread footings and stiffened slabs-on-grade are discussed in this chapter.

11.1 SPREAD FOOTING FOUNDATIONS

Shallow spread footings typically are not used in expansive soil applications unless the structures can tolerate some heave. Attempts have been made with various foundation alternatives to increase the bearing pressure of shallow footings to a point where the footing load approaches or exceeds the expansion potential of the bearing soil. These alternatives are generally not very successful. If such techniques are used, care must be taken not to exceed the bearing capacity of the bearing soil. Some modifications that have been proposed were shown in Figure 9.1. They include the following:

- Narrowing the width of the footing base.
- Placing the foundation wall or grade beam directly on grade without a footing.
- Providing void spaces within the supporting beam or wall to concentrate loads at isolated points.
- Increasing the reinforcement around the perimeter and into the floor slab to stiffen the foundation. This is essentially an unreinforced monolithic slab and wall foundation.

The problem with using a narrow footing in order to increase the bearing pressure is that the zone of influence of the applied stress beneath a narrow footing extends to a very limited depth. This is illustrated by considering the pressure bulb beneath a narrow footing, which is shown in Figure 11.1. If the load on the footing is as high as 5,000 psf (240 kPa), the applied stress at a depth of 3 ft (0.9 m) would be only 1,000 psf (48 kPa). At a depth of 6 ft (1.8 m), it would be less than 500 psf (24 kPa). Thus, even for an expansive soil with a very low swelling pressure the expansion would be reduced by the bearing pressure only to a depth of less than 3 ft to 6 ft (1 m to 2 m). This depth is much smaller than that of the active zone for most sites. Thus, increasing the bearing pressure has very little influence on the amount of heave. It is evident that the use of narrow spread footings in expansive soils should be restricted to soils having very low to low swell pressures or in situations where the amount of potential heave is within the tolerable limits of differential movement for the structure.

Where basements are used the foundation walls can be heavily reinforced so that they can span unsupported distances and resist cracking

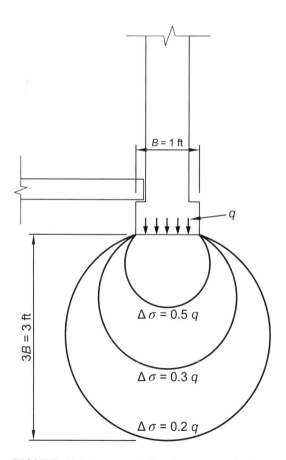

FIGURE 11.1. Pressure bulb below narrow footing.

due to differential movement. The structural configurations should be simple, and for split level construction the structure should be designed to consist of independently acting units connected by joints. Masonry brick and cinder block foundations are not suitable for this type of foundation.

Another method used to concentrate loads and increase bearing stress is to place shallow individual footings at several points under the foundation walls as shown in Figure 9.1c. This system has the same limitation as shown in Figure 11.1, and has only limited success at decreasing heave. Therefore, the superstructure should be relatively flexible. Load concentration on the pads is accomplished by providing a void space beneath the grade beam using the same technique as shown for pier and grade beam construction in Figure 9.1d. It is important to ensure that uplift pressures on the sides of the foundation walls are minimized or prevented. Care must be taken to ensure that by concentrating the loads on the pads the bearing capacity of the soil is not exceeded. The void space must be designed to accommodate the amount of expected heave.

11.1.1   Computation of Footing Heave

Computations for footing heave are similar to those presented in chapter 8 for free-field heave, except that the final net normal stress, σ''_f, is equal to the overburden stress, σ''_{vo}, plus the stress applied by the footing, $\Delta\sigma''_v$. The applied stress at any depth, z, below the footing can be calculated by any of several methods that have been published in the literature. Lambe and Whitman (1969) present charts similar to the one shown in Figure 11.1 that are based elastic theory. The charts can be used to calculate the total applied stress at depth beneath various footing shapes. A widely used simplified method that provides a good approximation for the elastic solution assumes that the applied total stress beneath a footing is distributed over an area that varies linearly with depth in the manner shown in Figure 11.2. For a strip footing of width B that is subjected to a load P, per lineal foot, the stress at depth z would be equal to

$$\Delta\sigma''_v = \frac{P}{(B+z)} \qquad (11\text{-}1)$$

where:

$\Delta\sigma''_v$ = total applied stress,
P = load on the footing, in pounds per lineal foot,
B = width of the footing, and
z = depth below the footing.

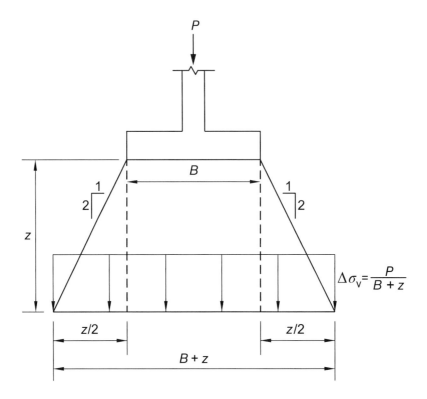

FIGURE 11.2. Approximation of applied stress distribution under strip footing.

Using the same manner of stress distribution for a rectangular footing with dimensions $B \times L$, the total applied stress at depth z would be,

$$\Delta\sigma_v = \frac{P}{(B+z)(L+z)} \qquad (11\text{-}2)$$

where:

P = total load on the footing, in pounds.

For a circular footing with diameter D, the total applied stress at depth z would be,

$$\Delta\sigma_v = \frac{4P}{\pi(D+z)^2} \qquad (11\text{-}3)$$

where:

P = total load on the footing, in pounds.

Shallow spread foundation systems are generally not considered suitable for expansive soil sites with a weighted risk factor, RF_w, greater than about 0.5 to 0.75. Of more concern than total movement of the footing is differential heave of different elements of the foundation system. Although differential movement of a foundation is commonly considered to be approximately equal to half of the total heave, there are many instances where the differential movement is equal to the total heave. The National Association of Home Builders (2010) states in its *Residential Construction Performance Guidelines* that the foundation levelness should be no more than $1/2$ inch (13 mm) higher or lower than any point within a distance of 20 ft (6 m). Differential heave beneath slabs-on-grade is discussed in more detail in Section 11.2.2.

11.1.2   Spread Footing Design Examples

The following examples demonstrate that although shallow foundations may be suitable for sites with low values of weighted risk factor, RF_w, they generally should not be used for sites with values of RF_w greater than about 0.5 to 0.75. They illustrate that the shallow foundation configurations shown in Figures 9.1a through 9.1c are generally not effective in reducing heave to tolerable amounts. The removal of expansive soil and replacement with nonexpansive soil has met with some success, but careful testing and analysis is essential even with that approach.

Examples 11.1 and 11.2 show the computation of free-field heave and heave of a footing constructed on two sites, one having a $RF_w = 0.5$ and the other with a $RF_w = 2.9$.

EXAMPLE 11.1

Given:

A site with a soil profile as given in Table E11.1-1. A single-story house will be constructed with a 16-in.-wide shallow strip footing foundation founded at a depth of 3 ft. The house will have a structural floor with a crawl space below. The dead load on the footing will be 2,100 lb/lin. ft.

TABLE E11.1-1   Soil Profile for Example 11.1

Depth (ft)	Soil Type	Water Content (%)	Dry Density (pcf)	CS % Swell, $\varepsilon_{s\%}$ @ 1,000 psf (%)	CV Swelling Pressure, σ''_{cv} (psf)
0–8	Silty clay	22	104.0	0.4	1,641
8–40	Sandy claystone	13	115.0	2.0	2,445

Find:

1. Compute the weighted risk factor, RF_w, for the site.
2. Compute the free-field heave for the site.
3. Compute the predicted footing heave.

Solution:

Part 1:

The weighted risk factor for the site is computed using equation (9-3).

The depth of potential heave, z_p, is computed by equating the overburden stress to the constant volume swelling pressure. It is assumed that it will extend to a depth that is in the claystone. Therefore,

$$(126.9 \times 8) + [130.0 \times (z_p - 8)] = 2{,}445$$

$$\Rightarrow z_p = 19.0 \text{ ft}$$

That depth is well within a depth to which it is expected that full wetting will progress within the design life of 100 years. Therefore, that depth is used as the design active depth.

For a design active depth of 19.0 ft, the weighted risk factor, RF_w, is computed using equation (9-3). Thus,

$$RF_w = \frac{8}{19}(0.4) \log\left(\frac{1{,}641}{1{,}000}\right) + \frac{11}{19}(2.0) \log\left(\frac{2{,}445}{1{,}000}\right) = 0.5$$

Part 2:

C_H is determined from equation (6-1).

For the silty clay:

$$C_H = \frac{0.004}{\log\left[\dfrac{1{,}641}{1{,}000}\right]} = 0.019$$

For the sandy claystone:

$$C_H = \frac{0.02}{\log\left[\dfrac{2{,}445}{1{,}000}\right]} = 0.052$$

The computations for free-field heave are presented in Table E11.1-2. The computed value of free-field heave is equal to 2.2 in. This amount of heave may present problems with flatwork and produce differential movement of slabs-on-grade such as a garage floor. Thus, maintenance of those elements must be expected.

TABLE E11.1-2 Free-Field Heave Computations for Example 11.1

Depth to Bottom of Layer (ft)	Depth Below Excavation (ft)	Interval (ft)	Soil Type	Total Unit Weight (pcf)	Incremental Stress @ Midlayer (psf)	Overburden Stress (psf)	Incremental Heave (in.)	Cumulative Heave (in.)	Cumulative Heave from z_p (in.)
0.00									2.23
1.00	1.00	1.00	Silty clay	126.9	63.4	63.4	0.32	0.32	1.91
2.00	2.00	1.00	Silty clay	126.9	63.4	109.3	0.21	0.53	1.70
3.00	3.00	1.00	Silty clay	126.9	63.4	317.2	0.16	0.69	1.54
4.00	4.00	1.00	Silty clay	126.9	63.4	444.1	0.13	0.82	1.41
5.00	5.00	1.00	Silty clay	126.9	63.4	571.0	0.10	0.92	1.31
6.00	6.00	1.00	Silty clay	126.9	63.4	697.8	0.08	1.00	1.23
7.00	7.00	1.00	Silty clay	126.9	63.4	824.7	0.07	1.07	1.16
8.00	8.00	1.00	Silty clay	126.9	63.4	951.6	0.05	1.12	1.11
9.00	9.00	1.00	Sandy claystone	129.9	65.0	1,080.0	0.22	1.34	0.89
10.00	10.00	1.00	Sandy claystone	129.9	65.0	1,210.0	0.19	1.53	0.70
11.00	11.00	1.00	Sandy claystone	129.9	65.0	1,339.9	0.16	1.69	0.54
12.00	12.00	1.00	Sandy claystone	129.9	65.0	1,469.9	0.14	1.83	0.40
13.00	13.00	1.00	Sandy claystone	129.9	65.0	1,599.8	0.11	1.94	0.29
14.00	14.00	1.00	Sandy claystone	129.9	65.0	1,729.8	0.09	2.03	0.20
15.00	15.00	1.00	Sandy claystone	129.9	65.0	1,859.7	0.07	2.10	0.12
16.00	16.00	1.00	Sandy claystone	129.9	65.0	1,989.7	0.06	2.16	0.07
17.00	17.00	1.00	Sandy claystone	129.9	65.0	2,119.6	0.04	2.20	0.03
18.00	18.00	1.00	Sandy claystone	129.9	65.0	2,249.6	0.02	2.22	0.03
19.00	19.00	1.00	Sandy claystone	129.9	65.0	2,379.5	0.01	2.23	

TABLE E11.1-3 Footing Heave Computations for Example 11.1

Depth to Bottom of Layer (ft)	Depth Below Excavation (ft)	Interval (ft)	Soil Type	Total Unit Weight (pcf)	Incremental Stress @ Midlayer (psf)	Overburden Stress (psf)	Footing Stress (psf)	Incremental Footing Heave (in.)	Cumulative Footing Heave (in.)
0.00									
1.00	0.00	1.00	Silty clay	126.9	63.4	63.4	0.00	0.00	0.00
2.00	0.00	1.00	Silty clay	126.9	63.4	109.3	0.00	0.00	0.00
3.00	0.00	1.00	Silty clay	126.9	63.4	317.2	0.00	0.00	0.00
4.00	1.00	1.00	Silty clay	126.9	63.4	444.1	1147.5	0.00	0.00
5.00	2.00	1.00	Silty clay	126.9	63.4	571.0	740.2	0.02	0.02
6.00	3.00	1.00	Silty clay	126.9	63.4	697.8	546.9	0.03	0.05
7.00	4.00	1.00	Silty clay	126.9	63.4	824.7	433.7	0.03	0.08
8.00	5.00	1.00	Silty clay	126.9	63.4	951.6	359.3	0.02	0.10
9.00	6.00	1.00	Sandy claystone	129.9	65.0	1,080.0	306.7	0.15	0.25
10.00	7.00	1.00	Sandy claystone	129.9	65.0	1,210.0	267.5	0.14	0.39
11.00	8.00	1.00	Sandy claystone	129.9	65.0	1,339.9	237.2	0.12	0.51
12.00	9.00	1.00	Sandy claystone	129.9	65.0	1,469.9	213.1	0.10	0.61
13.00	10.00	1.00	Sandy claystone	129.9	65.0	1,599.8	193.4	0.08	0.69
14.00	11.00	1.00	Sandy claystone	129.9	65.0	1,729.8	177.1	0.07	0.76
15.00	12.00	1.00	Sandy claystone	129.9	65.0	1,859.7	163.3	0.05	0.81
16.00	13.00	1.00	Sandy claystone	129.9	65.0	1,989.7	151.5	0.04	0.85
17.00	14.00	1.00	Sandy claystone	129.9	65.0	2,119.6	141.3	0.02	0.87
18.00	15.00	1.00	Sandy claystone	129.9	65.0	2,249.6	132.3	0.01	0.88
19.00	16.00	1.00	Sandy claystone	129.9	65.0	2,379.5	124.47	0.00	0.88

Part 3:

The calculation of footing heave follows the procedures presented above. The attenuation with depth of the applied footing stress is computed using equation (11-1). For purposes of illustration, computation of heave for the first soil layer below the footing will be shown here. The calculations for the footing heave are presented in Table E11.1-3. It is assumed that the footing is backfilled both inside and outside. The difference in unit weight between the concrete and the soil is neglected. The soil profile is divided into 19 layers, each having a thickness of 1.0 ft. At the midpoint of the first layer below the footing, the overburden stress plus the applied stress is

$$\sigma_f'' = (126.9 \times 3.5) + \frac{2{,}100}{(1.33 + 0.5)} = 1{,}592 \text{ psf}$$

The heave of that layer is computed to be

$$\rho_1 = C_H H_1 \log \frac{\sigma_{cv}''}{\sigma_f''} = 0.019 \, (12 \text{ in.}) \log \left(\frac{1{,}641}{1{,}592} \right) = 0.003 \text{ in.}$$

The computations for the entire soil profile are shown in Table E11.1-3. The slight difference between the computed value of applied stress shown in Table E11.1-3 and the value just computed is due to rounding off of decimal places.

The computed predicted footing heave shown in Table E11.1-3 is equal to 0.9 in. For this amount of heave, a shallow footing should be adequate. Even then, because this is an expansive soil site, some amount of maintenance should be expected over what would be normal for a nonexpansive site.

EXAMPLE 11.2

Given:

A site with a soil profile as shown in Table E11.2-1. A two-story residential house will be constructed having a continuous strip footing, 16 in. wide founded at a depth of 3 ft below the ground surface. The dead load on the footing, including the weight of the footing, is 2,100 lb/lin. ft.

TABLE E11.2-1 Soil Profile for Example 11.2

Depth (ft)	Soil Type	Water Content (%)	Dry Density (pcf)	CS % Swell, $\varepsilon_{s\%}$ @ 1,000 psf (%)	CV Swelling Pressure, σ_{cv}'' (psf)
0–8	Native clay	14	109.0	3.2	3,422
8–40	Claystone	12	116.1	4.6	4,976

Find:

1. Compute the weighted risk factor, RF_w, for the site.
2. Compute the expected free-field heave for the site.
3. Compute predicted heave of the continuous strip footing.

Solution:

Part 1:
The depth of potential heave, z_p, is computed by equating the overburden stress to the constant volume swelling pressure.

$$4,976 = 8(109.0 \times 1.14) + (z_p - 8)(116.1 \times 1.12)$$

$$\Rightarrow z_p = 38.6 \text{ ft}$$

That depth can be expected to be fully wetted within the design life of 100 years. Therefore, that depth is used as the design active depth. The weighted risk factor for the site is computed using equation (9-3).

$$RF_w = \frac{8}{38.6} (3.2) \log\left(\frac{3,442}{1,000}\right) + \frac{30.6}{38.6}(4.6) \log\left(\frac{4,976}{1,000}\right) = 2.9$$

Part 2:
Computation of the free-field heave follows the procedures outlined in Example 8.1.

For the native clay:

$$C_H = \frac{0.032}{\log\left[\frac{3,442}{1,000}\right]} = 0.060$$

For the claystone:

$$C_H = \frac{0.046}{\log\left[\frac{4,976}{1,000}\right]} = 0.066$$

The design active zone is considered to be equal to the depth of potential heave. The computations for free-field heave are similar to those shown in Table E11.1-2 but with different values. The computed value of free-field heave is equal to 11.7 in.

Part 3:
Table E11.2-2 shows computations for the footing heave in accordance with the procedures shown in Example 11.1. The maximum footing heave is calculated to be 7.2 in. It can be noted in Table E11.2-2 that because the layers into which the soil was divided are not even foot increments, the bottom of the footing and the boundary between the native clay and the claystone do not coincide exactly with the top of a layer. It was shown in chapter 8 that the effect of this is negligible.

TABLE E11.2-2 Footing Heave Computations for Example 11.2

Depth to Bottom of Layer (ft)	Depth Below Excavation (ft)	Interval (ft)	Soil Type	Total Unit Weight (pcf)	Incremental Stress @ Midlayer (psf)	Overburden Stress (psf)	Footing Stress (psf)	Incremental Footing Heave (in.)	Cumulative Footing Heave (in.)
0.00	0.00						0.00	0.00	0.00
1.29	0.00	1.29	Native clay	124.3	79.9	79.9	0.00	0.00	0.00
2.57	0.86	1.29	Native clay	124.3	79.9	239.8	1,190.2	0.21	0.21
3.86	2.15	1.29	Native clay	124.3	79.9	399.7	739.3	0.39	0.60
5.15	3.43	1.29	Native clay	124.3	79.9	559.6	508.4	0.41	1.01
6.43	4.72	1.29	Native clay	124.3	79.9	719.5	387.4	0.40	1.41
7.72	6.01	1.29	Native clay	124.3	79.9	879.3	312.9	0.58	1.99
9.01	7.29	1.29	Claystone	130.0	83.7	1,042.9	262.5	0.54	2.53
10.29	8.58	1.29	Claystone	130.0	83.7	1,210.2	226.0	0.50	3.03
11.58	9.87	1.29	Claystone	130.0	83.7	1,377.6	198.5	0.46	3.49
12.87	11.15	1.29	Claystone	130.0	83.7	1,544.9	176.9	0.43	3.92
14.15	12.44	1.29	Claystone	130.0	83.7	1,712.2	159.6	0.39	4.31
15.44	13.73	1.29	Claystone	130.0	83.7	1,879.5	145.3	0.36	4.67
16.73	15.01	1.29	Claystone	130.0	83.7	2,046.8	133.4	0.33	5.00
18.01	16.30	1.29	Claystone	130.0	83.7	2,214.1	123.3	0.30	5.30
19.30	17.59	1.29	Claystone	130.0	83.7	2,381.4	114.6	0.28	5.58
20.59	18.87	1.29	Claystone	130.0	83.7	2,548.7	107.1	0.25	5.83
21.87	20.16	1.29	Claystone	130.0	83.7	2,716.0	100.5	0.23	6.06
23.16	21.45	1.29	Claystone	130.0	83.7	2,883.3	94.6	0.20	6.26
24.45	22.73	1.29	Claystone	130.0	83.7	3,050.6	89.4	0.18	6.44
25.73	24.02	1.29	Claystone	130.0	83.7	3,217.9	84.8	0.16	6.60
27.02	25.31	1.29	Claystone	130.0	83.7	3,385.2	80.9	0.14	6.74
28.31	26.59	1.29	Claystone	130.0	83.7	3,552.6	76.8	0.12	6.86
29.59	27.88	1.29	Claystone	130.0	83.7	3,719.9	73.3	0.10	6.96
30.88	29.17	1.29	Claystone	130.0	83.7	3,887.2	70.2	0.08	7.04
32.17	30.45	1.29	Claystone	130.0	83.7	4,054.5	67.3	0.07	7.11
33.45	31.74	1.29	Claystone	130.0	83.7	4,221.8	64.6	0.05	7.16
34.74	33.03	1.29	Claystone	130.0	83.7	4,389.1	62.1	0.03	7.19
36.03	34.31	1.29	Claystone	130.0	83.7	4,556.4	59.9	0.02	7.21
37.31	35.60	1.29	Claystone	130.0	83.7	4,723.7	57.7	0.00	7.21
38.60		1.29	Claystone	130.0	83.7	4,891.0			

The predicted footing heave was calculated in Example 11.1 to be 0.9 in (23 mm). Although heave of that magnitude is generally considered to be tolerable for residential construction, the owner of the building should be made aware of the risk of some heave and the need for good landscaping and maintenance practices. Furthermore, although that magnitude of heave may be tolerable for some structures, it may not be for others, such as a safety critical building or a rigid structure not tolerant of differential movement.

Example 11.2 computed the heave for a site having a significantly higher value of RF_w. Footing heave on that site was computed to be 7.2 in. (183 mm), which is so high as to preclude the use of a spread footing foundation for almost any type of structure.

The same calculations were performed for an 8-in. (203-mm) -wide *stemwall* foundation constructed on the same soil profile as that in Example 11.2. For these calculations it was assumed that the foundation wall has no footing as shown in Figure 9.1b but without a void. The results of the computation are shown in Table 11.1, which compares the results of a number of different shallow foundation scenarios. The predicted heave of the foundation was calculated to be 7.0 in. (178 mm). This value of heave is almost the same as that for Example 11.2. Furthermore, the bearing pressure is so high as to pose potential bearing-capacity problems.

The same calculations were performed for a foundation wall supported on isolated square pads 16 in. (406 m) wide, as shown in Figure 9.1c. The pads were spaced to concentrate the dead load-bearing stress to 4,000 psf (190 kPa). The predicted foundation heave

TABLE 11.1   Comparison of Different Shallow Footing Scenarios

Footing	Fill	Depth of Overexcavation (ft)	Free-field Heave (in.)	Footing Heave (in.)
16 in. strip footing	No fill	0	11.7	7.2
8 in. footing	No fill	0	11.7	7.0
16 in. isolated pad footing	No fill	0	11.7	7.6
16 in. strip footing	Nonexpansive fill $\sigma''_{cv} = 0$ psf $\varepsilon_{s\%} = 0\%$	13 / 25	4.1 / 1.1	3.8 / 1.0
16 in. strip footing	Low-expansive fill $\sigma''_{cv} = 1{,}000$ psf $\varepsilon_{s\%} = 1\%$	13 / 25	5.4 / 2.4	3.8 / 1.0
16 in. strip footing	Expansive fill $\sigma''_{cv} = 2{,}200$ psf $\varepsilon_{s\%} = 1.75\%$	13 / 25	8.4 / 5.6	4.9 / 2.2

was computed to be 7.6 in. (193 mm). It seems counterintuitive that footing heave under the square pad with a higher bearing stress should be greater than that for the strip footing, which had a lower bearing stress. Comparison of equations (11-1) and (11-2) shows that the applied stress under the square pad attenuates faster with depth than that under the strip footing. Thus, although the heave in the upper 5 ft (1.5 m) or so is reduced, below that it actually increases compared to the strip footing. Thus, a shallow foundation system of any configuration would not be suitable for sites with high expansion potential.

Overexcavation of the upper several feet of expansive soil and replacement with nonexpansive fill was discussed in chapter 10 as a method to mitigate the expansion potential of a site. Thompson (1992) indicated that if there is 10 ft (3 m) or more of nonexpansive soil between the bottom of the footing and the top of the in situ expansive soil, the risk of damage is significantly reduced. The effectiveness of that method for the site used in Example 11.2 was analyzed by replacing the upper 13 ft (4 m) of soil with nonexpansive fill. Overexcavation to depths of this magnitude has been done on a subdivision-wide basis in Colorado. The overexcavation and replacement reduced the risk factor to a value of $RF_w = 2.13$. As shown in Table 11.1, the free-field heave was reduced to 4.1 in. (104 mm). The predicted heave of a 16-in. (406-m) -wide strip footing was calculated to be 3.8 in. (96 mm). That value of heave would still be considered intolerable for most structures.

It is of interest to calculate what depth of overexcavation would be required to limit the predicted footing heave to 1.0 in. (25 mm). The above calculations were repeated for several different depths and it was found that a depth of 25 ft (7.6 m) of nonexpansive fill beneath the footing would be needed to reduce the heave to 1.0 in. (25 mm). Overexcavation of that depth has been done in Colorado on a subdivision-wide basis in the Designated Dipping Bedrock Area (DDBA) that was discussed in Section 7.2.4. In those locations the footings were placed deeper so that basements could be constructed.

In many applications of this nature, instead of importing nonexpansive fill the overexcavated expansive soil is moisture-conditioned and recompacted. That practice is usually not successful in eliminating all expansion potential and even a low expansion potential in the fill can result in significant heave. An example of this is a site on which a forensic investigation was conducted. The moisture-conditioned and recompacted fill was tested and it exhibited expansion potential with $\sigma''_{cv} = 2,200$ psf (105 kPa) and $\varepsilon_{s\%} = 1.75\%$. The scenario analyzed above was repeated for a fill having those properties. The results are shown in Table 11.1 for the case labeled *expansive fill*. Even for the depth of fill of 25 ft (7.6 m) the footing heave is greater than what would be considered tolerable and the free-field heave is such that problems with flatwork and slabs-on-grade would be expected.

In the interest of investigating how much expansion potential can be tolerated and still consider the fill *nonexpansive,* the scenario was repeated for various degrees of expansivity of the fill. A *low-expansive* fill with $\sigma''_{cv} = 1{,}000$ psf (48 kPa) and $\varepsilon_{s\%} = 1.0\%$ was found to have the same footing heave as for the nonexpansive fill. However, it can be seen that the computed free-field heave was greater than for the nonexpansive fill. This shows that in cases where moisture-conditioned fill is being proposed, it is necessary for the engineer to perform additional testing and analysis on the fill to determine the effectiveness of the moisture-conditioning.

11.2 STIFFENED SLAB FOUNDATIONS

Stiffened slabs are a type of foundation that has been used for a variety of different types of problem soils. They have historically been used in areas with moderately expansive soils in many countries of the world. Stiffened slab construction is common in many areas in the southern and southwestern United States and in California. Stiffened slab foundations are also used extensively in Australia, Israel, and South Africa. Where the stiffened slab is placed at the level of the ground surface, it is typically referred to as a slab-on-grade foundation.

Stiffened slabs may be either conventionally reinforced or post-tensioned. Cross-sections of a conventional reinforced mat foundation and a post-tensioned slab foundation are shown in Figures 9.1e and 9.1f. The design procedures basically consist of determining bending moment, shear, and deflection for applied structural loads and expected heave patterns. Those aspects of the design are the responsibility of the structural engineer. The geotechnical engineer is responsible for providing the geotechnical input data. The required geotechnical parameters include the predicted free-field heave, the heave edge distance (mound geometry), and the stiffness modulus of the soil. The same general soil parameters are required for either conventionally reinforced slabs or post-tensioned slabs.

11.2.1 Edge Heave and Center Heave

Design of stiffened slab foundations on expansive soils is based on modeling the interaction of the soil with the slab subjected to the loads imposed by the structure. The slab and swelling soil are generally modeled as a loaded plate or beam resting on an elastic or elasto-plastic foundation. After construction of the slab, heave of the foundation soil can occur beginning from the outer edges, or it can be greater in the center. These two modes are depicted in Figures 11.3 and 11.4 and are termed *edge heave* or *center heave*. Edge heave represents a situation such as would occur due to excessive irrigation around the perimeter of the building. In this mode heave may also occur near the center of the slab but the heave at the edge is greater than at the center.

Under the center heave mode, the heave near the center of the slab is greater than at the edge of the slab. Center heave results most commonly from increased water content as a result of elimination of evapotranspiration. The increase in water content is greatest near the center of the structure and evapotranspirative losses around the edge cause a lesser

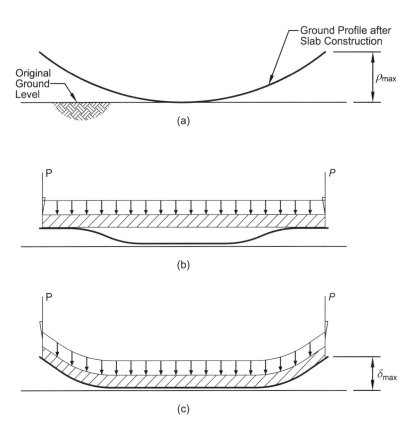

FIGURE 11.3. Edge heave mound profiles for slabs with varying stiffness: (a) mound profile after edge heave if no load is applied, that is, free-field heave with weightless slab (Case A); (b) mound profile for infinitely rigid slab with load (Case B); (c) mound profile for nonrigid stiffened slab with load (Case C).

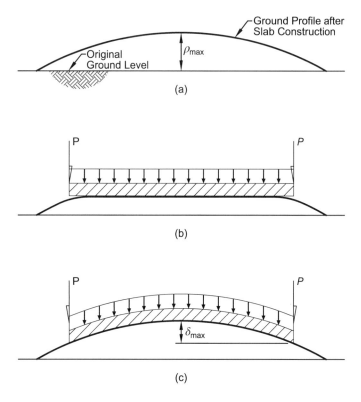

FIGURE 11.4. Center heave mound profiles for various cases of slabs-on-grade: (a) mound profile after center heave if no load is applied, that is, free-field heave with weightless slab (Case A); (b) mound profile for infinitely rigid slab with load (Case B); (c) mound profile for nonrigid stiffened slab with load (Case C).

degree of heave near the edges. In areas where climatic conditions cause shrink–swell conditions, shrinkage of the soil around the edges results in a mound profile tantamount to that shown for center heave conditions in Figure 11.4. This condition is also referred to as *edge drop,* and the design would be the same as for the center heave condition. Center heave represents the long-term most severe distortion condition (Lytton 1970).

Figures 11.3a and 11.4a show the mound that would occur beneath a weightless slab with no load applied. The mound is characterized by the maximum heave computed for conditions of no applied stress. This value of ρ_{max} would correspond to the free-field heave and is equal to the maximum mound height, as shown in Figures 11.3a and 11.4a.

Figures 11.3b and 11.4b show the mound that would result if a slab of infinite stiffness is placed on the mound with the applied structural loading. In this case, the entire slab would heave by the same amount with no differential displacement.

The situations depicted in Figures 11.3a or 11.4a and Figures 11.3b or 11.4b represent the two extremes. In reality, the stiffened slab will have some flexibility and the actual mound shape will be intermediate between those two degrees of flexibility. The actual shapes will be more like those depicted in Figures 11.3c and 11.4c.

The design procedure consists of predicting the mound shape and height and the relative stiffness of the soil and the slab. Analysis of the interaction between the soil and the structure allows the shear and bending moments to be determined for use in the structural design of the slab. A number of different design approaches have been developed, each prescribing a different combination of soil and structural design parameters. Several of these were discussed in Nelson and Miller (1992). Since then, various commercial software programs have made it possible to analyze the structural design requirements.

The required geotechnical parameters relate to maximum expected heave, the mound shape, and soil stiffness. The maximum heave of the soil can be computed using methods presented in chapter 8. The shape of the soil surface that will develop beneath a slab depends on expected variation of environmental conditions within the footprint of the slab. The variation of environmental conditions beneath the slab will depend on the overall environmental conditions of the site. Some design methods, such as the BRAB (1968) method or the PTI (2004) method base a form of *edge moisture variation distance* on climate. However, this applies only for nonlandscaped sites. Even xeriscape landscaping has some effect on the environment under the slab.

The form of the parameter used to express the soil stiffness will depend on the method of analysis being used by the design structural engineer. Most commonly, the parameters would include the modulus of elasticity, the modulus of subgrade reaction, or some similar elastic property. The structural engineer has the task of designing the slab so that under the loading conditions to be imposed on the slab, the slab is of sufficient stiffness that the differential movement of the slab is within tolerable limits.

11.2.2   Differential Heave

As noted above, the differential heave, δ_{max}, between one location on the foundation and another is more critical to the structure than the total maximum heave that occurs. The structural design of the slab is intended to result in a slab of sufficient stiffness such that the differential movement is within the limits that the structure can tolerate. Nelson, Chao, and Overton (2006) analyzed the variation of wetting profiles under slabs with time. The zone of soil that mostly contributes to differential soil heave is the zone of seasonal moisture fluctuation. Below that depth, the deeper soils contribute less to heave. The pattern of heave can change with time and the location of maximum differential movement can vary across the slab. Distortion of the foundation will be transmitted to the structure, and continued maintenance must be expected. The following case study illustrates the change in differential heave with time for a post-tensioned slab.

For landscaped sites the irrigation that is applied around the structure results in edge heave. Water initially infiltrates in a vertical direction. However, as water migrates vertically, it also spreads laterally, as discussed in chapter 7. Seepage analyses can be conducted to analyze the pattern of wetting over time under a foundation. The results of these analyses can be used to determine the pattern of differential heave during the design life of the structure.

CASE STUDY — DIFFERENTIAL HEAVE BENEATH SLAB-ON-GRADE FOUNDATION

A development consisting of 18 multiunit condominium buildings was constructed on a site north of Denver, Colorado. The buildings are three stories high and are supported on post-tensioned slab-on-grade foundations. The site is located on the Laramie Formation and has a RF_w rating of "very high."

Construction of the condominiums began around 2003 and continued through 2007. Beginning in about 2007, several of the buildings began to experience differential movement. Observed damage consisted of racked doors, cracked walls and ceilings, ceilings that were lifting off of the walls, and cracks in the post-tensioned slabs on the ground level. Elevation surveys of the buildings indicated that several of the buildings were tipping away from an adjacent golf course where a large amount of irrigation was being applied. The shape of the floor slabs at the ground level also showed that the slabs were being bowed or twisted. This was consistent with the damage observed in the structures. It was also noted that the damage of the buildings tended to be more severe at the third floor level than at the ground floor level. The soil profile consisted of up to about 10 ft of sandy clay underlain by silty claystone and claystone with interbedded layers of sandstone. Figure 11.5a shows the footprint of one of the buildings. Elevation surveys were conducted periodically through the east and west breezeways. The elevations were referenced to a deep benchmark that had been installed at the site at the beginning of the forensic investigation. The grading and drainage conditions at the site were very poor and there was evidence of ponding water at the site.

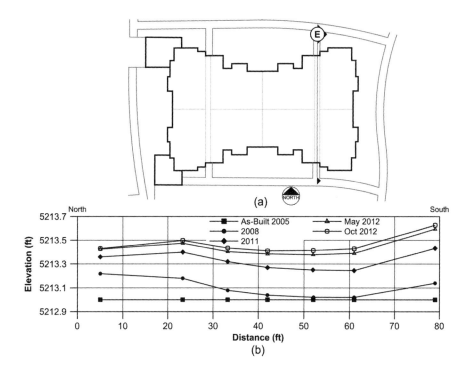

FIGURE 11.5. Results of elevation survey through building breezeway: (a) building footprint; (b) elevation measurements through east breezeway.

Figure 11.5b shows the elevation profiles measured at various times in the east breezeway. The total heave and differential heave are plotted against time in Figure 11.6. Even within a period of approximately 7 years after construction the heave is continuing. At the time of the investigation the post-tensioned slab in the east breezeway had experienced more than 7.4 in. of maximum heave and 2.5 in. of differential heave. The differential heave remains almost constant with time even though the total heave is continuing. However, the location of the maximum differential elevation changes with time. Remaining heave was calculated for the soil profile observed in the exploratory boring that was drilled closest to the building. The remaining heave was calculated to be 8.9 in.

The pattern of heave shown in Figure 11.5 is one of edge heave. This is typical for sites with poor drainage. It is evident that the stiffness of the slab was not adequate to limit the differential movement of the foundation to a tolerable amount. At a site with such highly expansive soils as this one, and where the predicted heave is as high as was calculated, a stiffened mat foundation is not recommended. Even for an infinitely rigid mat foundation the differential heave from one side, or one end, to the other could be of a magnitude that could cause severe tilting of the structure. At a site such as this one, a properly designed deep foundation would have performed much better.

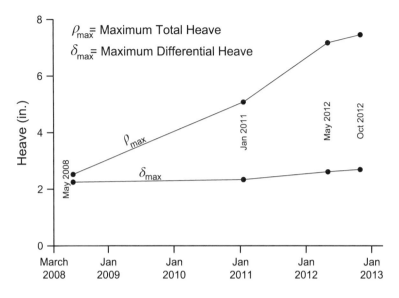

FIGURE 11.6. ρ_{max} and δ_{max} versus t for the east breezeway.

11.3 REMEDIAL MEASURES FOR SHALLOW FOUNDATIONS

Probably the most commonly used remediation plan for shallow foundations that have experienced distress due to heave of expansive soils is to underpin the foundation. Various methods of underpinning can be used depending on the soil and foundation conditions. Cosmetic repairs may be necessary for several months to years after a foundation has been releveled. When a foundation is releveled over a period of a few days, the superstructure, which took years to conform to the original foundation movement, will not spring back as quickly. It often takes a period of time for the superstructure to settle back onto the foundation after releveling.

11.3.1 Footing Foundations

Patented pier or pile systems such as micropiles, helical piles, or steel push piers are the most commonly used underpinning systems. In some cases, concrete drilled piers have been used, but access for the drilling can be problematical. Drilling for micropiles can use a small drilling machine that can be attached to the foundation wall. Micropiles are

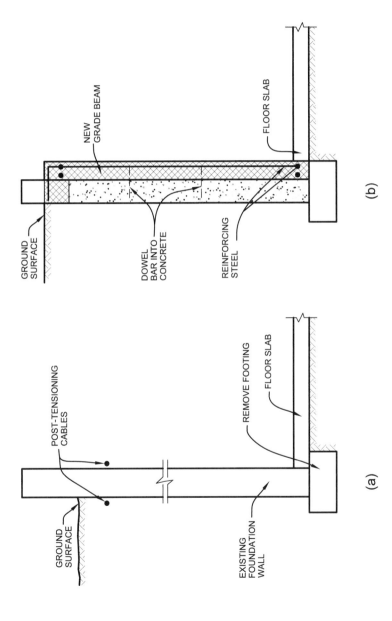

FIGURE 11.7. Stiffening of existing grade beam: (a) post-tensioning; (b) new grade beam (modified from Chen 1988).

315

typically 4 in. to 6 in. (100 mm to 150 mm) in diameter. Helical piles consist of one or more helices 8 in. to 14 in. (200 mm to 350 mm) in diameter attached to a steel shaft. The helical pile is turned into the ground using a form of power auger or hydraulic torque motor. The steel shaft can then be attached to the grade beam. Steel push piers are generally about 3 in. (80 mm) in diameter and are advanced by jacking against the foundation wall. Penetration of these piles is limited by the stiffness of the soil and the weight of the structure. This is demonstrated in the case study that follows.

When shallow foundations are underpinned the footing should be removed and voids should be excavated beneath the foundation wall. This may require stiffening of the foundation walls. Chen (1988) described stiffening of existing walls by means of post-tensioning or by construction of a new grade beam adjacent to the existing wall. Repair of foundation grade beams by these methods is shown in Figure 11.7.

The following case study demonstrates two interesting points (Nelson, Fox, and Nelson 2014). It demonstrates the use of an economical and innovative geophysical method to determine the length of existing push piers and helical piles. It also demonstrates the need to use an appropriate remediation method and the advantage of using micropiles over push piers and helical piles in highly overconsolidated clays.

CASE STUDY—USE OF INADEQUATE UNDERPINNING SYSTEMS

A single-story residence was originally constructed on spread footings on a site located in northern Colorado that was known to contain highly expansive clay-stone to great depth. Shortly after construction the house began to experience significant movement of the foundation and basement slab. The house was involved in litigation and was subsequently purchased by a salvage buyer who hired engineers to recommend repairs to the foundation and basement areas. The recommended repairs consisted of steel push piers to stabilize the perimeter foundation, helical piles to stabilize the interior footings, and structural repairs to the foundation walls. The new foundation system was subsequently installed, the house was releveled to the extent possible, and the interior cosmetic damages were repaired. The specifications called for the push piers and helical piles to be installed to a depth of 30 ft below the foundation. Following the supposed repair of the house, it was sold to a new owner who finished the basement and occupied the house for approximately 9 years. After 7 years in the house, the new owner noticed that the main floor of the house was out of level, and that there were a number of cracks in the drywall and exterior brick veneer.

Elevation surveys of the house indicated that at least 3 in. of additional movement had occurred since the house was repaired (Nelson, Fox, and Nelson 2014). An investigation into the repair logs for the house indicated that there were inconsistencies between the original repair design and the actual repair. The repair contractor indicated that the helical piles and push piers had been installed to a depth of at least 30 ft. However, the geotechnical data from previous investigations indicated that the stiffness of the bedrock would prevent advancement of the push piers or helical piles to that depth.

Several methods of measuring the length of the piles were investigated. Most of these methods involved costs that were well above the available resources. A local geologist suggested an innovative and economical geophysical method that could be used to determine the depth to which the steel push piers and the helical piles were actually installed. This method was based on measurement of the magnetic characteristics at depth. To provide access for the geophysical probes, borings were drilled adjacent to three of the steel push piers and between two helical piles. These borings were advanced using conventional drilling equipment typically used for underpinning by micropiles. Borings were drilled through the slab to a depth of 35 ft.

A probe was lowered into the borehole that measured the magnetic response of the soil around the boring. This probe was able to detect the adjacent pile and indicated the depth to which each pile was installed. The data from this probe was able to confirm that all of the piles were installed well short of the design depth. Depths of between 8 and 20 ft were measured.

It was recommended that the existing piles be disconnected from the house and the foundation be underpinned using cased or sleeved micropiles to a depth of approximately 65 ft with the upper 45 ft being cased. The new micropiles would be adequately anchored below the active zone such that upward movement of the piles would not occur. It was recommended that the existing slab-on-grade be removed and replaced with a structural floor. Furthermore, it was recommended that a void greater than the free-field heave value be installed below the structural floor to prevent future expansion of the surficial soils from affecting the slab.

This case demonstrates the advantage of micropiles over push piers or helical piles in soil or bedrock of the nature encountered in this case. The micropile drilling equipment was able to penetrate to the specified depth and the depth was easily verified. Clearly, the push piers or the helical piles could not penetrate to the specified depth. If the preferred method was push piers and or helical piles, predrilling in the stiff claystone would have been required.

11.3.2   Stiffened Slab-on-Grade

Stiffened slab-on-grade foundations are intended to provide a rigid support that can withstand soil heave without resulting in large differential movements. Therefore, when damage occurs it indicates that the slab is not sufficiently stiff. Remedial measures to correct this problem are diffi-cult. Supporting the slab at isolated points will likely be ineffective if the slab is not stiff enough to carry the loads between supports. Increasing the stiffness of the slab is difficult to accomplish at this point.

If access beneath the slab is possible, beams can be constructed under the slab and underpinned to provide additional support. If the load-bearing elements of the structure can be isolated, load can be transferred to outside piers or other forms of support. In some cases innovative techniques to increase the stiffness of the slab-on-grade by post-tensioning the stiffened beams and/or superstructure have been successful. Where future heave is predicted to be large it may be necessary to move the building off the slab, remove the slab, and construct a new foundation.

11.3.3   Other Methods

Other remedial measures can include drainage and moisture control methods, and soil treatment. Drainage improvements can help to reduce further volume changes where damage is a result of an increase in the water content in foundation soils. Interceptor drains may be necessary if water intrusion is due to gravity flow of free water in a subsurface per-vious layer. Surface drainage can be improved by regrading, altering the gutter and downspout system, and providing a moisture barrier or subdrain. A graded swale or ditch, preferably lined, can be constructed to divert surface water runoff away from the structure.

In some areas, chemical injection is used as a remedial method. This method typically involves injecting a low pH or potassium-based solution into the expansive soil in an attempt to eliminate the remaining expansion potential. This method might be effective in more permeable soils but not in areas where low permeability soils or expansive bedrock are present.

References

Building Research Advisory Board (BRAB). 1968. "Criteria for Selec-tion and Design of Residential Slabs-On-Ground." Publication 1571, National Academy of Sciences Report Number 33 to Federal Housing Administration, Washington, DC.

Chen, F. H. 1988. *Foundations on Expansive Soils*. New York: Elsevier Science.

Lambe, T. W., and R. V. Whitman. 1969. *Soil Mechanics*. New York: John Wiley and Sons.

Lytton, R. L. 1970. "Design Criteria for Residential Slabs and Grillage Rafts on Expansive Clay." Report for Australian Commonwealth Scientific and Industrial Research Organization (CSIRO), Melbourne, Australia.

National Association of Home Builders. 2010. *Residential Construction Performance Guidelines—Contractor Reference* (4th ed.). Washington, DC: Builder Books.

Nelson, J. D., and D. J. Miller. 1992. *Expansive Soils: Problems and Practice in Foundation and Pavement Engineering*. New York: John Wiley and Sons.

Nelson, J. D., K. C. Chao, and D. D. Overton. 2006. "Design Parameters for Slab-on-Grade Foundations." *Proceedings of the 4th International Conference on Unsaturated Soils*, Carefree, AZ, 2110–2120.

Nelson, J. D., Z. P. Fox, and E. J. Nelson. 2014. "In Situ Measurement of Steel Push-Piles and Helical Piers in Expansive Soils." *Proceedings of the UNSAT2014 Conference on Unsaturated Soils*, Sydney, Australia.

Post-Tensioning Institute (PTI). 2004. *Design of Post-Tensioned Slabs-on-Ground* (3rd ed.). Phoenix, AZ: Post-Tensioning Institute.

Thompson, R. W. 1992. "Performance of Foundations on Steeply Dipping Claystone." *Proceedings of the 7th International Conference on Expansive Soils*, 1, Dallas, TX, 438–442.

12

Design Methods for Deep Foundations

It was shown in chapter 11 that shallow spread foundation systems are generally not adequate for expansive soil sites except for sites with low risk factor, *RF*, ratings. Stiffened mat foundations may be adequate for some structures on sites with moderate *RF* ratings, but even then careful design and highly stiffened systems are needed. For sites that exhibit potential free-field heave of several inches, the most reliable system is a deep foundation with a structural floor. This chapter presents the methods for computation of pier heave and design procedures for deep foundations.

12.1 PIER AND GRADE BEAM FOUNDATION

The drilled pier and grade beam foundation system shown in Figure 12.1 can be used for both lightly and heavily loaded structures on expansive soils. The grade beam is designed to support the structural load between the piers. The grade beam may be a reinforced concrete basement wall or a stiff beam supported by the piers. It must be designed to mitigate the effects of differential pier movement on the superstructure. A void space must be maintained beneath the grade beam in order to isolate the structure from the soil and prevent soil swelling pressures from producing uplift forces on the grade beams. The void space also helps to concentrate the structural load on the piers to assist in counteracting uplift pressures.

The piers are typically reinforced concrete shafts. For residential structures, 10 in. or 12 in. (250 mm or 300 mm) diameter piers are commonly used. For heavier commercial structures, larger diameter piers are used. Pier diameters less than 10 in. (250 mm) are considered micropiles, and require particular attention to detail during construction to allow for proper placement of concrete along the entire length. Regardless of diameter, concrete should be placed using a tremie chute to prevent void spaces, honeycombed concrete, or excessive mixing with soil from the sides of the holes.

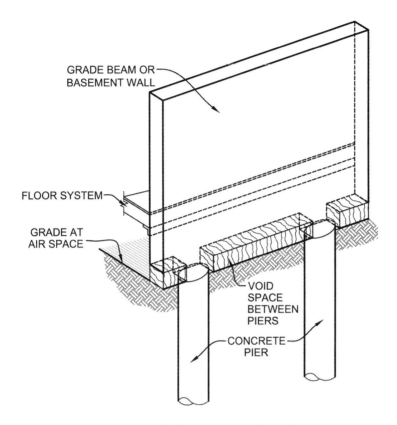

FIGURE 12.1. Typical detail of pier and grade beam foundation system.

The top several feet of the pier should be poured within a form completely to the top of the soil surface to prevent sloughing of soil at the top of the pier, which would cause a *mushrooming* effect on the pier. Figure 12.2 shows a drawing of a mushroomed pier and the forces acting on the top. The effect of mushrooming, along with uplift skin friction, has been seen to cause large uplift forces on the piers. Void spaces have been observed under the bottom of the piers that have experienced excessive heave due to uplift forces on such mushroomed tops.

Design of the pier and grade beam system consists primarily of calculating the predicted free-field heave and the predicted pier heave. The pier heave is a function of the free-field heave.

The depth of the design active zone is one of the most important parameters in pier design. The design active zone was defined in chapter 1, but it must be recognized that the presence of the pier can influence the depth of wetting. If the interface between the soil

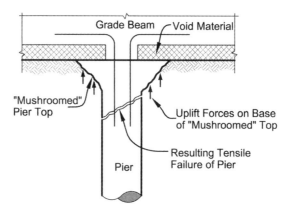

FIGURE 12.2. Incorrect construction of pier top, resulting in pier failure.

and the pier shaft is not sealed well, it can provide access for water to penetrate to depths greater than would be expected without this pathway, thereby causing deep-seated heave. This can be particularly critical if water-bearing strata or perched water tables are intersected. It may be appropriate to consider such deep wetting when defining the design active zone.

Assessing the performance of a pier requires a period of time for monitoring. Pier heave generally does not begin until some time after construction, whereas slab heave can begin almost immediately. Pier heave can lag slab heave by several years, resulting in problems with coverage by warranty programs. The onset of pier heave will depend on the rate at which the soil is wetted. Heave of different elements of the structure will occur at different times. Figures 12.3 and 12.4 demonstrate the mechanism responsible for the lag time between slab and pier heave. Figure 12.3 shows a cross-section of a pier and grade beam foundation system with a slab-on-grade concrete basement floor. Progression of a hypothetical wetting front at the site is shown in Figure 12.4a.

For the system shown in Figure 12.3, wetting of the soil under the slab, and hence, heave of the slab will begin almost immediately after construction for reasons discussed in chapter 7. Thus, progression of slab heave will occur as shown in the upper curve in Figure 12.4b. At some time, such as $t_1 = 4$ years, the wetting front will have progressed to the depth of $z_{A1} = 10$ ft as shown in Figure 12.4a.

Two different pier lengths will be considered for this example, one at 20 ft (6.1 m) and another at 35 ft (10.7 m). Pier heave generally does

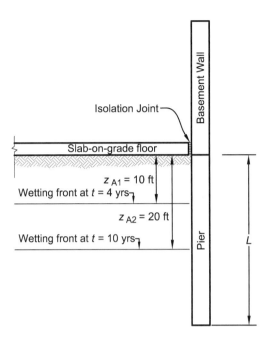

FIGURE 12.3. Foundation and floor system for analysis of hypothetical site (Nelson, Overton, and Durkee 2001).

not begin until wetting has progressed to a depth of about half the length of the pier. At the end of 4 years, the depth of wetting would be 10 ft (3.0 m), and slab heave of 2.5 in. (64 mm) would have occurred as shown in Figures 12.4a and 12.4b. Because the depth of wetting has reached about half its length, the 20 ft (6.1 m) pier would be starting to experience movement. The 35 ft (10.7 m) pier would not have experienced any.

At the end of 8 years, the depth of wetting would be 17.5 ft (5.3 m), and slab heave of 3.75 in. (95 mm) would have occurred, as shown in Figures 12.4a and 12.4b, respectively. The 20-ft (6.1-m) pier would have experienced over 1.5 in. (38 mm) of movement and the 35-ft (10.7-m) pier would be starting to experience movement.

This phenomenon can have serious financial consequences. Although the slab heave becomes evident within a few years, the pier heave may not become evident until warranties have expired. The illustration shown in Figure 12.4 points out the need for careful design of the foundation system, and for consideration of the building element interactions.

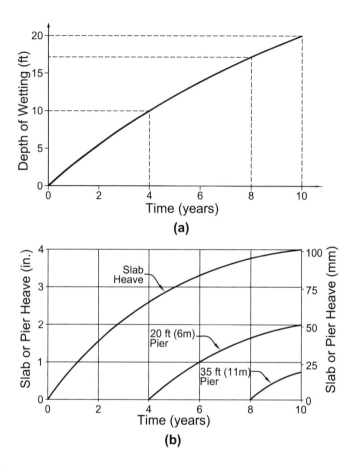

FIGURE 12.4. Lag time between onset of pier heave relative to slab movement: (a) progression of wetting front; (b) slab and pier heave (modified from Nelson, Overton, and Durkee 2001).

12.1.1   Design Methods

The principle on which pier design is based is to found the piers in a sound stratum at sufficient depth that will provide anchorage to minimize movement due to uplift forces exerted by the expansive soils. If a stable nonexpansive stratum or the depth of potential heave exists at a shallow depth, the pier may be designed as a rigid member anchored sufficiently deep in that stratum so as to prevent movement. This method is termed the *rigid pier method*. However, if the design active zone is deep, the required pier length calculated by the rigid pier method may be too long to be practical. The pier may then be designed in such a manner as to allow for some amount of tolerable movement. To do this, the predicted pier heave must be computed and the pier length designed accordingly so that it is within the amount of movement that the structure can accommodate. Nelson and Miller (1992) presented such a method to calculate pier heave. That method has been termed the *elastic pier method*. The elastic pier method was developed for uniform soil profiles and piers with limited length to diameter ratios and is difficult to apply to more complex soil profiles and pier geometry. More recently, a finite element method of analysis was developed to compute pier movement in expansive soils having variable soil profiles, complex wetting profiles, large length-to-diameter ratios, and complex pier configurations and materials. The model has been named APEX (for Analysis of Piers in EXpansive soils). Details of the method are presented in Nelson, Chao et al. (2012) and Nelson, Thompson et al. (2012) and are summarized in Section 12.1.1.2.

12.1.1.1 *Rigid Pier Method*

In the rigid pier design method, the uplift forces are equated to the anchorage forces, and it is assumed that there is no heave of the pier. The forces assumed to be acting on a rigid pier are shown in Figure 12.5. The principle of the design is that the negative skin friction below the design active zone plus the dead load, P_{dl}, must resist the uplift pressures exerted on the pier by the expansive soil. This assumes that the bottom of the pier is founded below the zone where soil expansion can occur.

Chen (1988) and O'Neill (1988) presented similar methods of analysis for rigid piers. Chen (1988) assumed that the uplift skin friction is constant throughout the design depth of wetting. O'Neill (1988) considered that for a short interval at the bottom of the design active zone there will be a transition zone where the uplift skin friction increases from zero (at the bottom) to a limiting constant value that exists throughout the upper part of the design active zone. The length of this transition zone has been discussed in chapter 7, and it can be investigated on a project-specific basis if an analysis of soil wetting is performed. If such an analysis is not performed, it is conservative to assume that the length of the transition zone is very small or zero.

The skin friction in the uplift and anchorage zones is treated simply as Coulomb skin friction. The friction force is taken as being equal to the net normal stress acting on the side of the pier times a coefficient of friction (Chen 1988; Nelson and Miller 1992). The net normal stress acting on the pier will be equal to the swelling pressure of the soil. Thus, the uplift skin friction, f_u, is a function of the CV swelling pressure of the

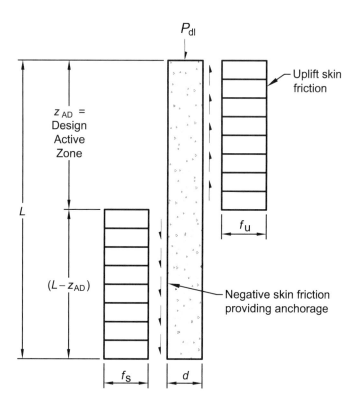

FIGURE 12.5. Forces acting on a rigid pier in expansive soil.

soil and can be expressed by the form shown in equation (12-1).

$$f_u = \alpha_1 \sigma''_{cv} \tag{12-1}$$

where:

α_1 = coefficient of uplift between the pier and the soil, and
σ''_{cv} = the CV swelling pressure in terms of net normal stress.

In the anchorage zone below the design active depth one of two possible conditions could exist. If the soil below that zone is not expansive, the lateral stress would be the lateral earth pressure at rest. In that case, the anchorage skin friction, f_s, can be computed as

$$f_s = \alpha_2 K_0 \sigma''_{vo} \tag{12-2}$$

where:

α_2 = coefficient of anchorage between the pier and the soil,
K_0 = coefficient of lateral earth pressure at rest, and
σ''_{vo} = overburden stress.

If the design active zone is the depth of potential heave, and if expansive soil exists below that depth, and if that soil can be wetted, the anchorage skin friction would be given by an equation similar to equation (12-1). However, the critical condition would be at the time when the design active zone has been wetted but the soil in the anchorage zone has not. In that case, equation (12-2) would apply. Therefore, for design purposes equation (12-2) should be used.

Model studies by Chen (1988) indicated that the value of α_1 should range from 0.09 to 0.18. Taking into consideration Chen's value, and work by O'Neill (1988), Nelson and Miller (1992) indicated that the values of α_1 and α_2 would range from 0.10 to 0.25. Some practicing engineers in the Front Range area of Colorado have frequently used a value of 0.15 for α_1 (CAGE 1999).

Long-term measurements on drilled piers were conducted at the Colorado State University (CSU) Expansive Soils Field Test Site. Full-scale drilled concrete piers, 14 in. (360 mm) in diameter and 25 ft (7.6 m) long, were constructed in claystone of the Pierre Shale. They were instrumented with embedded vibrating-wire strain gauges for concrete. Readings were taken over a period of 10 years. The results indicated that reasonable values for α_1 and α_2 range from about 0.4 to 0.6 (Benvenga 2005).

The values obtained at CSU appear to be more consistent with values normally assumed for adhesion factors in the design of piles in ordinary soils. Thus, it is believed that a value of α between 0.4 and 0.6 is the more reasonable range, mainly because this is based on actual measurements on full-scale concrete piers.

The value of σ''_{cv} can be determined from oedometer tests, as discussed in chapter 6. This should actually be determined on a horizontally oriented sample, but if that is not possible the vertical swelling pressure can be used. Normally, the vertical swelling pressure will be greater that the horizontal swelling pressure. However, in steeply dipping beds, the opposite may be true.

The total uplift force, U, can be computed by integration of the skin friction, f_u, over the area of the pier within the design active zone depth, z_{AD}. For a uniform distribution of uplift skin friction as shown in Figure 12.5, U would be computed as,

$$U = \pi d f_u z_{AD} \qquad (12\text{-}3)$$

where:

d = pier shaft diameter.

The uplift force must be resisted by the applied dead load and skin friction in the anchorage zone beneath the design active zone. For a straight shaft pier with a uniformly distributed anchorage skin friction as shown in Figure 12.5, the resistance force, R, would be given as

$$R = P_{dl} + \pi d f_s (L - z_{AD}) \qquad (12\text{-}4)$$

where:

R = resistance force,
P_{dl} = dead load,
f_s = anchorage skin friction below the design active zone,
L = length of pier, and
z_{AD} = depth of design active zone.

By setting equations (12-3) and (12-4) equal, Nelson and Miller (1992) presented the equation for required length of a rigid straight shaft pier in a single soil layer as

$$L = z_{AD} + \frac{1}{f_s}\left[\alpha_1 \sigma_{cv}'' z_{AD} - \frac{P_{dl}}{\pi d}\right] \qquad (12\text{-}5)$$

The pier must be reinforced over its entire length to resist the tensile forces that are developed. The tensile force can be computed using equation (12-6).

$$F_t = U - P_{dl} \qquad (12\text{-}6)$$

where F_t is the maximum interior tensile force in the pier, and other terms have already been defined.

12.1.1.2   APEX Method

In many cases, the pier may extend through various strata of expansive soil that exhibit widely varying properties. Also, micropiles are finding wider use in expansive soil applications. These piles exhibit L/d ratios well in excess of 20. A finite element method of analysis was developed that is capable of analyzing complex soil heave profiles and nonuniform soil-to-pier interface conditions. It is capable of analyzing piers that are cased or sleeved over a portion of their length. As mentioned earlier, that method has been termed the APEX method. The development of that code is summarized below. A more detailed description of the code and its validation is presented in Nelson, Chao et al. (2012) and Nelson, Thompson et al. (2012).

APEX Field Equations
Material properties can be specified point-wise throughout the analysis domain. The swell is assumed to be isotropic, and is accounted for by writing the constitutive equations as

$$\varepsilon_{rr} = \frac{1}{E_s}[\sigma_{rr} - v(\sigma_{\theta\theta} + \sigma_{zz})] + \varepsilon_{iso} \tag{12-7}$$

$$\varepsilon_{\theta\theta} = \frac{1}{E_s}[\sigma_{\theta\theta} - v(\sigma_{zz} + \sigma_{rr})] + \varepsilon_{iso} \tag{12-8}$$

$$\varepsilon_{zz} = \frac{1}{E_s}[\sigma_{zz} - v(\sigma_{zz} + \sigma_{\theta\theta})] + \varepsilon_{iso} \tag{12-9}$$

where:

$\varepsilon_{rr}, \varepsilon_{\theta\theta}, \varepsilon_{zz}$ = components of strain in cylindrical coordinates,
E_s = modulus of elasticity of the soil,
$\sigma_{rr}, \sigma_{\theta\theta}, \sigma_{zz}$ = components of stress in cylindrical coordinates,
v = Poisson's ratio, and
ε_{iso} = isotropic swelling strain. This is the same as $\varepsilon_{s\%}$ for an isotropic soil.

The Boundary Conditions
The soil-to-pier interface is modeled such that either slip between the soil and the pier (Coulomb friction) or failure within the soil adjacent to the pier (Mohr–Coulomb failure) can take place. Axial strain in the pier is assumed to be negligible relative to strain in the soil.

   The boundary conditions at the soil-to-pier interface specify a relationship between the nodal displacement and nodal force. This specification is used for the vertical component of displacement at nodes where the soil would be in contact with the pier. The interface element is shown in Figure 12.6. The force at the interface is given by equation (12-10).

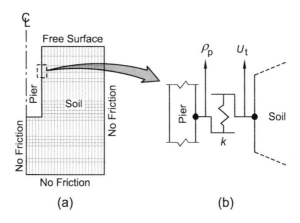

FIGURE 12.6. Boundary conditions: (a) soil boundary conditions; (b) pier–soil boundary conditions.

$$F_t = k(\rho_p - U_t) \tag{12-10}$$

where:

F_t = the nodal force tangent to pier,
ρ_p = the pier heave,
U_t = the nodal displacement tangent to the pier, and
k = a parameter used to adjust shear stress.

The parameter k is considered to be similar to a spring constant of the connection between U_t and ρ_p.

Development of Pier Heave
The pier must be in equilibrium, and therefore, the tangential forces exerted on the pier by the soil must equal the total external load on the pier. Figure 12.7 illustrates how ρ_p is adjusted in order to bring about equilibrium. Figure 12.7a illustrates the boundary conditions before the soil swells. In this state there is no uplift force on the pier. Figure 12.7b illustrates the state of the boundary conditions after the swelling has taken place but before any pier heave has occurred. The shear forces create an upward force on the pier, and the pier is no longer in equilibrium. In order to maintain equilibrium, the pier must move upward by the appropriate amount to create both upward and downward forces acting on the pier, as illustrated in Figure 12.7c.

Interface Stress
Limiting values for slip failure are defined by Coulomb friction or shear failure within the soil next to the pier, as defined by Mohr–Coulomb

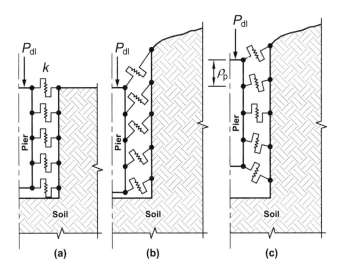

FIGURE 12.7. Pier and soil interface conditions: (a) initial conditions of no pier heave; (b) soil heave with no pier heave; (c) pier heave to maintain equilibrium of pier (Nelson, Chao et al. 2012).

shear strength. In either case, they are expressed in terms of normal and tangential components of stress. For the Mohr–Coulomb failure theory, the soil begins to fail at a peak strength value and is adjusted to decrease to the residual strength as strain decreases.

If the stress in the soil adjacent to the pier exceeds that which is necessary to cause soil failure by the Mohr–Coulomb theory, and there has been no slip of the soil based on Coulomb friction, a small zone of soil adjacent to the pier will fail in shear and the shear stress on the side of the pier will equal the shear strength of the soil.

Coulomb friction is defined by a linear relationship between normal force and skin friction, as was discussed in Section 12.2.1.1. Shear strength is defined by the conventional shear strength equation. Figure 12.8 shows the two relationships. Interface failure will occur by the lower value of the two mechanisms as demonstrated by the solid line in Figure 12.8.

Input Parameters for Pier Heave Computation
The APEX code requires that input parameters be specified on a layer-by-layer basis for the entire soil profile. The primary elastic properties used in APEX analysis are the Young's modulus of the soil, E_s, Poisson's ratio, v, and the coefficient of earth pressure at rest, K_0. Other parameters are the pier–soil adhesion factor, α, the shear

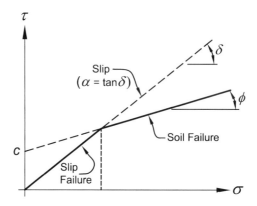

FIGURE 12.8. Strength envelopes for slip and soil failure modes.

strength of the soil including the peak and residual angle of internal friction, ϕ, and the cohesion, c.

Young's modulus of the soil, E_s, can be measured in the laboratory by means of unconfined compression or triaxial shear tests. It can also be determined from oedometer test results by correlation with the constrained modulus, D, measured during the loading portion of the test. In the design charts the value of E_s is expressed in units of bars. When that is the case it is denoted by E_A.

Poisson's ratio, υ, will vary with the stiffness of the soil, and hence, its water content. For stiff clays and claystones a typical value will be from 0.2 to 0.3 and for soft clays it can vary up to values around 0.4.

The coefficient of earth pressure at rest, K_0, can vary from values below 1.0 for softer soils up to values of 3.0 or more for highly overconsolidated clays (Lambe and Whitman 1969).

The soil-to-pier adhesion factor, α, was discussed in Section 12.1.1.1. Research at Colorado State University has shown that reasonable values of α range from about 0.4 to 0.6 (Benvenga 2005).

Pier Design Charts
Design charts were developed using the APEX program for use in facilitating pier design for sites where the soil conditions can be represented by a simplified heave profile. Charts were prepared for two types of cumulative free-field heave distributions. The shapes of the two types of distributions are shown in the insets on Figures 12.9a and 12.9b.

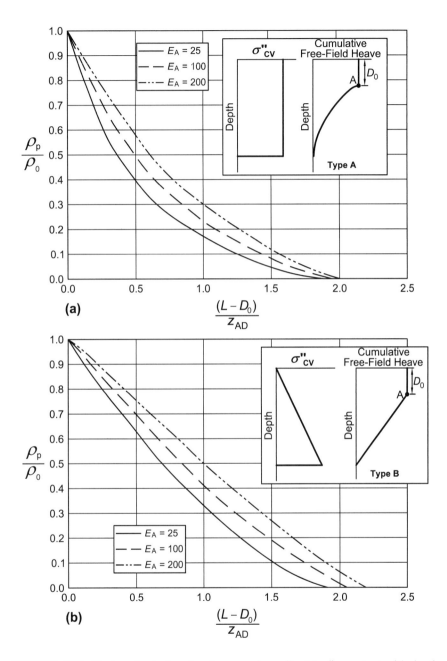

FIGURE 12.9. Design charts for piers in expansive soils: (a) σ_{cv}'' constant with depth; (b) σ_{cv}'' increasing with depth.

The top segment of the two distributions shown in the insets is a vertical line of depth D_0. This represents the depth of a layer of nonexpansive fill such as was considered in Table 11.1 in chapter 11. If such a layer does not exist the value of D_0 is zero. The Type A profile was prepared for the case where the expansion potential of the soil below the depth D_0 is constant with depth. In that case the cumulative heave profile has a logarithmic shape, as shown in Figure 12.9a. The Type B profile shown in Figure 12.9b was prepared for the case where nonexpansive soil exists at the top of the soil layer below which the expansion potential below the depth D_0 increases with depth.

Figure 12.9 presents the ratio of pier heave to free-field heave, ρ_p/ρ_0, as a function of pier length relative to the depth of the design active zone, $(L - D_0)/z_{AD}$. The design curves were developed for a zero dead load condition. The value of α was held constant at 0.4 over the entire depth. The value of v was taken as 0.4, and K_0 was assumed to be 1.0. In Figure 12.9, Young's modulus of the soil is expressed as E_A, which is E_s in units of bars. Because nonexpansive or remolded soil would likely be not as stiff as the expansive soil, the value of E_A for nonexpansive fill above a depth of D_0 was assumed to be equal to one-third that of the expansive soil.

The design charts must be used with caution. All parameters could not be kept in a dimensionless form when using the APEX program to develop the charts. They are expected to represent conservative conditions. Nevertheless, the charts can be used as a guide to the general magnitude of the required pier length. A more accurate method of analysis would be to use the APEX program and actual soil conditions.

12.1.2   Load-Bearing Capacity

Heave of the pier foundation is generally the governing potential mode of failure to be considered when designing the pier foundation on expansive bedrock. However, that may not always be true if the bedrock is in a badly fractured state such that significant relative slip between bedrock fragments can occur. The load carried by the pier foundation must be borne by friction and/or end bearing. Furthermore, prior to wetting, or if the anticipated wetting does not occur, the expansive bedrock must be capable of carrying the load placed upon the foundation without a shear failure and with the resulting settlements being tolerable for the foundation. Methods for design of the pier under normal conditions, prior to heave taking place, can be found in various textbooks on foundation engineering.

12.2 PATENTED PIERS

Patented piers, such as helical piles, micropiles, and push piers, are often used for remediation of foundations in expansive soils that have experienced intolerable heave. They are also used for new construction, but not as frequently as conventional drilled piers. The small size and versatility of the drilling equipment with low headroom requirements make patented piers well-suited for applications in confined areas such as crawl spaces or garages. The design and application of patented piers are discussed in the following sections.

12.2.1 Helical Piles

Helical piles derived their origin as anchors for structures such as power poles or transmission lines. They began to be used fairly frequently for foundation systems in the early 1990s. The original helical pile system had little resistance to lateral loads. More recently, helical pile systems have been developed that have a large diameter for the upper portion of the helical pile, such that they have the ability to provide some resistance to lateral force (Perko 2009). Figure 12.10 shows a diagram of a helical pile.

   Helical piles should be advanced until both (1) the required minimum length is obtained, and (2) the required final installation torque is achieved. One of the disadvantages of the use of helical piles in expansive bedrock is the difficulty in penetrating hard expansive bedrock with a SPT blow count value of 50 blows per ft or greater.

   The helical pile must be designed so that the predicted heave is within tolerable limits. The design must also provide for adequate pullout capacity to resist the uplift forces imparted on the foundation. When a helical pile system is used in expansive soils, it is important that no "slack" or looseness exist in the connections between rod sections.

   The principle of the design is that the pullout capacity of the helical bearing plates plus the dead load must resist the total uplift force exerted on the pier. Uplift forces are produced by the swelling pressures acting on the pier above the design active zone, and uplift forces produced by heave of the expansive soil acting on other parts of the foundation system. As indicated in equation (12-3), the uplift force acting on the shaft of the helical pile can be computed by integration of the skin friction, f_u, over the area of the pier within the design active zone depth, z_{AD}. This force is normally small due to the small diameter of the shaft and disturbance of the adjacent soil during installation. Other contributions to uplift force are foundation specific. The pullout capacity of

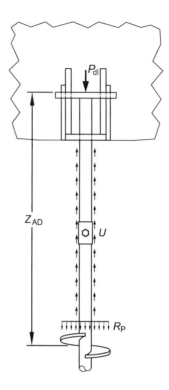

FIGURE 12.10. Helical pile.

the helical bearing plates, R_p, plus the dead load, P_{dl}, can be equated to the total uplift force, U, as given by equation (12-3). Pullout capacity can be computed using methods presented in Perko (2009).

For the design of helical piles, the value of α_1 in equation (12-1) is different from the value of α_1 for the design of ordinary drilled concrete piers. Perko (2009) indicated that a value of $\alpha_1 = 0.1$ is appropriate for design of the helical piles in order to account for the reduced adhesion along the steel shaft. In fact, the actual α_1 value may be less than 0.1, because as the pier is advanced into the soil, the material in the annular space above the helix, in which the shaft is centered, is disturbed. It is not recompacted and the swell potential in this area is reduced.

The skin friction along the shaft is generally low due to the small diameter of the shaft and the remolding and straining of soil that occurs immediately around the shaft during construction of the piers. In predicting heave of a helical pile, it is assumed that heaving of the soils immediately around the shaft above the depth of the helix has no influence on the pier. The heave of the pier is limited by the amount by which the helix can move, which is the amount that the soil at the depth of helix will heave. This value can be determined from the calculations of free-field heave.

When the depth of the design active zone is not deep, the minimum depth of the bearing plates of a helical pile may be placed greater than the design active zone depth, z_{AD}. Perko (2009) indicated that some practitioners received good design results without any additional penetration beyond z_{AD}, whereas some practitioners use a distance of 3 ft to 5 ft (1 m to 1.5 m) below z_{AD} to provide some margin for error.

12.2.2 Micropiles

Micropiles have been used to underpin foundations since the early 1950s, and they are increasingly being used for underpinning foundations experiencing heave due to expansive soils. The small surface area of the micropile reduces the ability of the expansive soils to transfer load to the micropile. Further reduction of the uplift forces can be accomplished by the use of a casing in the upper parts of the shaft. If the upper zone is cased, care must be taken to seal the annular space around the drilled shaft.

A typical micropile is constructed by first drilling a small diameter boring, generally 4 in. to 6 in. (100 mm to 150 mm) in diameter. A steel reinforcing bar is inserted and grout is tremied into the hole. If a casing is used it is inserted into the hole for the upper several feet sufficient to minimize uplift forces. The capacity of a micropile to support a structure and resist uplift in expansive soil is provided primarily through friction at the interface of the shaft and the surrounding soil. Uplift shear stresses create tensile stresses within the micropile, the magnitude of which depends on the frictional characteristics along the soil-to-pile interface. These frictional characteristics are design variables that must be incorporated into the design of the micropile.

Details of a typical micropile are shown in Figure 12.11. The upper portion of the micropile is cased with a PVC sleeve while the bottom portion has grout in direct contact with the soil. Depending on the method of construction and the fit between the PVC casing and the drilled hole, grout can flow up in the annulus between the PVC and the side of the boring. An "all-thread rod" is typically used to reinforce the micropile and provide a means for attachment to the foundation (Nelson et al. 2013).

When a micropile is tremie grouted from the bottom of the hole, the grout flows up the inside of the casing in the upper cased section and is in direct contact with the casing material. Usually, the grout can flow

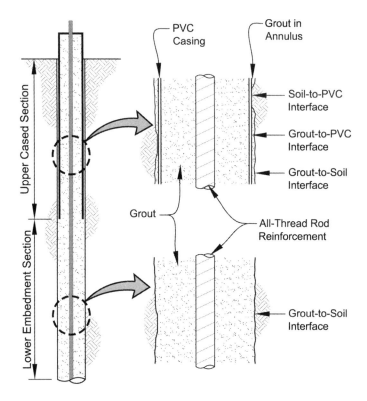

FIGURE 12.11. Typical configuration of micropile installed in expansive soil (Schaut et al. 2011).

up into the annulus formed between the outside of the casing and the soil. Grout may also flow down into the annulus from grout overflow at the ground surface. Good practice is to seal off the annulus to prevent deep-wetting of the soil. Figure 12.11 shows examples of the different interface conditions.

Figure 12.12 shows the PVC casing placed in the drilled hole prior to grout placement. Figure 12.13 shows the tremie grouting of the micropile. After curing of the grout, the micropile is connected to the foundation wall by means of a bracket bolted to the wall. Such a bracket can be seen in Figure 12.13. It is very important that the micropile be attached to the bracket on the wall or grade beam so that it can resist uplift forces due to potential uneven movement of the micropiles or uplift forces exerted on the foundation walls.

Either the rigid pier method or the APEX method, as described in Sections 12.1.1.1 and 12.1.1.2, are suitable for design of the micropiles. Because of the large L/d ratio of a typical micropile, and the fact that

FIGURE 12.12. Micropile prior to grouting with annulus around the outside of casing (Schaut et al. 2011).

FIGURE 12.13. Installation of micropile during grouting.

the interface conditions may vary along the length of the micropile, case-specific analysis using the APEX program is recommended instead of relying on the design charts in Figure 12.9.

Regardless of the method of analysis used, a thorough under-standing of the frictional characteristics of the pertinent interfaces

TABLE 12.1   Summary of Micropile Interface Shear Strength Test Data (Schaut et al. 2011)

Interface	Direct Shear Test				Modified Triaxial Test			
	ϕ_p	C_p (psf)	ϕ_r	C_r (psf)	ϕ_p	C_p (psf)	ϕ_r	C_r (psf)
Grout-to-PVC (smooth, dry)	16.3°	0	11.7°	0	12.8°	0	10.1°	0
Grout-to-PVC (rough, dry)	–	–	–	–	20.1°	0	15.3°	0
Remolded soil-to-PVC (in situ water content)	16.6°	923	15.1°	700	–	–	–	–
Remolded soil-to-PVC (inundated)	18.6°	0	15.4°	0	–	–	–	–
Soil-to-PVC (in situ water content)	–	–	–	–	11.2°	0	9.7°	0
Soil-to-grout (in situ water content)	20.0°	2,231	16.4°	1,296	–	–	–	–
Soil-to-grout (inundated)	19.8°	1,314	14.5°	851	–	–	–	–

Note: ϕ_p = peak angle of internal friction, C_p = peak cohesion, ϕ_r = residual angle of internal friction, and C_r = residual cohesion.

is required. Schaut et al. (2011) performed conventional direct shear tests and modified triaxial tests to evaluate the interface friction characteristics for each of the interfaces described above. Table 12.1 summarizes the interface shear strength results. As shown in Table 12.1, the soil-to-grout interface has the highest interface shear strength, whereas the soil-to-PVC interface has the lowest shear strength.

12.2.3   Push Piers

Push piers are another form of underpinning that is used for reme-diation of distressed foundation systems. They can be installed using relatively compact equipment with low headroom requirements. Push piers consist of short sections of steel tube that are coupled together by an internal sleeve. Figure 12.14 shows the typical components of the push piers. For remediation of a foundation system on expansive soils, a heavy-duty bracket is attached to the foundation wall or grade beam after the foundation has been exposed. The steel shaft is pushed into the ground using a hydraulic ram. The weight of the structure provides the resistance necessary to force the push pier into the ground. Therefore, the reaction force available to advance the pier is limited by the weight of the structure and the strength of the foundation wall or grade beam. This was illustrated in the case study presented in chapter 11.

FIGURE 12.14. Components of push piers (courtesy of Magnum Piering, Inc.).

The inside of the push pier may be grouted using a tremie method to increase the capacity of the pier. In some cases, a reinforcing bar may also be inserted into the shaft. The design methods previously described are suitable for design of the push piers. The frictional characteristics at the interfaces involved, as discussed in Section 12.2.2, also apply to the design of the push piers.

Push piers are used extensively in regions like the central United States where seasonal moisture content changes cause heaving of near surface clay till and cause distress to shallow foundations. In these areas, the soils are not as overconsolidated as in the western United States, and predrilling often is not necessary. A friction reduction ring slightly larger in diameter than the pier shaft is used on the bottom end of the push pier to create an annulus during pier driving. The small annulus allows the pier to penetrate two to three times deeper than would be achieved without the friction reduction ring. This depth is important in resisting shallow expansive soil heave.

Drilling equipment can be used to create a pilot hole for push piers and allow for penetration into very stiff clays, mudstone, and shale. The depth of the mini-drilling devices that attach to the push pier bracket are typically limited to about 40 ft. The drilled push pier is a hybrid of the push pier and the micropile; it offers the advantages of low-cost push pier casing and the penetration capability of micropile drilling techniques.

12.3 DEEP FOUNDATION DESIGN EXAMPLES

Examples of deep foundation design are illustrated in this section. Examples 12.1 through 12.4 all use the soil profile presented in Table E12.1. In these examples, the design active zone is assumed to be equal to the depth of potential heave, and the deep foundation is designed assuming the soils/bedrock will be fully wetted to that depth. If the design active zone depth were shallower than the depth of potential heave, the foundation design would be similar, but the calculated design depth and degree of wetting would be determined from a water migration analysis, as discussed in chapter 7. In that case, the primary difference in the design analysis would be the effect that partial wetting would have on the heave profile as was presented in Example 8.4.

Chapter 7 discussed how the soil profile can become wetted to the extent necessary to develop the full expansion potential of the soil. If the depth of potential heave is not overly deep, the design active zone may consist of the fully wetted depth of potential heave. That is the assumption made in the examples presented in this section.

12.3.1 Rigid Pier Design Example

The rigid pier design method assumes that the uplift and resistance forces are in equilibrium, and therefore, that there is no pier movement. Example 12.1 illustrates the computations. The following steps should be noted.

1. The design active zone is considered to be the depth of potential heave, which is computed by setting the overburden stress equal to the CV swelling pressure, σ''_{cv}. The overburden stress at the bottom of the tan claystone is seen to be less than σ''_{cv} for the tan claystone but greater than that for the gray claystone. Thus, the interface between these two strata represents the depth of potential heave.
2. The uplift force on the pier is computed from the friction acting in the design active zone. This computation uses the values of σ''_{cv} in the design active zone and a value for α based on measurements performed at Colorado State University.
3. The depth of the required anchorage zone is computed by equating the anchorage force to the uplift force less the dead load. This computation necessitates an evaluation of the lateral stress acting on the lower portion of the pier. For this example, it is assumed that the lateral stress is equal to the lateral earth pressure at rest, with a value of $K_0 = 1$.

EXAMPLE 12.1

Given:

The soil profile and properties are shown in Table E12.1. Drilled piers 12 in. in diameter will be constructed with the top of the pier at the ground surface. The dead load on the pier will be 11 kips.

TABLE E12.1   Soil Profile for Example 12.1

Depth (ft)	Soil Type	Water Content (%)	Dry Density (pcf)	CS % Swell, $\varepsilon_{s\%}$ @ 1,000 psf (%)	CV Swelling Pressure, σ_{cv}'' (psf)
0–26	Tan claystone	12.3	114	4.3	4,527
26–50	Gray claystone	11.0	108	3.6	2,008

Find:

Using the rigid pier design method, compute

1. Required length, L_{reqd}, of a straight shaft pier.
2. Tensile force for the straight shaft pier.

Solution:

Part 1:
The depth of potential heave is computed first. At the bottom of the tan claystone, i.e., at depth = 26 ft, the overburden stress is

$$\sigma_{vo}'' = 128 \times 26 = 3,328 \text{ psf}$$

This value is less than σ_{cv}'' in the tan claystone, but it is greater than that of the gray claystone. Thus, the depth of potential heave is at the interface between the tan and gray claystones. As noted above, this will be taken as the depth of the design active zone. Thus, $z_{AD} = 26$ ft.

The required rigid pier length can be computed from equation (12-5) if the soil profile consists of a single uniform soil. However, in this example the soil profile comprises two different soil strata. The required length of a rigid pier is calculated by equating the uplift forces as depicted in Figure 12.5 to the sum of the negative (anchorage) skin friction forces and the dead load. Based on the values of α measured by Benvenga (2005), reasonable values of α_1 and α_2 are assumed to be 0.4 for both uplift and anchorage.

From equation (12-3), the uplift force is calculated to be

$$U = \pi d\alpha_1 \sigma_{cv}'' z_{AD}$$

$$U = \pi \times \left(\frac{12}{12}\right) \times 0.4 \times 4,527 \times 26 = 147,909 \text{ lb}$$

From equation (12-4), the resistance force is calculated to be

$$R = P_{dl} + \pi d f_s (L - z_{AD})$$

To compute the anchorage skin friction, f_s, the lateral stress acting on the pier below the design active zone must be computed. For this computation, the critical condition exists when the zone of soil above the design active zone has been wetted but the soil below that has not. Thus, it may be assumed that the lateral stress, σ_h'', acting on the pier is equal to the earth pressure at rest. Thus,

$$\sigma_h'' = K_0 \sigma_{vo}''$$

The value of K_0 is conservatively estimated to be 1.0 for an overconsolidated clay. The lateral stress will vary with depth in the anchorage zone. The overburden stress at the depth of the design active zone was computed above to be 3,328 psf. Thus, at a depth of 26 ft

$$(\sigma_h'')_{26ft} = 1.0 \times 3{,}328 \text{ psf}$$

At the bottom of the pier the lateral stress acting on the pier is equal to

$$(\sigma_h'')_L = 1.0[3{,}328 \text{ psf} + 120(L - 26)]$$

The average lateral stress in the anchorage zone is

$$(\sigma_h'')_L = \frac{3{,}328 + [3{,}328 + 120(L - 26)]}{2} = 3{,}328 + 60(L - 26)$$

and the average skin friction in that zone is

$$f_s = 0.4[3{,}328 + 60(L - 26)] = (24L + 707.2)$$

Thus, the total resistance force is

$$R = 11{,}000 + \pi \left(\frac{12}{12}\right)(24L + 707.2)(L - 26)$$

Setting $R = U$ gives a quadratic equation that can be solved to give

$$L_{reqd} = 49 \text{ ft}$$

Part 2:
The maximum tensile force, P_{max}, will occur at the depth of the design active zone and will be equal to

$$P_{max} = P_{dl} - U = 11{,}000 - 147{,}909 = -136{,}909 \text{ lb} = -136.9 \text{ kips}$$

The negative sign indicates that the force is tensile.

Depending on the potential for cracking of the concrete in the pier and the consequential potential for corrosion of the steel, it may be desired to increase the amount of steel to provide for some sacrificial reinforcement. That decision must be made at the discretion of the designer.

12.3.2   APEX Design Example

Examples 12.2 and 12.3 illustrate the use of the design charts presented in Figure 12.9 for two different soil profiles. Comparison of the required pier length computed in Examples 12.1 and 12.2 shows that if a tolerable movement of 1 in. is acceptable for the structure, the length of the pier was reduced by a significant amount.

EXAMPLE 12.2

Given:

The same soil profile and properties as for Example 12.1. The elastic modulus, E_s, of both the native clay and claystone was measured to be 93,000 psf. The cumulative free-field heave profile was computed using the method demonstrated in Example 8.1 and is shown in Figure E12.2. The allowable pier movement is 1.0 in.

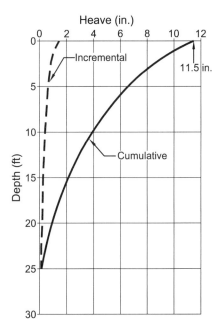

FIGURE E12.2. Cumulative free-field heave profile determined from Example 12.1.

Find:

The required pier length using the design charts in Figure 12.9.

Solution:

The cumulative free-field heave curve shown in Figure E12.2 has a shape similar to the one shown in Figure 12.9a. Because the soil has expansion potential up to the ground surface, the value of D_0 is zero. The pier design chart shown in Figure 12.9a will be used to determine the required pier length.

The elastic modulus of the native clay and claystone was measured to be 93,000 psf. To convert this unit to bars, it is convenient to remember that 1 bar is almost exactly equal to 1 atmosphere, or 2,116 psf.

Thus,

$$E_A = \frac{93,000}{2,116} = 44 \text{ bar}$$

For the value of free-field heave shown in Figure E12.2, the allowable normalized pier heave was $\rho_p/\rho_0 = (1.0/11.5) = 0.09$. Interpolating between the curves for $E_A = 25$ and $E_A = 100$ bar, $(L - D_0)/z_{AD}$ is read off the design chart shown in Figure 12.9a to be 1.38.

Remembering that $D_0 = 0$,

$$L_{reqd} = 1.38 \times 26 = 36 \text{ ft}$$

The exact soil profile was also analyzed using the APEX computer program. The required pier length was computed to be 38 ft, which agrees closely with the length determined using Figure 12.9a.

EXAMPLE 12.3

Given:

The same soil profile and properties as in Example 12.1 except that the upper 10 ft of tan claystone was excavated and replaced with nonexpansive fill. The cumulative free-field heave was computed as in Example 8.1 and is shown in Figure E12.3. The allowable pier movement is 1.0 in. The water content and dry density of the fill were 17.2 percent and 104.6 pcf, respectively.

Find:

The required length of a straight shaft pier using the design charts developed from the APEX program.

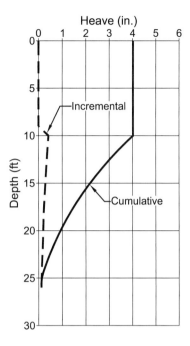

FIGURE E12.3. Cumulative free-field heave profile determined for Example 12.3.

Solution:

The cumulative free-field heave was computed to be 4.2 in. Thus, the pier will be designed for a value of ρ_p/ρ_0 equal to

$$\frac{\rho_p}{\rho_0} = \frac{1.0}{4.2} = 0.24$$

The cumulative heave profile shown in Figure E12.3 has a shape similar to that shown in Figure 12.9a. The depth D_0 is equal to the depth of overexcavation of 10 ft. The value of E_A for the claystone was measured in Example 12.2 to be 44.

Interpolating between the curves for $E_A = 25$ bar and 100 bar, the required value of $(L - D_0)/z_{AD}$ is read from the pier design chart in Figure 12.9a to be 0.82.

$$\frac{L - D_0}{z_{AD}} = \frac{L - 10}{26} = 0.82$$

$$L_{reqd} = 31.3 \text{ ft (say, 32 ft)}$$

The exact soil profile was also analyzed using the APEX computer program. The required pier length was computed to be 29 ft.

12.3.3   Helical Pile Design Example

The design of helical piles for heave in expansive soil is illustrated in Example 12.4. The solution consists of determining at what depth the free-field heave is equal to the design tolerable heave.

EXAMPLE 12.4

Given:

The same soil profile and properties as in Example 12.1.

Find:

Required length of a helical pile for a maximum tolerable movement of 1 in.

Solution:

As discussed in Section 12.2.1, the helix will move by the same amount that the soil at the depth equal to the length of the helical pier will heave. This value can be determined from the calculations of free-field heave. The free-field heave profile is shown in Figure E12.2. In that figure, the depth at which the cumulative heave is equal to 1.0 in. is 20 ft. Thus, a helical pile 20 ft long can be expected to heave about 1 in.

12.4   REMEDIAL MEASURES FOR DEEP FOUNDATIONS

Distress to structures founded on deep foundations on expansive soils can be caused by a variety of factors.

- Uplift of the foundation through skin friction
- Uplift on the base of grade beams or basement walls
- Loads from uplift or shrinkage under a floor slab connected to the foundation system
- Improper pier design, (e.g., pier length too short, or pier reinforcement inadequate)
- Improper construction of void space under grade beams
- Improper pier construction (e.g., allowing a mushroom to form at the top of the pier; see Figure 12.2), not anchoring the pier in stable bedrock, poor concrete quality, or void in pier shaft
- Excessive lateral pressure on foundation walls

12.4.1   Pier and Grade Beam Foundation

Remedial measures for an existing pier and grade beam foundation will depend on the mechanism causing the failure. If the failure is due to insufficient void space beneath grade beams, floor slabs, or mushroomed piers, it may be possible to simply excavate beneath those structural members to create enough void space to accommodate the future heave. This, of course, assumes that the foundation system is still structurally sound and can be repaired without being replaced.

In other cases, replacement of foundation elements, along with a full program of underpinning, may be needed. Depending on the nature of the cause of distress, some factors that should be considered include the following:

- Underpin the structure with new piers to replace improperly placed, damaged, or short piers.
- Remove "mushroom" at the top of piers.
- If climate is prone to periods of low precipitation, a horizontal or vertical membrane can stabilize moisture conditions.
- Construct interceptor drains to prevent ponding. Make sure wood products are well ventilated to avoid rot.
- Install sump pumps to remove perched water.

12.4.2   Underpinning

Underpinning may be necessary in cases where the existing deep foundation system has been damaged by the expansive soils, where the piers are too short, or where other structural defects are present that would require replacement. Partial underpinning of any foundation is not recommended. Such repairs often result in variable bearing conditions and cause excessive differential movement as the unrepaired portion of the foundation continues to move.

Underpinning consists of installing new foundation elements to provide additional support or to transfer the structural loads to different soil or rock strata. The new piers can be used to support the old foundation walls or grade beams. Alternatively, new grade beams can be installed. It is also important that the new grade beam be solidly attached to the piers to prevent heaving soils from lifting the structure off of the piers.

Underpinning methods that have been used vary from one locale to another. Most commonly, they consist of micropiles, helical piles, or push piers. In some areas, large-diameter replacement piers have been drilled next to the foundation and the foundation supported on brackets or beams that extend under the existing grade beams or basement walls.

References

Benvenga, M. M. 2005. "Pier-Soil Adhesion Factor for Drilled Shaft Piers in Expansive Soil." Master's thesis, Colorado State University, Fort Collins, CO.

Chen, F. H. 1988. *Foundations on Expansive Soils*. New York: Elsevier Science.

Colorado Association of Geotechnical Engineers (CAGE). 1996. "Guideline for Slab Performance Risk Evaluation and Residential Basement Floor System Recommendations (Denver Metropolitan Area) Plus Guideline Commentary." CAGE Professional Practice Committee, Denver, CO.

Lambe, T. W., and R. V. Whitman. 1969. *Soil Mechanics*. New York: John Wiley and Sons.

Nelson, J. D., K. C. Chao, Z. P. Fox, and J. S. Dunham-Friel. 2013. "Grouted Micropiles for Foundation Remediation in Expansive Soil." *Proceedings of the Center for Innovative Grouting Materials and Technology (CIGMAT) Conference*, Houston, TX, 20–42.

Nelson, J. D., K. C. Chao, D. D. Overton, and R. W. Schaut. 2012. "Calculation of Heave of Deep Pier Foundations." *Geotechnical Engineering Journal of the Southeast Asian Geotechnical Society and Association of Geotechnical Societies in Southeast Asia* 43(1): 12–25.

Nelson, J. D., and D. J. Miller. 1992. *Expansive Soils: Problems and Practice in Foundation and Pavement Engineering*. New York: John Wiley and Sons.

Nelson, J. D., D. D, Overton, and D. B. Durkee. 2001. "Depth of Wetting and the Active Zone." *Expansive Clay Soils and Vegetative Influence on Shallow Foundations*, ASCE, Houston, TX, 95–109.

Nelson, J. D., E. G. Thompson, R. W. Schaut, K. C. Chao, D. D. Overton, and J. S. Dunham-Friel. 2012. "Design Procedure and Considerations for Piers in Expansive Soils." *Journal of Geotechnical and Geoenvironmental Engineering* 138(8): 945–956.

O'Neill, M. W. 1988. "Adaptive Model for Drilled Shafts in Expansive Clay." In *Special Topics in Foundations*, Geotechnical Special Publication No. 16, edited by B. M. Doas, ASCE, 1-20.

Perko, H. A. 2009. *Helical Piles—A Practical Guide to Design and Installation*. Hoboken, NJ: John Wiley and Sons.

Schaut, R. W., J. D. Nelson, D. D. Overton, J. A. H. Carraro, and Z. P. Fox. 2011. "Interface Testing for the Design of Micropiles in Expansive Soils." *Proceedings of the 36th Annual Conference on Deep Foundations*, Deep Foundations Institute, Boston, MA, 137–145.

13

Floors and Exterior Flatwork

Floor slabs that are commonly used in areas with expansive soils include slabs-on-grade, stiffened slabs, and structural floors. In the context of this chapter, the slab-on-grade, sometimes referred to as a slab-on-ground, consists of a nonreinforced or lightly reinforced concrete slab that rests directly on the ground with little consideration given to its structural capacity. The stiffened slab is similar to the slab-on-grade but is constructed with intensive reinforcement. This would include post-tensioned slab systems, which oftentimes also serve as the foundation. The design of stiffened slab foundations was discussed in chapter 11. The structural floor consists of a flooring system supported by the grade beams, foundation walls, or other internal supports. A void space beneath the floor prevents contact between the soil and the floor. An example of each floor type is presented in Figure 13.1. The pros and cons for each floor type are presented in Table 13.1. These considerations are also discussed by the Foundation Performance Association (2004). The design considerations for each floor type are discussed in the following sections.

13.1 SLABS-ON-GRADE

Prior to the 1980s, most residential houses, school buildings, and industrial and warehouse structures used nonreinforced or lightly reinforced slab-on-grade construction. A slab-on-grade is illustrated in Figure 13.1a. Nonreinforced slabs-on-grade may be constructed for structures on nonexpansive soils. Lightly reinforced slabs may be used for structures on expansive soils with low expansion potential and where differential floor movement will not hinder the function of the space above the floor. A lightly reinforced slab that is only reinforced with wire mesh is usually not adequate. The use of a grid of deformed bar reinforcing is becoming more common. A lightly reinforced slab placed on expansive soils must be isolated from the foundation walls by a slip joint so that it can move independently from the foundation walls as the soil below the slab swells or shrinks. This type of floor system is

351

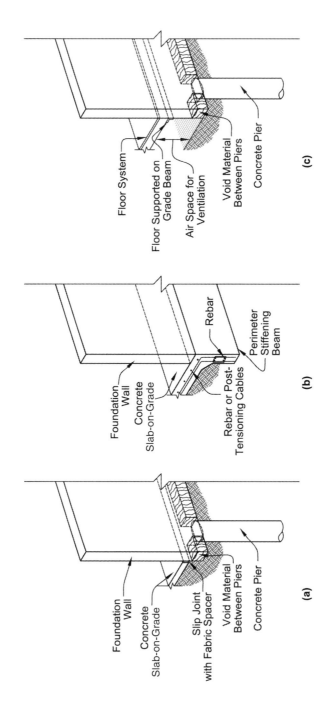

FIGURE 13.1. Typical configurations of various floor slabs for structures on expansive soils: (a) slab-on-grade; (b) stiffened slab; (c) structural floor.

TABLE 13.1   Comparison of Floor Slabs Constructed for Structures on Expansive Soils

Floor Type	Advantages	Disadvantages	Comments
Slab-on-Grade	Comparatively easy and quick to construct. Construction cost is lower than those of stiffened slabs and structural floors. Construction joints and isolation joints can be used with this system to allow separate concrete placements.	Does not reduce amount of differential vertical heave that can occur. Differential movement between the slab and foundation elements is more likely to be transmitted to the superstructure resulting in damage.	Select structural fill should be used beneath slab to reduce total and differential vertical movements. Subgrade and fill, if used, should be field-verified for conformance to geotechnical specifications. This system is not designed to prevent foundation heave.
Stiffened Slab	May be less expensive than structural floors. Slab thickness and reinforcing is usually less than that of structural floors. Differential movement is usually less than that of slab-on-grade.	Must be heavily reinforced to be effective in reducing differential heave in highly expansive soils. Does not significantly reduce the amount of total heave.	Stiffened grade beams should be continued across slab. Slab and stiffened beams must be tied together to act as a single reinforced unit. Select structural fill should be used to reduce potential soil movement. Subgrade and fill, if used, should be field-verified for conformance to geotechnical specifications. This system is not designed to prevent foundation tilt.
Structural Floor	Reduces vertical movements experienced by the floor due to expansive soils, provided that sufficient void space is maintained under the slab. Results in less differential movement of superstructure. Select structural fill is not needed beneath the slab. Fill can consist of expansive or nonexpansive soil. Increases dead load on the foundation.	Usually results in higher construction cost than those of slabs-on-grade and stiffened slabs. Requires more engineering design effort than a slab-on-grade. Extra time required for construction. Wood deterioration and mold growth issues can occur in structural wood floor system.	Most reliable method to prevent differential floor movement. Slab is more heavily reinforced than nonstructural slab. Adequate ventilation must be provided in the void space to prevent mold and radon gas accumulation. Design must account for adequate void space beneath slab after free-field heave has taken place. Steel or concrete floor systems are superior to wooden floor systems in order to avoid mold growth.

commonly referred to as a *floating slab*. Chen (1988) noted that there is no such thing as a truly floating slab. As long as the basement slab is used as a reaction member for the basement wall to resist lateral earth pressure, the concept of a truly floating slab cannot be realized.

Slabs-on-grade placed on expansive soils always pose a risk of cracking and heaving. This design must allow the floor to undergo expected heave movement without adversely affecting the function of the floor slab, and without causing damage to the rest of the structure. Floating slabs-on-grade may perform adequately when constructed on non- to low-swelling soils. They are sometimes used in conjunction with overexcavation and replacement where non- to low-swelling fills have been placed below the slab. The design of the slab-on-grade is not appropriate for moderate to highly expansive soils or steeply dipping bedrock areas where the slab may undergo intolerable movement due to heaving of the underlying soils.

It is important that the slab be isolated entirely from the superstructure above it in order to avoid the risk of transferring movement to the upper stories. Slab isolation must be provided at several locations. One location is where the slab abuts the grade beam or foundation wall. Standard expansion joint material is often used as the slab isolation material but experience has shown that this practice is not very effective. An appropriate slab isolation system should not be capable of transmitting shear stresses. Use of a multilayered low-friction material or a void space that is sealed by a durable long-lasting caulking material is preferred. As noted above, true isolation cannot be achieved if the floor must support the basement wall against lateral earth pressure. Figure 13.2 shows the results of using an inadequate isolation system. In addition to causing slab distress, the upward forces that are transmitted at the interface between the slab and the grade beam or foundation wall will exacerbate heave of the foundation. Also, interior load bearing columns should be supported on separate foundation elements that are isolated from the slab.

Non–load-bearing interior walls must be isolated from potential slab movement as well. This is commonly accomplished by constructing the partition walls with a gap or void at the bottom of the wall. The wall is supported by the joists of the first floor such that it is suspended a specified distance above the slab. Figure 13.3 shows a detail of this construction. The amount of the gap distance should be designed based on expected free-field heave. Figure 13.4 shows a partition wall that was damaged by slab heave. It can be seen that the bottom wall gap was completely closed by slab heave. The wall studs were cut to avoid further damage to the superstructure. Figure 13.5 shows a steel stud that was buckled due to heave of the slab-on-grade floor.

FIGURE 13.2. Slab-on-grade floor subjected to heave. Friction between the wall and the slab restricted slab movement.

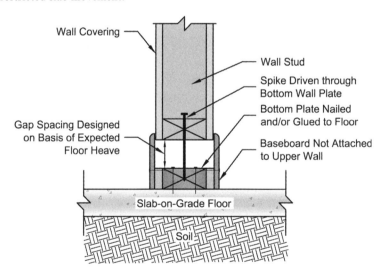

FIGURE 13.3. Suspended partition wall.

Conventional slab construction on nonexpansive soil often includes several inches of gravel beneath the concrete slabs. The use of gravel beneath the slab allows the uniform distribution of slab load and the uniform curing of concrete, thus reducing shrinkage cracks. The gravel layer along with plastic sheeting can also act as a vapor barrier from water in the soil below.

FIGURE 13.4. Closed gap of suspended non–load-bearing interior wall. Note the studs that were cut to prevent transmittal of forces to upper floors.

On nonexpansive soil, the advantages of the gravel layer on nonexpansive soil are outweighed by disadvantages. If water intrudes beneath the slab, the gravel layer provides a path for it to travel over the entire footprint of the house resulting in a higher risk of both slab and foundation heave for the entire structure.

It has been common for soils reports to include language indicating that there is some risk of movement of the slab-on-grade floor, and provide recommendations for construction of the floor providing the owner is "willing to accept the risk of some movement of the floor." Unfortunately, the soils reports generally do not quantify the risk of movement. A more appropriate approach to quantify the risk of movement is to perform calculations of expected free-field heave so as to inform the owner of what amount of risk is involved. If the owner is advised that there is a risk of several inches (or hundreds of millimeters) of movement, they are usually more willing to accept the extra cost of a structural floor.

13.2 STIFFENED SLABS

Stiffened slabs can be used for floors in areas with moderate to highly expansive soils. Attempts have been made to devise a stiffened slab system that can be economically built to withstand reasonable movement

FIGURE 13.5. Steel stud buckled due to floor heave.

of the expansive soils. A typical configuration of the stiffened slab is shown in Figure 13.1b and Figure 9.1. The stiffened slab system usually serves as the foundation as well. Potential free-field heave and maximum tolerable movement need to be taken into account when designing this type of the system. The design of stiffened slabs was discussed in chapter 11.

13.3 STRUCTURAL FLOORS

In the 1960s it was recognized that slab-on-grade floors in basements were heaving, and designs began to incorporate a structural floor over a crawl space. However, during this time period basement floors were most commonly used for storage and mechanical rooms and thus, slab-on-grade floors were typically designed and constructed. With the popularity of walkout and garden level basements, basements were more commonly finished for living space in the 1980s. Heave of the slab-on-grade floors in the living spaces was more problematic than if the space was not finished, and therefore structural floors became more common (Nelson, Chao, and Overton 2006). At that time, soil reports began to note that the most reliable method to avoid such problems was the use of structural floors.

The structural floor is supported by the foundation walls and other interior foundation elements such that it is isolated from the expansive soil. A structural floor system is illustrated in Figure 13.1c. One advantage of the structural floor is that the weight of the floor system is transferred directly to the foundation, thus increasing the dead load pressure on the foundation. Initially, structural floor systems were constructed using wood or composite decking supported on wood floor joists. Many building codes require a minimum clearance beneath wood floor systems of 18 in. (0.5 m). Thus, the actual constructed space must be equal to that amount plus the expected free-field heave.

The crawl space beneath the floors must have adequate ventilation to avoid the accumulation of moisture that could lead to mold development and deterioration of the wood. Ventilation of the crawl space also facilitates exhausting of radon gas. Problems with wood floor systems have led to the more frequent use of concrete floor systems. A concrete deck supported by steel beams and joists is a common system. These systems require a smaller clearance beneath the floor, but they also must allow for the expected free-field heave. The amount of void space must be designed based on predicted free-field heave.

Another system for structural floors consists of precast, prestressed panels spanning between the grade beams. Figure 13.6 shows a detail of such a floor system. The advantage of that system is that it eliminates the need for a separate beam and joist system. This system has been used for floor systems in buildings constructed of concrete and masonry for many years, but has yet to be used for basement floors.

The initial cost of construction of a structural floor may be higher than that of a slab-on-grade or even a stiffened slab, but the higher initial cost will be offset by better long-term performance. Because of greater awareness of expansive soil issues by the public, there have been instances where the market value of a house with a structural floor was enhanced. The cost of repairing a slab-on-grade damaged by heaving of expansive soils can far exceed the initial cost of a structural floor.

13.4 EXTERIOR SLABS AND FLATWORK

Exterior slabs and flatwork such as sidewalks or driveways are particularly susceptible to heave. However, they can tolerate a greater amount of movement than floor systems and still fulfill their function as hardscaping. Because they are easily accessible and the cost of replacement is not excessive, it is generally not economical to isolate them from the soil or to provide for a stiffened slab system.

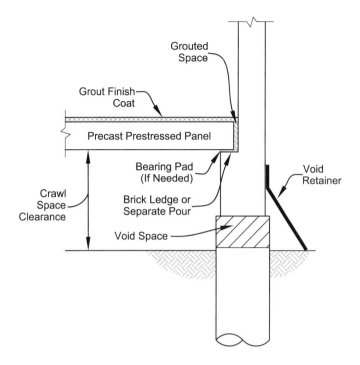

FIGURE 13.6. Precast prestressed panel floor system.

Exterior slabs should not be connected to the foundation of the main structure. They should be allowed to move as an isolated member to avoid damaging the structure or creating uplift forces on the foundation. Figure 13.7 shows an example of the damage that can occur when slabs are not isolated. The most common method of isolating the exterior flatwork from the foundation is to use common expansion joint material. Care must be taken to provide sufficient slope of the slab away from the building. Sufficient separation from the interior floor elevation must be provided so that water is not directed back into the foundation or building interior after heave occurs. An effective means of providing a transition to the structure is to construct a section of the slab or flatwork over a void space, as shown in Figure 13.8.

13.5   REMEDIATION TECHNIQUES

Many repair methods have been proposed to repair damaged slabs. This section will present an introduction to some methods that have been used successfully. Regardless of the method selected, it is necessary for

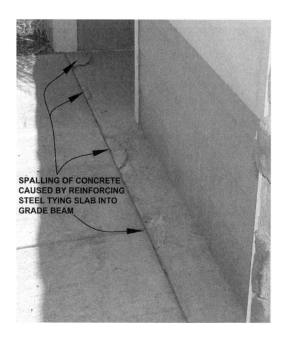

FIGURE 13.7. Damage caused by tying exterior slab into grade beam.

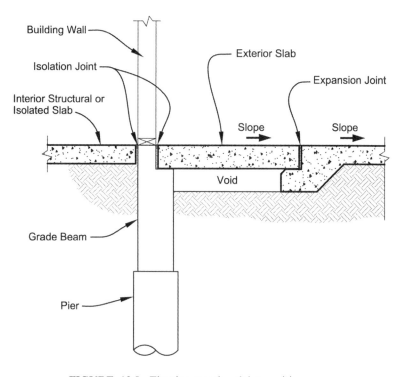

FIGURE 13.8. Floating exterior slab transition zone.

the engineer to evaluate the remaining future heave and evaluate the effectiveness and future risks of using a particular repair method.

13.5.1 Structural Floor Systems

Replacement of a slab-on-grade with a structural floor system is often the most effective method of repairing a slab if the expected remaining heave of the slab is intolerable. This method requires the removal of the slab-on-grade and replacement with a slab or flooring system that is supported on the building foundation. This type of repair increases the loads on the foundation. Engineering analyses must be conducted to determine if additional foundation elements are needed to support the additional loads. The repair plan must also ensure that sufficient void space is constructed beneath the slab.

13.5.2 Moisture Control

Moisture control may provide an effective repair method for slabs, provided that the amount of remaining heave is tolerable. Moisture control may mitigate the fluctuations in movement if the slab is in an area where shrinking and swelling types of soil are present. Moisture control has been effective in some cases, such as garage slabs, by removing the slab, removing and replacing some depth of the subgrade soil with moisture-conditioned soil, and then replacing the slab.

13.5.3 Chemical Injection

Chemical injection has been used with varying degrees of success. The application of this technique was discussed in chapter 10. The effectiveness of this technique depends on the permeability of the soil being treated. A relatively even distribution of the chemical agent may be achieved if the soil is of a uniform nature, such as remolded clay. However, in many native soils such as those with fractured claystone, it is very difficult to achieve an even distribution of the chemical in the expansive soil or bedrock.

The depth of application is also somewhat limited. One advantage this method does have is that the chemicals can be injected through small ports drilled in the slab so that the entire slab does not have to be removed.

13.5.4 Isolation of the Slab

In some cases the amount of remaining heave is very limited. In those cases, the owner may choose to tolerate the movement and not spend money and effort on remediation of the slab. In such cases, it is advisable to isolate the slab from the rest of the structure. Installation, or in some cases reinstallation, of the gap at the bottom of walls is advisable so that the slab movement will not be transmitted to the structure.

It may also be necessary to isolate the slab from the foundation elements as well. If the slabs-on-grade have been doweled into the stem walls of the structure, the slab should be cut as close to the edge of the foundation as possible and a small perimeter strip of the slab removed. The reinforcing steel dowels can then be cut out and the perimeter strip of the slab replaced. An isolation zone should be placed between the slab and the wall.

13.5.5 Exterior Slabs

Exterior slabs can be treated using any of the methods previously discussed. Most commonly the damaged slab is removed, the subgrade is replaced or reconditioned, and the slab is replaced. At sites with highly expansive soils, it may be necessary to replace the slab several times. At locations where the slab abuts the structure, a detail like that shown in Figure 13.8 can be constructed.

References

Chen, F. H. 1988. *Foundations on Expansive Soils*. New York: Elsevier Science.

Foundation Performance Association (FPA). 2004. "Foundation Design Options for Residential and Other Low-Rise Buildings on Expansive Soils." Document No. FPA-SC-01-0, FPA, Houston, TX.

Nelson, J. D., K. C. Chao, and D. D. Overton. 2006. "Design Parameters for Slab-on-Grade Foundations." *Proceedings of the 4th International Conference on Unsaturated Soils*, Carefree, AZ, 2110–2120.

14

Lateral Pressure on Earth Retaining Structures

Lateral pressure[1] acting on basement walls or retaining walls must be resisted by structural means incorporated into the walls. Various methods of analysis have been developed to calculate the lateral forces exerted by the soil on the wall. The most commonly used equations for lateral earth pressure are some form of Rankine's equations of lateral earth pressure or the Coulomb method, which considers equilibrium of the soil mass next to the wall. In the simplest, and perhaps the most common form, the force is expressed as an equivalent fluid pressure acting against the wall. This means that as the depth, or height, of soil that is supported increases, the pressure also increases in a linear fashion. There also exist other, more rigorous methods of calculating earth pressure that take into account surface slopes, shape of the wall face, and so on. All of these methods result in calculation of lateral forces that are considered acceptable for purposes of design when applied in the appropriate fashion. These methods have been covered extensively in other texts on soil behavior and will not be restated here.

In the case of expansive soils, the distribution of forces becomes more complicated. The forces generated by expansive soils are generally additive to the forces generated by equivalent fluid pressure calculations. There are some limitations to the amount of the lateral swelling pressure that can be developed, and there are some methods that can be used to reduce the swelling pressure.

14.1 COMPUTATION OF LATERAL PRESSURE FROM EXPANSIVE SOILS

The lateral earth pressure experienced by a retaining structure is made up of several components. Under a typical nonexpansive soil-loading

---

[1]In classical continuum mechanics, the term *pressure* usually refers to an isotropic stress in which the stress is the same in all directions. However, in foundation design the term *pressure* has historically been used to refer to the stress acting on the structure, whether it is isotropic or not.

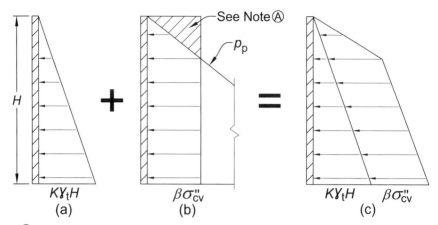

Ⓐ This part of the lateral swelling pressure exceeds the passive pressure. Therefore, the swelling pressure is limited by the passive pressure.

FIGURE 14.1. Lateral earth pressure distribution in expansive soil: (a) earth pressure from nonexpansive soil; (b) earth pressure from expansive soil; (c) total earth pressure for expansive soil.

scenario, the lateral earth pressure would consist of the lateral earth pressure as calculated by conventional lateral earth pressure equations (e.g., Rankine or Coulomb theory). Additional pressure caused by surcharge and hydrostatic loads would also be added to the lateral pressure that must be resisted by the structure.

Figure 14.1a shows the pressure distribution of a conventional nonexpansive soil without hydrostatic pressure from free water. In a scenario with nonexpansive backfill, the pressure acting against the wall would equal the unit weight of the backfill multiplied by a coefficient of earth pressure, K. For a conventional foundation wall, this may be the coefficient of active earth pressure, K_a, or the coefficient of earth pressure at rest, K_0, or some value between those.

If the backfill is expansive, additional pressure is exerted due to the tendency for the soil to swell as it becomes wetted. If the structure was infinitely rigid and the soil was totally confined, the additional lateral stress due to the tendency to swell would be equal to the CV swelling pressure, σ''_{cv}. This would rarely be the case. Most structures will exhibit some deflection, the amount of which would depend on the rigidity of the structure. Furthermore, the soil is normally orthotropic in nature such that for a flat-lying deposit the swelling pressure in the lateral (i.e., horizontal) direction is usually less than that in the vertical direction.

If the deposit is steeply dipping, the swelling pressure in the lateral direction could be even greater than that in the vertical direction.

The effect of orthotropy, however, relates primarily to undisturbed natural deposits. Backfill soils typically are remolded and compacted in place. That process destroys the natural structure of the soil and results in a more isotropic structure. In that case, the swelling pressure would tend to be independent of direction. Some research has suggested that for purposes of computing lateral earth pressure, the lateral swelling pressure can be considered to vary from approximately 0.7 (Sapaz 2004) to 1 (Katti, Katti, and Katti 2002) times the vertical swelling pressure. That swelling pressure would be added directly to the conventional earth pressure applied to the wall. Those stresses are shown by the value $\beta\sigma_{cv}''$ in Figure 14.1b. The coefficient β is a reduction factor to account for the difference between horizontal and vertical swelling pressures.

The lateral pressure exerted on the wall by the swelling of the expansive soil cannot exceed the passive pressure of the soil. If the swelling pressure is greater than the passive pressure of the soil, p_p, the soil will fail along a passive wedge, resulting in a limiting value of the lateral swelling pressure for that portion of the wall, as shown in Figure 14.1b. Figure 14.1c shows the resultant lateral pressure acting on a wall, taking into consideration both the lateral earth pressure and the lateral swelling pressure.

14.2 TESTING FOR MEASURING LATERAL SWELLING PRESSURE

Measurements of lateral swelling pressure have been made using various laboratory test methods such as oedometer equipment, triaxial equipment, in situ instrumentation, and large-scale laboratory tests. In practice, the most commonly used method of estimating lateral earth pressure of expansive soil is to apply a correction factor to the vertical swelling pressure, σ_{cv}''. That was accounted for in the coefficient β shown in Figure E14.1.

Results of lateral earth pressure measurements have been reported in technical papers by Ofer (1982); Chen (1988); Fourie (1989); Aytekin, Wray, and Vallabhan (1993); Katti, Katti, and Katti (2002); and Sapaz (2004). Their results showed that lateral swelling pressures due to the wetting of expansive backfill soil can be many times greater than the lateral earth pressure that would be exerted by nonexpansive backfill. Their results also showed that even small lateral strain during soil expansion resulted in a significant reduction in the swelling pressure (Katti, Katti, and Katti 2002).

Regardless of which test method is used to measure the lateral swelling pressure of a soil, the resulting value can still be large. In most actual situations, it is unlikely that the soils will be completely restrained from any lateral strain. Nevertheless, the addition of even relatively low swelling pressure can have a very significant impact on the design requirements of an earth retaining structure. For example, lateral swelling pressure of as little as 300 psf (14 kPa) will double the total lateral load exerted on a 10-ft (3-m) -high retaining wall. Most expansive soil construction sites have soils with expansion pressures far exceeding 300 psf (14 kPa). It is evident that expansive soils should not be used as retaining wall backfill. In fact, Section 1610 of the International Building Code (2012) states the following: "As expansive soils swell, they are capable of exerting large forces on soil-retaining structures; thus, these types of soils are not to be used as backfill."

14.3   REDUCTION OF LATERAL SWELLING PRESSURE

Various methods of reducing the lateral swelling pressure have been proposed and implemented. The simplest method of reducing lateral swelling pressure is simply to backfill entirely with nonexpansive soil. However, depending on the nature, stiffness, and thickness of the backfill material, lateral swelling pressures can still be exerted on the walls.

Moisture-conditioning has been suggested as a method to reduce the expansion potential of the soil (Chen 1988). With this method, the backfill soil is mixed to a water content generally 1 to 4 percent above optimum water content during compaction. The concept behind this method is that the expansive backfill will be preswelled during placement, thereby eliminating the future swell potential of the soil. Although this method may appear to be somewhat effective, at least initially, it is very difficult, if not impossible, to permanently maintain the proper water content of the soil. During dry periods, the soil will tend to dry out, causing it to shrink or settle, particularly near the surface. This results in cracking at the surface and pulling away of the backfill soil from the retaining structure. When wet conditions return, the surface cracks provide a pathway for water to enter into the backfill soil, thereby increasing the potential for lateral swelling pressure to negatively affect the structure.

Another method suggested by Katti, Katti, and Katti (2002) is to place a layer of cohesive nonswelling (CNS) material between the expansive soil and the retaining structure. Katti's research showed that up to 95 percent reduction of the applied swelling pressure can be achieved by installing 2 ft (50 cm) or more of CNS as backfill next to the wall. Katti's research was performed using only one specific type of soil, which was black cotton clay from the Malaprabha River Valley in India. Testing should be done with other soils to properly use this method. Nevertheless, Katti's results suggest a direction for future research and provide a basis for estimating design guidelines for backfill in expansive soil.

14.4 DESIGN FOR LATERAL EARTH PRESSURE

The following examples demonstrate the magnitude of design earth pressure that would result from the use of expansive backfill soil.

EXAMPLE 14.1

Given:

A basement wall is to be constructed in an area with expansive clay soil. The angle of internal friction, ϕ, of the clay soil is equal to 22 degrees. The backfill will be compacted to a dry density, γ_d, of 102.3 pcf at a water content of 22 percent. The CV swelling pressure for this soil was measured in the laboratory to be 1,000 psf. The basement will have a structural floor. The height of the backfill behind the wall will be 10 ft.

Find:

1. Lateral earth pressure acting on the wall with nonexpansive soil backfill.
2. Lateral earth pressure acting on the wall, considering the expansion potential of the backfill.

Solution:

Part 1:
Since this is a basement wall, it will be restrained at the bottom by the structural floor and at the top by the first floor joists. Thus, this wall cannot tolerate lateral movement or rotation and must be designed for the at-rest condition. The coefficient of earth pressure at rest can be estimated from a simplified version of Jaky's equation (Jaky 1948; Bowles 1996):

$$K_0 = (1 - \sin \phi) = 0.625$$

$$\gamma_t = \gamma_d(1 + w\%) = 124.8 \text{ pcf}$$

The lateral earth pressure expressed as an equivalent fluid pressure, p_0, is,

$$p_0 = 124.8 \times 0.625 \times z = 78.0z \text{ psf}$$

where z is the depth below the top of the backfill.

The earth pressure distribution and total load, P_0, are shown in Figure E14.1a.

$$P_0 = \frac{1}{2}78.0H^2 = 3,900 \text{ lb}$$

where H is the height of the wall.

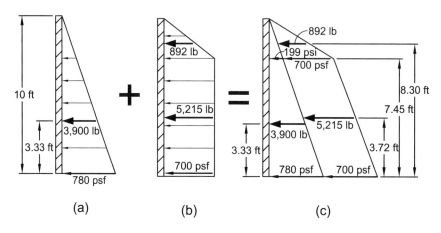

(a)   (b)   (c)

FIGURE E14.1. Lateral earth pressure distribution with expansive soil contribution: (a) earth pressure from nonexpansive soil; (b) contribution due to expansive soil; (c) total pressure distribution.

Part 2.

On the basis of results that were just discussed, the value for β is assumed to be 0.7. The lateral swelling pressure is equal to $0.7 \times 1,000 = 700$ psf, as shown in Figure E14.1b. As noted above, the expansive swelling pressure will be limited to the passive earth pressure of the soil, K_p. The maximum passive pressure, p_p, of the soil can be estimated from Rankine's equation.

$$K_p = \tan^2\left(45 + \frac{\phi}{2}\right) = 2.198$$

$$p_p = 124.8 \times 2.198 \times z = 274.3 \times z \text{ psf}$$

The depth at which the passive earth pressure is equal to the lateral expansive earth pressure is equal to $700/274.3 = 2.55$ ft.

The resulting expansive earth pressure distribution is shown in Figure E14.1b and the total resultant pressure distribution is shown in Figure E14.1c. The total load distribution from the fill would be only 3,900 lb per lineal foot if it were nonexpansive. The total load contributed by the expansive earth pressure is 6,107 lb per lineal foot, for a total load of 10,007 lb per lineal foot. Thus, even a relatively low swelling pressure in the backfill has resulted in an increase of more than 250 percent.

Example 14.1 has shown that the lateral earth pressure will be increased manyfold by using expansive soil backfill. In addition to that, the use of expansive soil backfill can have unintended consequences in that it will also produce uplift skin friction forces on the foundation wall. Example 14.2 illustrates the effects of expansive soil backfill on uplift forces on the foundation wall.

EXAMPLE 14.2

Given:

The same foundation wall system as in Example 14.1.

Find:

Foundation uplift from skin friction due to expansive backfill on the basement wall.

Solution:

The friction angle, δ, for a clay soil with an angle of internal friction, ϕ, of 22 degrees placed against concrete can be estimated from the following equation as being between 13.2 and 17.6 degrees.

$$\delta = 0.6 \ to \ 0.8 \times \phi$$

Thus, assuming $\delta = 17$ degrees, the coefficient of friction, $\tan \delta$, for the soil against the wall is

$$\tan \delta = \tan 17° = 0.306$$

Using the lateral earth pressures calculated in Example 14.1, the potential uplift friction on the wall is shown in Figure E14.2. The maximum uplift capacity per lineal foot of wall, acting at the base of the wall, p_v, is equal to the lateral earth pressure at the base of the wall multiplied by the coefficient of friction previously calculated:

$$p_v = 1,480 \ lb \times 0.306 = 453 \ lb$$

This value is below the vertical swelling pressure of the soil, 1,000 psf, which would be the limiting factor for uplift due to skin friction against the foundation. Therefore, all of the soil friction capacity of the soil if available for uplift. Using this example, the total available uplift capacity per lineal foot, P_v, acting on the foundation walls is calculated as

$$P_v = 10,007 \ lb \times 0.306 = 3,062 \ lb$$

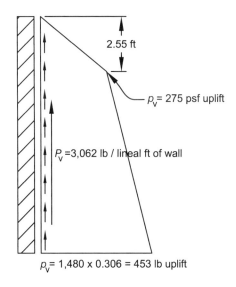

FIGURE E14.2. Distribution of uplift forces acting on wall from skin friction.

This uplift force represents a significant component of the uplift that must be resisted by the structure. Considering that this value is the result of a soil with a horizontal swelling pressure of only 700 psf of pressure, the potential exists for significant damage from soils with higher swelling pressures.

References

Aytekin, M., W. K. Wray, and C. V. G. Vallabhan. 1993. "Transmitted Swelling Pressures on Retaining Structures." *Unsaturated Soils, Proceedings of the 1993 ASCE National Convention and Exposition*, ASCE, Dallas, TX, 32–43.

Bowles, J. E. 1996. *Foundation Analysis and Design* (5th ed.). New York: McGraw-Hill.

Chen, F. H. 1988. *Foundations on Expansive Soils*. New York: Elsevier Science.

Fourie, A. B. 1989. "Laboratory Evaluation of Lateral Swelling Pressure." *Journal of Geotechnical Engineering* 115(10): 1481–1486.

International Building Code (IBC). 2012. *International Building Code.* Falls Church, VA: International Code Council.

Jaky, J. 1948. "Pressure in Silos." *Proceedings of the 2nd International Conference on Soil Mechanics and Foundation Engineering*, The Netherlands, 1, 103–107.

Katti, R. K., D. R. Katti, and A. R. Katti. 2002. *Behaviour of Saturated Expansive Soil and Control Methods*, Revised and enlarged edition, The Netherlands: Balkema.

Ofer, Z. 1982. "Laboratory Instrument for Measuring Lateral Soil Pressure and Swelling Pressure." *Geotechnical Testing Journal*, ASTM 1(4): 177–182.

Sapaz, B. 2004. *Lateral versus Vertical Swell Pressures in Expansive Soils.* Master's thesis, Middle East Technical University, Ankara, Turkey.

Index